Sustainable Agriculture Reviews

Volume 17

Series editor
Eric Lichtfouse

Other Publications by Dr. Eric Lichtfouse

Books

Scientific Writing for Impact Factor Journals
Nova Publishers 2013

Sustainable Agriculture
Springer 2009

Sustainable Agriculture Volume 2
Springer 2011

Environmental Chemistry. Green Chemistry and Pollutants in Ecosystems
Springer 2005

Rédiger pour être publié ! Conseils pratiques pour les scientifiques
Springer 2012, 2e édition.

Journals and Series

Agronomy for Sustainable Development
www.springer.com/journal/13593

Sustainable Agriculture Reviews
www.springer.com/series/8380

Environmental Chemistry Letters
www.springer.com/journal/10311

Environmental Chemistry for a Sustainable World
www.springer.com/series/11480

Blog

Agronomy blog
http://www1.montpellier.inra.fr/agronomy-blog

Magazine

Publier La Science
https://listes.inra.fr/sympa/d_read/veillecaps/

More information about this series at http://www.springer.com/series/8380

Eric Lichtfouse
Editor

Sustainable Agriculture Reviews

Volume 17

Springer

Editor
Eric Lichtfouse
INRA, UMR1347 Agroécologie
Dijon, France

ISSN 2210-4410 ISSN 2210-4429 (electronic)
Sustainable Agriculture Reviews
ISBN 978-3-319-16741-1 ISBN 978-3-319-16742-8 (eBook)
DOI 10.1007/978-3-319-16742-8

Springer Cham Heidelberg New York Dordrecht London

Printed on acid-free paper

Springer International Publishing AG Switzerland is part of Springer Science+Business Media (www.springer.com)

Contents

Chapter 1
Research and Development Priorities in the Face of Climate Change and Rapidly Evolving Pests

Marco Barzman, Jay Ram Lamichhane, Kees Booij, Piet Boonekamp, Nicolas Desneux, Laurent Huber, Per Kudsk, Stephen R.H. Langrell, Alain Ratnadass, Pierre Ricci, Jean-Louis Sarah, and Antoine Messean

Abstract Agriculture faces the challenge of meeting increasing food demands whilst simultaneously satisfying ever stringent sustainability goals. Taken together with the ever increasing rate of integrated globalisation and other anthropogenic impacts, this challenge is further complicated by climate change. Climate change is indeed increasingly recognised as a considerable risk to agriculture in the European Union, particularly with respect to direct impacts on crop production and yield stability. A major impact threat is the further risk from new and emerging invasive alien species, and potential novel pathogenically aggressive adaptations in existing indigenous pests and pathogens, which, hitherto, have been managed with conventional practices and approaches.

The introduction of several exotic pests such as *Tuta absoluta*, *Bemisia tabaci*, and *Bactrocera* fruit flies in Europe points out the changing trend in

---

The views expressed by Stephen R. H. Langrell are purely his own and may not in any circumstances be regarded as stating an official position of the European Commission.

M. Barzman • J.R. Lamichhane • A. Messean (✉)
French National Institute for Agricultural Research (INRA), Avenue Lucien Brétignières, 78850 Thiverval-Grignon, France
e-mail: Marco.Barzman@grignon.inra.fr; Jay-Ram.Lamichhane@grignon.inra.fr; Antoine.Messean@grignon.inra.fr

K. Booij • P. Boonekamp
Wageningen University, Postbus 69, 6700AB Wageningen, The Netherlands
e-mail: kees.booij@wur.nl; piet.boonekamp@wur.nl

N. Desneux
French National Institute for Agricultural Research (INRA), UMR1355-ISA 400 route des chappes, BP 167, 06903 Sophia Antipolis, France
e-mail: Nicolas.Desneux@sophia.inra.fr

L. Huber
French National Institute for Agricultural Research (INRA), Avenue Lucien Brétignières, 78850 Thiverval-Grignon, France
e-mail: Laurent.Huber@grignon.inra.fr

E. Lichtfouse (ed.), *Sustainable Agriculture Reviews*, Sustainable Agriculture Reviews 17, DOI 10.1007/978-3-319-16742-8_1

pathogen adaptation to new regions due to climate change thereby threatening the viability of European crop production. Likewise, slight increases in temperature heighten disease severity caused by indigenous pathogens such as *Leptosphaeria maculans*, *Fusarium graminearum* and *Dickeya* spp. on oilseed rape, cereals and potato, respectively in Europe. Over the last century, there has been an increased global mean temperature by 0.74 °C which is projected to rise by 3.4 °C by the end of twenty-first century. This raise in temperature has resulted in increased pest pressure in European agriculture through a shift from lower latitudes pole-wards and from lower to higher altitudes. In view of this, the development of anticipatory adaptive strategies, resulting in more resilient cropping systems, is the only alternative to tackle evolving pests under changing climate in order to ensure food security for a global population estimated to reach 9.6 billion by 2050.

Keywords Climate change • European network • Pest evolution • Research priority • Sustainable agriculture • Plant health measures

1.1 Introduction

The twenty-first century began with the double challenge of meeting food demands and satisfying sustainability goals. The incidence of emerging new and more aggressive pests (pests refers to all organisms considered harmful to plants, including

P. Kudsk
Department of Agroecology – Crop Health, Aarhus University,
Forsøgsvej 1, 4200 Slagelse, Denmark
e-mail: Per.Kudsk@agrsci.dk

S.R.H. Langrell
Agriculture and Life Sciences in the Economy Unit, Institute for Prospective Technological Studies (IPTS), European Commission-Joint Research Centre (JRC),
Edificio EXPO, E-41092 Seville, Spain
e-mail: Stephen.LANGRELL@ec.europa.eu

A. Ratnadass
Agricultural Research Centre for International Development (CIRAD), HortSys Research Unit, TA B-103/PS4, F-34398 Montpellier Cedex 5, France
e-mail: alain.ratnadass@cirad.fr

P. Ricci
French National Institute for Agricultural Research (INRA),
400 route des chappes, BP 167, 06903 Sophia Antipolis, France
e-mail: Pierre.Ricci@sophia.inra.fr

J.-L. Sarah
Agricultural Research Centre for International Development (CIRAD),
TA 179/04 Avenue Agropolis, 34398 Montpellier cedex 5, France
e-mail: jean-louis.sarah@cirad.fr

insects, pathogens and weeds; Figs. 1.1 and 1.2) affecting yield levels and stability, and thereby threatening food security, has increased over the past number of years. Although climate change will certainly affect the distribution, in addition to anthropogenic impacts, and severity of pest incidence it is not possible to make simple and sweeping assumptions regarding the potential overall crop losses. This is in part because of the unpredictable adaptation of certain pests to a changing environment.

Climate change may indeed have dramatic effects in specific regions of the world. In some regions, excessively dry, wet or warmer conditions may directly hamper crop production, and current crop protection methods may become ineffective. Recent examples of pest adaptation to warmer temperatures have given rise to escalating disease epidemics in new areas where, in retrospect, current systems were excessively vulnerable to unforeseen threats. Less direct effects, such as modifications of cropping systems with the introduction of new crops or changes in crop or pest phenology can also have significant consequences. To deal with the

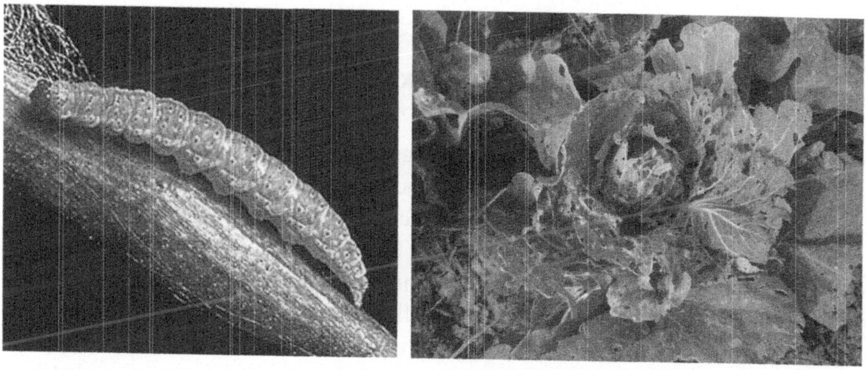

Fig. 1.1 Diamondback moth caterpillar (*Plutella xylostella*) and the damage it causes on cabbage. The infestation often results in complete crop destruction

Fig. 1.2 The white fly (*Bemisia tabaci*) and the sooty mold caused by this pest on cabbage (Photographs courtesy of Philippe Ryckewaert)

interactions between the environment, crops, cropping systems and pest in order to work out manageable adaptive strategies within such scenarios, a new multi-disciplinary and integrative knowledge base is imperative.

As pests are, and will continue to be, responsible for crop losses (which may amount to more than 40 % worldwide (Oerke 2006)), the reduction of their impact is more relevant than ever, not only to produce sufficient food and commodities, but also to reduce excessive input use and unnecessary CO_2 emissions. In spite of improving quarantine policies, we can be confident that increasing global transport of plants, plant commodities and products, as well as other anthropogenic activities, and climate change will further accelerate the spread of pests. Mitigation and adaptation strategies are therefore needed to reduce the impact of new pests as well as existing but evolving pest populations in plant production systems. To develop such anticipatory adaptive strategies resulting in more resilient cropping systems requires new insights, innovative knowledge and appropriate measures adapted to our current socio-economic constraints framed against a clear and informative risk context and realities at field level.

1.2 Policy and Regulatory Context

Climate change is increasingly recognised as a considerable risk to agriculture in the European Union, in particular with respect to direct impacts on crop production and the stability of yields. It is becoming increasingly central within the European policy agenda, mainly with respect to EU food security. Potential adaptation of existing pest populations to new environments, as well as exotic incursions of new, better adapted pest genotypes or species, will further add to the lack of predictability and stability of current cropping systems. Further indirect losses in cropping systems due to climate change, through the probable changing impact of pest and diseases, have, in relative terms, received very little attention or recognition compared to human or animal health issues.

Although further integration of the roles of plant health and crop protection expertise for the creation of more resilient cropping systems appears to be required, such recognition, although still rather tentative and subject to further policy integration and elaboration, is consistent within the framework of adaptive strategy needs as outlined in the European Commission White Paper, Adapting to climate change: Towards a European framework for action (COM(2009) 147/4)(European Commission 2009a), although no specific actions are outlined.

The impact of climate change, globalisation and increased intra-Community movement on the spread and introduction of existing and potential invasive alien species, respectively, is seen as one of the central considerations behind the current revision of the Common Plant Health Regime,[1] with an emphasis on sustainability (with regards to potential increased pressures on cropping systems and yields) and

[1] See http://ec.europa.eu/food/plant/strategy/index_en.htm

overall food security. Against this background, recent, and potentially future, changes to the legislative topography, responding to increasingly stringent environmental and human health concerns, for example, Regulation (EC) No 1107/2009 (European Commission 2009b) and Directive 2009/128/EC (commonly referred to as the Framework Directive) (European Commission 2009c), and their respective implications, will add further practical and operational challenges and complications in maintaining the sustainability of cropping yields.

Taken together, it is unclear how exactly both pest evolution and legislative scenario changes, notwithstanding uncertainty regarding their medium and longer term individual and interactive impacts, will impact on cropping system resilience, or how such changes and implications can be effectively captured within an effective anticipatory adaptive policy context.

1.3 Scope

The main consequence of climate change and accelerated globalisation is a heightened level of unpredictability regarding spatial and temporal interactions between weather, cropping systems and pests. This is the ongoing crop protection challenge that farmers, advisers, researchers and policy makers face. Within this context, this assessment aims to begin to identify the knowledge, tools and approaches needed to manage a continuously changing situation in terms of how cropping systems and legislatively required Integrated Pest Management (IPM) strategies can be designed to better deal with invasive, as well as existing but evolving, pest populations in Europe (Fig. 1.3). For invasive alien pest species that manage to become established, we seek approaches to reduce their impact. For indigenous pests, we seek prevention, escape or control strategies to manage plant-pest evolutionary processes. Our aim is to identify relevant research needs and policy implications.

Fig. 1.3 The tomato leaf miner (*Tuta absoluta*) and the damage it causes on tomato. This pathogen is originating from South America and was first detected in eastern Spain in 2006. Since then, this pest rapidly invaded other European countries. In absence of effective control measures, it can cause up to 80–100 % yield losses in tomato (Photographs courtesy of Judit Arnó)

1.4 What Is Climate Change in the Context of Pests?

Climate change is generally associated with increased CO_2 levels, higher seasonal temperature profiles and more precipitation. These global climate change effects will affect the distribution and severity of pests, but whether such changes will cause more devastating outbreaks cannot be answered in general terms.

Higher levels of CO_2 might increase photosynthesis and crop yields, but will also affect plant morphology, canopy structure and hence micro-climate and the micro-organisms that live on crops. Thus far, different studies have given conflicting results, showing that elevated CO_2 levels may have negative, neutral or positive effects on fungal growth. Indeed, the Climapest project (http://www.macroprograma1.cnptia.embrapa.br/climapest/english-version), covering 20 different crops throughout Brazil, concluded that climate change can "decrease, increase or have no impact on plant diseases, pests, or weeds depending on the region and the time period." For example, experiments have shown that in an environment with higher CO_2 levels, some rust-caused diseases seemed to increase while others decreased in severity.

The effects of temperature change are manifold and will affect crops and plant diseases in many different direct and indirect ways. Higher temperatures are often associated with faster rates of development, growth and reproduction and hence the fear that plant diseases will become more severe makes intuitive sense, particularly because in warmer climates crops such as those found in the tropics tend to suffer greater pest problems than crops in more temperate regions. Yet it is not easy to predict which combinations of crops and pests will tend to become problematic in the future and in which areas (Shaw and Osborne 2011).

Precipitation is very important for crops to grow. However, rainfall and moisture are also vital for the occurrence of fungal and bacterial species. Infection, sporulation and spread of many fungi are moisture-dependent and outbreaks are often triggered by long periods of wetness. Leaf wetness, dew formation and splashing have long been studied as key factors for many fungal species.

In general, climate change models ignore possible effects on dynamics and infectivity of pests and diseases. The most important reason is that we do not have long term monitoring data or an empirical approach to feed modelling systems that might be used to predict impacts and mitigation scenarios with sufficient levels of certainty (Scherm 2004; Shaw and Osborne 2011). In addition, this pattern of climate change factors is not equally distributed over the globe. It is anticipated that a general shift of a milder climate towards the poles will improve the potential of crop production but, in turn, hotter and drier conditions in many already semi-arid areas of the world will limit the possibilities for agriculture. Therefore it is unlikely that general models can ever be developed.

Nevertheless, all climate models predict a higher frequency of extreme and fluctuating weather conditions, which would influence the interactions between crops, pests and diseases in an unpredictable way, and consequently increase the risk of complete failure of current, and possibly future, crop protection strategies.

The body of knowledge on the effects of climate change, new exotic pest incursions and globalisation on crop protection points to a higher level of unpredictability regarding future crop-pest-climate interactions and increased frequency and amplitude of climatic fluctuations. It can be assumed that there will be acceleration in the rate of exotic pests entering and establishing in Europe as well as an accelerated rate of evolution of existing pest populations. In such a situation, the priority may shift from maximising yields to achieving more stable levels of production and avoiding total crop failures. It will be important to develop IPM strategies that are dynamic and internally diversified to locally adapted cropping systems more resilient to fluctuating weather conditions and new and evolving pests.

1.5 What Is the Effect of Climate Change on Pests?

A wide range of pests including insects, fungal and bacterial pathogens and weed may benefit from climate change. The way in which climate change impacts their evolution can be summarised as below.

1.5.1 Insects

Climate and weather patterns are of primary importance for the distribution, development and population dynamics of insects. Although insects may regulate body temperatures in certain ranges, many species – especially smaller herbivores – are almost exclusively ectotherms (Willmer et al. 2004). Insect physiology is therefore primarily driven by temperature which means that phenology, reproduction and developmental rates significantly change when populations are exposed to other climatic regimes. Climatic envelopes (the range and pattern of climatic parameters) and day length largely determine the potential distribution of insect species when host plants are available. On the other hand, insects are constantly adapting their life history patterns to the available climatic regimes or season lengths.

Due to their development, evolutionary adaptations in life cycles occur in very short time periods so that ongoing climatic changes are already reflected in according adaptations, for example daylight signals that induce diapause (Bradshaw and Holzapfel 2011). Increased pest outbreaks are judged ‘virtually certain’ in the IPCC Synthesis Report (IPCC 2007). Aside from population dynamics *per se*, current climatic projections predict that the distribution of insect species will shift from lower latitudes polewards and from lower to higher altitudes. These range shifts in insect distributions can already be observed in nature as a response to global warming (Walther et al. 2002; Battisti et al. 2005; Parmesan 2006; Gutierrez et al. 2009). For example, the stinkbug *Acrosternum hilare* in England and Japan has shifted its range by more than 300 km northward with a temperature increase of only 2 °C.

Lower latitudes usually have more complex pest assemblages; i.e. more pest species and antagonists per crop than higher latitudes towards the poles. Climatic shifts are therefore expected to increase the number of harmful species in temperate regions, especially in those of the northern hemisphere. For example, higher temperatures may extend the distribution range of the European corn borer to maize areas previously free of this species (Fig. 1.4). This may not only lead to crop losses but also to more secondary infections by Fusarium moulds and consequent mycotoxin problems in this crop. Pests may also spread or invade into areas and find new host plants that were not infested before. Being the driving factor for establishment in new distribution ranges, global warming may also increase the travelling speed of invasive pests by decreasing the function of mountain ranges as cold barriers (Aluja et al. 2011).

Concerning phenological changes, the earlier onset and increased length of the growing season also means that indigenous polyvoltine species may be able to complete more generations per year leading to higher population levels at the season's end. Formerly univoltine species may also become bi- or multivoltine. This is potentially the case with codling moth, a key pest in apple throughout several parts of Europe (Stoeckli et al. 2012).

Fig. 1.4 European corn borer (*Ostrinia nubilalis*) caterpillar infestation on corn. This pest causes devastating losses on corn ears, as well as the stalks, by chewing tunnels, which cause the plants to fall over

Besides higher population pressure, the potential for damage is expected to last longer within the growing season and crops will have to be accordingly protected. As a consequence, established management strategies that are focused on low pesticide residues will have to be largely adapted to avoid the possible resistance build-up to plant protection chemicals and thereby ensure their sustainability in future use (Samietz et al. 2015). Further knowledge about pest biology, especially the potential adaptation of insect pests to climatic changes, is necessary and needs to be combined in simulation studies in order to prepare management systems that can withstand the challenges of global warming.

Shifts in phenology due to changing temperatures may also change the synchrony between host plants, pest species and their natural enemies. This may lead to unexpected interactions the outcome of which, in terms of more or less damage, is hard to predict. Furthermore, invasive species may have more opportunity to become established in areas that were formerly unsuitable as a habitat due to the climatic conditions (Fig. 1.5).

Besides higher temperatures during the growing season, higher winter temperatures are equally significant. The distribution of many species is limited by their low temperature tolerance. Given milder winters, more species are able to survive and colonise crops. Some species may skip their sexual stage and have new asexual generations throughout the year which may lead to early high population levels. There are a number of instances worldwide where aphids have shifted from holocyclic (primary and secondary hosts) to anholocyclic (only secondary hosts, i.e. only the crop plants) (Radcliffe and Ragsdale 2002).

These processes are likely to have an impact on crops and increase pest management costs. One well documented instance, which although not directly linked to

Fig. 1.5 Damage caused by fruit fly (*Ceratitis cosyra*) on mango. Infestation of this pest makes the fruit unmarketable cull resulting in severe economic losses

increased temperatures still reveals the importance of such a shifting process in aphid phenology, is in Scotland. Oilseed rape production was introduced with farmers growing vernalized oilseed rape (autumn planted with the crop remaining green during the winter). Green peach aphids could overwinter as anholocyclic forms (asexual generations) on oilseed rape and colonise potato earlier than those aphids that overwintered on the primary host (Fenton et al. 1998; Woodford 1998). If increased temperatures allow aphids to remain on secondary hosts (either crop or non-crop), it allows for the rapid colonisation of crops early in the season.

The phenology of pests will also be modified by increased temperatures. A 20-year study in the UK demonstrated that winter temperature was the dominant factor affecting aphid phenology and that a mere 1 °C increase in winter temperature advanced the migration phenology by 19 days (Zhou et al. 1995). Overall, higher temperatures would allow insect populations to colonise crops earlier and develop faster. In this way, crop damage will occur more rapidly than what is currently observed. Finally, as several of these insect species are also vectors of viruses, crops may also become more vulnerable due to earlier and more severe infections.

Besides temperature, rising carbon dioxide levels will increase the carbon-nitrogen balance in crop plants and hence their structure and palatability for leaf chewing insects. Each species may respond differently to all these changes and this will affect concentrations of constitutive and induced defensive chemicals in plants, insect feeding behaviour, competition between pests, interactions among pests and natural enemies and ultimately the damage to crops (Trumble and Butler 2009). However, our current knowledge on these aspects remains largely fragmentary.

Over recent decades there has been constant growth in the number of reports of invasive species reaching new areas. Invasions are primarily due to the rapid movement of people and agricultural commodities. However, warmer temperatures mean that insects which could not previously survive in potential new areas are able to establish and colonise. A case in point is the potato psyllid, which has migrated into California several times over the past century, the populations never lasting more than a year. Cool winter temperatures meant it could only survive in Mexico and southern Texas during this period of the year. However, it migrated again into California in 2000 and successfully established (Liu and Trumble 2007), inducing substantial losses in tomato, potato and pepper crops.

In addition, new emerging species often spread into completely new ecological settings where most of their natural enemies are missing. An example is the introduction of the B-biotype (*Bemisia tabaci*) in Brazil which was responsible of vectoring of viruses present in native plants only onto cultured tomato crops, leading to new virus diseases in this crop (Fernandes et al. 2008). Whether antagonists will also extend their ranges, thereby following the herbivores, is unknown for the moment, especially in the case of introduction via globalisation/international trade rather than the mere extension of an area of distribution via climate change. In-depth ecological knowledge about both pests and their natural enemies in the country of origin may help to be better prepared and develop more robust cropping systems. This may apply to both already introduced pests and vectored diseases such as *Tuta absoluta* (Desneux et al. 2010; Zappalà et al. 2013), *Bemisia tabaci* (Tahiri et al. 2006), tomato yellow leaf curl virus (TYLCV), *Plutella xylostella* and its parasitoids (Sarfraz et al. 2005), and to potential invasive species such as *Bactrocera* fruit flies (Stephens et al. 2007) as well as to robust cropping systems vis-à-vis these pests currently being developed and tested in the tropics (Licciardi et al. 2008; Vayssieres et al. 2009).

1.5.2 Fungal Diseases

The incidence of plant disease outbreaks depends on complex interactions between many factors. Overall, the development of a virulent strain within a diverse population, the presence of host plants missing appropriate resistance to a specific virulent strain, plant architecture, a uniform cropping system allowing spread, weather conditions and limited antagonistic activities of non-pathogenic populations play an important role in disease outbreaks. The mode of nutrition also has an important role. For example, biotrophic fungi deriving nutrition only from living host tissue are more successfully controlled than necrotrophs which derive nutrition from both live and dead plant materials (Beed et al. 2011). Taking all these factors together, it is a challenge to speculate on the effects of climate change, particularly when long-term datasets from the past are missing to develop and test predictive models for the future. Nevertheless, our knowledge of the phenology of plant-disease interactions, population genetics of pathogens as well as host crops, and examples of overwhelming establishment of new diseases in a region, provides insights into how climate change may affect incidence of disease.

First, some features of climate change will influence disease phenology. Higher temperatures and/or elevated CO_2 will speed up the life cycle of some pathogenic fungi thereby increasing the development of inoculum. The latter can lead to high levels of infection and accelerated evolution of new strains (Chakraborty and Dutta 2003). Crop protection must be timely and effective and if not precise and complete, a new wave of infection can cause a new disease epidemic thereby overcoming crop resistance. Prolonged generations of pests will be able to infect crops at a later stage then at present. For example, it has been estimated that phoma stem canker could cause a 10–50 % decrease in the total yield of oilseed rape in the UK depending on future climate conditions (Barnes et al. 2010). Another effect of climate change is that anthesis in, for example, wheat in the UK will be earlier in the season, which is a more favourable time for *Fusarium* ear blight infection and consequently increases the chances of finding mycotoxins in cereal produce (Madgwick et al. 2011). In addition, if the crop can be infected later in the season, it is more mature or even in senescence, losing its natural resistance, and therefore the pest has a higher chance to overwinter in leftover stubble or storage organs.

Second, climate change may affect the expression of plant resistance traits in a positive or negative way. Breeding for resistance takes a long time and today's resistant varieties are made for present agricultural conditions. It has been experimentally demonstrated that the expression of quantitative resistance against phoma stem canker in oilseed rape dropped dramatically when the temperature was raised from 20 to 25 °C (Huang et al. 2009). Indeed, the authors showed that the percentage of the leaf area infected increased from 5 to 50 %. Taken together, the expression of resistance genes in the host plant and their efficacy may decrease dramatically because of climate change. In addition, we could envisage that the increased generation cycles of pests triggered by climate change might select more aggressive pathogen populations. Such a selected population combined with hampered resistance in the host may lead to unpredicted, or unprecedented, epidemiological outbreaks.

Thirdly, when the genetic variation of a crop is low over a large area, and a new or adapted strain of a pathogen has taken over the population or become dominant, the effects can be dramatic. A classic example is wheat production in the Central, West Asian and North African regions which feeds more than one billion people. Although many wheat varieties are grown in this huge area, all have a similar genetic background. With a slightly increased temperature and decreased rainfall, as observed over recent decades, a new type of yellow rust (*Puccinia striiformis*, referred to as *Ug99*) emerged and was able to extend from Africa to India within 15 years, leading to widespread epidemics thereby highlighting how vulnerable an area can be when slight climate changes occur (Hovmøller et al. 2011).

Looking more closely at new strains of *Puccinia striiformis*, the causal agent of wheat rust disease (Hovmøller et al. 2011), it has been observed that new strains have adapted to produce spores at warmer temperatures than usual for this fungus. This may have probably increased the rate of disease expansion on a global scale (Milus et al. 2009). This new strain spread to at least three new continents within 3 years, faster than previously reported for any crop pathogen (Hovmøller et al. 2008). This gave rise to severe epidemics in warm wheat production areas (Hovmøller et al. 2010) where yellow rust was previously absent or infrequent, implying that current cropping systems were not prepared for the new situation (Fig. 1.6). At a temperature regime typical for these areas, isolates of the new strain produced three to four times more spores per day than strains found previously. The rapid spread of the new strain is probably the cumulative result of increased pathogen fitness, successful colonisation of new, warm environments, an increase in the number of spores in the atmosphere, long-distance dispersal of these by wind, and increasing anthropogenic impacts such as increased regional and global travel and commerce. Similar chains of events should be expected for other pathogens, reinforcing the need for coordinated, international action to ensure sustainable crop disease management under climate change.

Another example is *Phytophthora ramorum*, a pathogen native and restricted to California and not found in Europe (Fig. 1.7). However, with the ever increasing trade and planting of foreign and exotic plant species in Europe (such as rhododendron), in combination with humid summers and milder winters, this pathogen was not only able to cross the ocean by way of a hitherto unrecognised introduction pathway (and which it most probably has done on multiple previous occasions) but was also able to establish in our region. Today, this fungal pathogen is not only considered established and endemic in Europe, but is continuing to extend its geographical as well as host range (Grunwald et al. 2012).

To the three issues examined in detail above, we can add a raft of others. Besides the changes in host-pathogen interactions, which climate change will induce (of which phenology and host resistance are components), we can note other changes

Fig. 1.6 Yellow rust (*Puccinia striiformis*) caused by the fungal pathogen on winter wheat. New strains of this fungus have adapted to produce spores at warmer temperatures than usual thereby increasing the rate of disease expansion on a global scale

including fecundity, canopy size, as well as long and short-distance dispersal, etc. Changes in the geographical distribution of hosts and pathogens, new pathogens, new remnant vegetation and increased plant stress will all be important issues in the development of fungal diseases in changing climatic conditions.

As effects of climate change on fungal diseases remain difficult to predict in general terms, a practical method might be detailed modelling of each individual crop-pathogen situation in a projected climate change situation of a certain geographic area. Looking at north-west Europe with predicted warmer and more humid winters, and warmer and drier summers, polycyclic and monocyclic fungi can be classified in seven 'ecotypes', based on dissemination method of spores, infection condition requirements, latent period to weather conditions etc. (West et al. 2012). Being classified in a specific ecotype one can predict if climate change in that region will support an increase or a decrease infection of a certain fungus-crop combination. This methodology might be useful to test in other regions of Europe.

In conclusion, the complex interplay of changing resistance expression, new virulence of pathogens, other host plants and, over-arching all these, more favourable infection conditions due to climate change, warrants a careful inventory of the new knowledge needed to prevent unprecedented disease outbreaks in the future and their ultimate control.

Fig. 1.7 Sudden oak death caused by the oomycete pathogen *Phytophthora ramorum*. The pathogen can cause the disease on a wide range of plant species (Photograph courtesy of Yana Valachovic)

1.5.3 Bacterial Diseases

It is expected that plant pathogenic bacteria could become a more serious threat to plant health. One reason is that bacteria together with plant or plant products can easily migrate around the globe. Another reason is that new strains emerge from warmer regions and easily adapt to our more temperate regions. A telling example in Europe can be found in potato crops, where *Pectobacterium* spp. were the principle bacterial diseases. However, over the past decade *Dickeya* spp. have taken over. The latter have been identified as pathogens from tropical and sub-tropical regions and have become established in Europe due to increasing milder climatic conditions. Compared to the *Pectobacterium* spp., *Dickeya* spp. are more aggressive and virulent and are now responsible for 50–100 % of field infections, causing significant damage in potato (Czajkowski et al. 2011).

As climate change may enhance plant pathogenic bacteria threats it is important to realize that bacteria maintain the possible ability to behave as 'kingdom hoppers': moving from human to animal to plant to environment, and *vice versa*. Examples include the Shiga-toxin producing *Escherichia coli* (*STEC*) in Germany in 2011, where infection probably took place abroad at the production site of seeds. Such a situation is not new and many incidents have been reported earlier. An example is

190 outbreaks in the United States between 1973 and 1997 through the consumption of *E. coli* or *Salmonella* contaminated fresh products (Slayton et al. 2013). Human pathogenic *E. coli* and *Salmonella* grow extremely well in the intestines of pigs and cows and survive well in manure, even if mixed with soil as a fertiliser. Experiments with fluorescent-labelled *Salmonella* showed that these human pathogenic bacteria are easily taken up by the roots of lettuce seedlings and transported to above-ground lettuce leaves, surviving as endophytes in the tissue (Klerks et al. 2007). Recently it had been reviewed that bacterial species, especially those belonging to the large family of *Enterobaceriaceae*, can easily be adapted in human, animal and plant environments. Horizontal gene transfer between species can easily occur in a joint habitat as soil/manure and the plant rhizosphere, leading to bacteria that are well adapted to live on plant as well as human hosts. In an analogy to zoonosis (bacteria transmitted from animal to human), this new class of plant bacteria is termed 'phytonosis' (Van Overbeek et al. 2014). One of the most important threats of phytonosic bacteria are acquired genes for antibiotic resistance, known for being abundantly present in the bacterial community of the plant rhizosphere (Van Overbeek et al. 2014).

1.5.4 Weeds

Weeds compete with crops for light, nutrients and water and it is likely that the effect of environmental disturbances such as rising CO_2 or increasing temperatures will be manifested as a change in the competitiveness between crops and weeds. Being inherently adapted to the prevailing conditions, it is likely that weeds will respond more favourably to climate change than crops. This has been confirmed in the majority of studies examining the response of crops and weeds to increasing levels of CO_2 (although the physiological characteristic of crop and weeds being either a C_3 or C_4 plant will also determine the response to CO_2) (Hatfield et al. 2011; Ziska 2011). While CO_2 is mainly expected to influence crop-weed competition, the most likely effect of a rising temperature is the northwards expansion of native and invasive weed species (Hatfield et al. 2011).

In comparison to many diseases and insect pests, weeds are relatively non-mobile and their spread and establishment into new regions is expected to take longer than for other pest groups. However, as the projected climatic warming may exceed maximum rates of plant migration in postglacial periods (Malcolm et al. 2002) it could favour the most mobile weed species. Many agricultural weeds have characteristics associated with long-distance dispersal (Rejmanek 1996) and a wide geographic range such as small seeds, phenotypic plasticity and short juvenile period suggesting that agricultural weeds may be among the fastest to spread (Dukes and Mooney 2000).

Most of the current information on the impact of climate changes on future weed distribution is based on predictions performed using bioclimatic envelope models (Hyvönen et al. 2012). These models, however, have a limited value for predicting the long term changes in the composition of weed communities (Thuiller et al. 2008). Firstly, it is assumed that climatic conditions alone can predict range shifts without considering biotic and management practices such as soil type, tillage practice and cropping sequence. Furthermore, the bioclimatic envelope models assume that weed species will only gain a foothold in areas that have a climate similar to its native range but this assumption has recently been challenged by studies revealing that the range expansion of alien species exceeded that predicted by the models (Clements and DiTommaso 2011). Rapid genetic evolution and/or phenotypic plasticity may explain these observations (Maron et al. 2004; Clements and DiTommaso 2011).

For an agricultural weed to be successful it must adapt not only to the climate but also to the constraints of the cropping system. Thus, the range shift of a weed species should not only be studied in the context of climate but should also include biotic parameters but this has rarely been the case so far. In the context of range shifts the term *niche* is often used. According to Hutchinson (1957), the abiotic factors and most notable climate determine the fundamental *niche* of a species while the interactions with the biotic factors together with the dispersal properties determines the realised *niche* of a species. Although the concepts of fundamental and realised *niches* are useful also for understanding range shifts of agricultural weeds they do not take into account the economic impact of agricultural weeds. To overcome this shortcoming, McDonald et al. (2009) introduced the concept of the damage *niche*, which they defined as 'the environmental conditions that make specific weeds abundant, competitive, and therefore damaging the production of particular crops'. Determining the damage *niche* of a weed species requires information on the crop-weed interactions and is thus region and cropping system specific.

Despite its relevance to the impact of climate change on the distribution and impact of agricultural weeds the "damage niche" concept has only been applied in a few other studies (Stratonovitch et al. 2012; Bradley 2013). Using the damage niche concept based and the HadCM3 projections for the periods 2046–2065 and 2080–2099 Stratonovitch et al. (2012) found a possible northward shift in the range of *Alopecurus myosuroides* (Fig. 1.8) in winter wheat but also a local-scale variation due to variation in soil type and water holding capacity. The competitive balance was predicted to shift in favor of the crop due its deeper root system making the crop less prone than the weed to the more frequent drought stress events predicted, i.e. the damage niche was predicted to reduce under climate change.

1.6 Climate Change Is Also Affected by Crop Protection Strategies

Adapting crop protection strategies is important to stabilising food availability but crop protection also has an important role to play in mitigating climate change. Indeed, crop protection can contribute to the reduction of greenhouse gas (GHG) emissions. In the UK, arable production systems account for more than 7 % of the total GHG emission and in a recent study it was found that effective and proper crop protection was by far the most important factor for decreasing GHG emissions in the future (Carlton et al. 2012). Reducing crop damage minimises losses in invested

Fig. 1.8 Black grass (*Alopecurus myosuroides*) infestation in winter wheat crop. This weed is subjected to a possible northward shift in Europe and also to a local-scale variation

nitrogen, a significant producer of GHG. A quantified example with regards to fungicide use against diseases of UK oilseed rape shows that yield benefits from fungicide use significantly outweigh GHG emissions associated with production and application of fungicide. Reduced GHG emissions associated with more efficient use of nitrogen per tonne yield are considerable. For UK wheat, barley and oilseed rape use, such benefits are >1.5 Mt CO_2 equivalent per year. With regards to forest systems, crop protection might be the most important factor in enhancing sequestration of CO_2 by the "green lungs" of the world.

It is surprising that in most climate reports, and climate change debates, the role of plant pests and diseases in mitigating the effects of climate change are at best only marginally covered. Therefore it is perhaps timely that we elaborate new crop protection strategies that are robust enough for the changing and challenging climate conditions ahead. Within this context, we face a new challenge: can diversifying crop protection help mitigate climate change?

1.7 Effect on Crop Protection Strategies

Current agricultural practices may need to be revisited because of climate change. IPM is a widely used strategy for insect control, for example, integrating cultural, biological and chemical controls to reduce insect pest populations below a threshold which will cause economic losses. Researchers and growers will need to revise these current IPM programmes and approaches to address several important effects of increasing temperatures and climate change more broadly. Indeed, pest development is often more rapid at higher temperatures and will therefore develop faster, with the effect that crop damage could occur more rapidly than expected. A recent study also suggests that herbicide performance may be reduced at elevated CO_2 levels (Ziska 2010). In addition, as temperatures increase, the frequency of spring frosts will decline and the resulting extended frost-free periods may increase the duration and intensity of insect outbreaks. Farmers may be tempted to take advantage of the changing climate by planting crops earlier in the season. However, plants will then be available for insect pests earlier, in turn allowing populations an earlier opportunity to infect and potentially add additional generations during a typical growing season. For many pests, it means bigger populations by the end of the season and higher associated yield losses.

Indirect effects of climate change may occur due to shifts in cropping patterns and crop distributions. New combinations of crops may be grown in particular regions and in different growing conditions with changing husbandry practices. If ecological knowledge and farmer experience is lacking for these new situations, unexpected outbreaks may become more frequent or more difficult to manage.

In some cases, small differences in temperature and humidity can determine the preferred protection strategy. A study in the Central Valley of California (Daane and Caltagirone 1989) showed that an IPM strategy based on the use of parasitoids against the black scale *Saissetia oleae* on olive (Fig. 1.9) could work in the northern part of the Valley. In the southern part, however, the slightly warmer and drier

Fig. 1.9 Infestation of black scale (*Saissetia oleae*) on olive which can lead to severe damage to this crop reducing drastically the photosynthetic ability of leaves

conditions were not favourable to the parasitoids. In this area, the authors recommended heavy pruning to exacerbate the hot and dry conditions to increase summer mortality of the pest.

Taking into account the uncertainty of local and temporal climate change effects in different regions of the world, its effects on plant characteristics and crop production as well as on pest and disease epidemiology and infection probabilities, designing appropriate and reliable crop protection strategies present considerable agricultural research and innovative challenges.

Earlier we have described the possible effects of climate change on the behaviour of pest and diseases, the occurrence of new invasive pathogens, emergence of novel strains from existing indigenous populations, plant resistance characteristics, new hosts for pest and diseases and more favourable climatic conditions for infections. What are the consequences for our current crop protection tools? Can we rely in the future on chemical pesticides as we overwhelmingly do today? And if not, are our current IPM tools suitable to be used in the future?

These questions need to be addressed to guarantee resilient crop protection strategies for the specific climate change impacts in each region. We need to study new resistance genes with optimal expression under variable climate change situations. We have to collect more data on the behaviour of pests and diseases in future circumstances and the efficacy of biological control organisms. We have to assess the role for alternative plant hosts, intercropping and landscape management for pests and diseases as well as for beneficial organisms. And we need to develop Pest Risk Analysis models to predict, not only the risk of new invasive species of pathogens, but novel indigenous emergent, their chances of establishment in new cropping areas, and how to counteract disease disasters through local adaptive and resilient cropping and crop protection systems. In other words, all the questions we currently address in order to develop effective and robust IPM approaches with less reliance on chemical crop protection can, *mutatis mutandis,* be raised to develop locally adaptive and diversified IPM systems which are resilient enough to meet extreme weather fluctuations as a result of climate change.

1.8 Recommendations and Action Points

Faced with the uncertainty regarding the effects of temperature, CO_2 and other changes in climatic patterns on crop protection, a number of action points have been identified in terms of research areas and the organisation of research and extension, to face the challenges. While projected changes don't automatically translate to 'doom and gloom' scenarios, the level of uncertainty is such that policy, research and extension should be prepared for potential worst case scenarios or extreme events following a 'no regrets' approach. Preparation, in turn, should produce increased resilience in cropping or horticultural systems. This high level of uncertainty also needs to be borne in mind when considering models, which are perceived as essential tools to support decision making but whose limits constantly need to be repeated and explained. This is true for models driven either by weather changes or those tackling exotic pest establishment. The predictive power of modelling on the effects of climate change on pest and disease occurrence is seriously hindered by biotic and abiotic uncertainty. This means that when using model outputs for policy or farm management, this uncertainty should be taken into account in order to develop flexible responses and be continuously adaptive.

1.8.1 Human Resources

Historically, agricultural systems have adapted to face major changes, and this capacity to change requires human resources. For example, 2–3 years after the initial arrival of the invasive pest *Tuta absoluta*, the severity of damage was drastically reduced due in large part to farmers having learned how to manage and reduce the impact of this pest. This was particularly true in areas that had experienced the earlier *Bemisia tabaci* invasion and had subsequently developed IPM responses to it.

This increased uncertainty and the accelerated rate and intensity of disturbances we are now witnessing means this capacity for change needs to be pro-actively accelerated. Paradoxically, in several sectors the appropriate expertise and financial support are declining. This can be seen in the area of extension services, where, for example, in the USA long-term reliance on herbicide-resistant crops has translated into a reduced number of advisers with weed management skills. Lack of the required expertise can also be seen in the area of plant health. Regarding the threat posed by new and significantly more virulent yellow rust strains, for example, identifying strains requires highly skilled staff, the expertise for which, and in other areas of plant health and protection, is being continuously eroded across Europe. Although recognised, action needs to be taken to reverse the decline in plant health expertise and skilled crop protection specialists in Europe.

One approach which could be usefully employed to address shortcomings in human and financial resources is to join forces, pool resources and develop new collaborations for pest management. For example, cooperation between the historically compartmentalised experts in plant health and crop protection could help to draw on shared expertise regarding the anticipation of invasive and alien species and modelling, particularly with regards to understanding the processes leading to establishment (though it should be borne in mind that distinctions between invasive species and more aggressive strains of existing ones may not always be pertinent when considering their spread and impact). Large research groups such as Climapest, which brings together 134 researchers from Brazil and Argentina, which are already working on this issue, could be linked to other groups, although conditions in Europe will, of course, differ from those in South America. The authors strongly recommend the promotion of collaborative efforts within the research community between the historically compartmentalised plant health and crop protection spheres.

Another useful avenue is to develop participatory approaches involving new stakeholders, including farmers, inspired by Web 2.0 and taking advantage of the new technologies which make it easier to exchange information. This type of approach could, for example, improve worldwide tracking of global wheat rust. We can look to existing participatory systems as possible models for this approach, including the Finnish web application EnviObserver, which enables farmers to report pests with mobile phones, and the French 'Epidémiosurveillance' regional networks, which offer weekly bulletins. It should be noted, however, that farmers will often refrain from reporting the presence of regulated pests on their farm.

1.8.2 Resilient Cropping Systems

From the point of view of cropping systems, 'preparing for the worst case scenario' translates into uncovering what currently makes cropping systems resilient to extreme, variable and unpredictable situations. Promising avenues to achieve resilience include genetic diversification and combining crop protection systems. An example which has been developed for late blight in potato crops relies on varietal resistance coupled with new methods to monitor loss of resistance. Another example, mentioned earlier, is the combination of the opportunistic use of native natural enemies with entomopathogens against the invasive species *Tuta absoluta* in tomato crops. However, the literature review on this subject reveals a notable scarcity of knowledge and references regarding what constitutes robust or resilient cropping systems and, more generally, on how cropping systems can be manipulated to reduce the risks associated with invasive and alien species and rapidly evolving pest species within the context of climate change. This reflects a serious knowledge gap within a priority research area and is one which needs urgent attention. The authors strongly recommend this new field of study be examined as a priority. We note also the need to enlist social and economic researchers to work on these issues to ensure that the solutions proposed are both socially and economically appropriate for stakeholders.

1.8.3 Crop-Weed Competition

Weeds have a much greater potential to cause crop losses than many insect pests and pathogens, but the study of climate change impact on crop-weed competition has been overlooked. In some cases, differential responses between weeds and crops could be a source of opportunity. For example, red or weedy rice shows a much stronger response to rising CO_2 than cultivated rice and consequently could serve as a unique source of genes which may potentially help adapt cultivated rice to climatic uncertainties. It is suggested to adopt the concept of the damage niche in future studies to take into account the impact on, for example, management practices on the impact of weeds on crop yields in a changing climate.

1.8.4 Anticipation/International Monitoring

As has been made clear throughout the production of this report (preliminary meetings, workshop), strategies that identify the vulnerabilities of crop systems prior to pest introduction and establishment are a worthwhile and feasible approach. An example can be seen in the international collaboration between the UK and China to help the latter decrease the risk of invasion from *Leptosphaeria maculans*, a *Phoma-causing* agent that can lead to devastating effects. The collaboration included support on pest risk analysis to determine the risk of entry and establishment, training on symptom recognition, PCR diagnosis, epidemiology, cultivar resistance screening techniques, conducting surveys on the distribution of *Phoma* agents, and workshops with policy makers on risks and strategies to prevent spread.

International, global coordination has also been identified as necessary to monitor both invasive alien species and the appearance of new and more aggressive strains of existing pest species. The Borlaug Global Rust Initiative, covering 20 countries in several continents, is an example of coordination on new highly virulent strains of rust that have the potential to induce widespread wheat crop failures. It operates as a worldwide warning system for potential rust outbreaks. The creation of an international database on harmful organisms accessible by all countries would be another useful step, and could fall within the remit of the European and Mediterranean Plant Protection Organisation.

Likewise, the existence of the European Commission's EUROPHYTE rapid alert system for Europe (http://ec.europa.eu/food/plant/plant_health_biosafety/europhyt/index_en.htm) and the development for the EPPO pest reporting system currently in development (pestreporting.eppo.int) are of particular importance in this regard. Lastly, for prevention, good synergies and information exchange should be encouraged to promote the most rapid response to new and emerging risks and incursions where elements of cooperation and information sharing should be encouraged focusing on promoting vigilance and international information exchange from across the globe, with particular emphasis on principle trading partners.

1.8.5 Adaptation Strategies Including Breeding for and Sustaining Resistance

A priority area should be breeding for resistance and devising strategies to sustain that resistance. It should follow an assessment of how existing farming systems, crop rotation practices and varieties will perform under changing climatic conditions. It requires breeding for varieties that are both adapted to changing climates and those containing temperature-stable resistance genes. Deployment strategies or the combination of new varieties with other methods will be needed to ensure that resistance to pest is sustained.

1.8.6 Biological Control

Research and development in biological control is another priority area which will help to adapt cropping systems to the appearance of new pests and changing climatic conditions. Making use of opportunistic native natural enemies against new exotic pests can also be seen as a promising avenue. Looking beyond Europe, in North Africa and West and Central Asia, for example, there is a need for biological control agents specifically adapted to warmer and drier conditions.

1.8.7 Pest Risk Analysis

Even though there are recognised limits to pest risk analysis in terms of its capacity to predict establishment, it remains an important early warning tool. It can help to identify those pest species which present the highest risk for European cropping systems. With pest risk analysis, it becomes possible to propose adaptations to cropping systems in advance or to conduct research on the ecological requirements and/or control options related to pest origins. Greater progress could be made with PRA if it could be made more spatially explicit or based on 'hotspots' or case studies so that more focused, site-specific conditions affecting the likelihood of establishment are taken into account.

1.8.8 Extension

Advisory services currently focus on optimising existing systems and, faced with climate change, this approach will have to change in order to concentrate on making systems more resilient. Diversifying cropping systems in general with strategies such as more varied rotations (as is recommended for tackling *Diabrotica* in maize, for example) can be regarded as major leverage points which can be addressed via extension services. Changes will also be required in the way extension is considered in the innovation process. The involvement of multiple stakeholders, including public-private partnerships, and new, more collective approaches to extension are required to meet the challenges of building more robust cropping systems.

1.9 Conclusion

Increasing global trade of plant and plant products fosters the introduction and dissemination of exotic pests from one region to another while climate change favors their establishment throughout new regions which were previously unfavourable for their adaptation. In addition, the life cycle of indigenous pests within a given region is likely impacted under climate change leading to higher number of generations per season. It means that the pressure of pests in agriculture will increase both in space and time due to both native and invasive species. Hence, durable plant protection strategies are needed to tackle such challenges. In Europe, further legislation restrictions in the use of plant protection products, increasing awareness in consumer health and environmental issues add additional pressure to develop sustainable solutions, and to identify the knowledge, tools and approaches to address the challenge of food security and safety.

Although climate change is expected to have a scientific heavy impact on European agriculture by influencing the stability of crop yield, scientific information on the consequences is lacking and studies in this regard are still in their infancy. Research based on broader collaborative approach should be made in order to develop anticipatory adaptive strategies resulting in more resilient cropping systems. To this aim, interregional, transnational and global networking of researchers and stakeholders is an effective way to better use the limited resources needed to address this central twenty-first century challenge.

Acknowledgements This report is the fruit of a joint initiative of the Institute for Prospective Technological Studies, one of the institutes in the European Commission's Joint Research Centre, and the ENDURE European Research Group, which brings together some of Europe's leading agricultural research, teaching and extension institutes with a special interest in IPM. It is built on a technical report and a workshop of international experts, and the authors would like to thank all those who contributed to what we believe is a valuable and much-needed insight into the research and policy implications of meeting this twenty-first century challenge. The authors wish to thank Andrew Lewer for his initial collaboration in the construction of this manuscript.

References

Aluja M, Guillén L, Rull J, Höhn H, Frey J, Graf B, Samietz J (2011) Is the alpine divide becoming more permeable to biological invasions? – insights on the invasion and establishment of the Walnut Husk Fly, Rhagoletis completa (Diptera: *Tephritidae*) in Switzerland. Bull Entomol Res 101:451–465. doi:10.1017/S0007485311000010

Barnes AP, Wreford A, Butterworth MH, Semenov MA, Moran D, Evans N, Fitt BDL (2010) Adaptation to increasing severity of phoma stem canker on winter oilseed rape in the UK under climate change. J Agric Sci 148:683–694. doi:org/10.1017/S002185961000064X

Battisti A, Stastny M, Netherer S, Robinet C, Schopf A, Roques A, Larsson S (2005) Expansion of geographic range in the pine processionary moth caused by increased winter temperatures. Ecol Appl 15:2084–2096. doi:org/10.1890/04-1903

Beed F, Benedetti A, Cardinali G, Chakraborty S, Dubois T, Garrett K, Halewood M (2011) Climate change and micro-organism genetic resources for food and agriculture: state of knowledge, risks and opportunities. Commission on genetic resources for food and agriculture, FAO. doi:http://www.fao.org/docrep/meeting/022/mb392e.pdf

Bradley BA (2013) Distribution models of invasive plants over-estimate potential impact. Biol Invasions 15:1417–1429. doi:10.1007/s10530-012-0380-0

Bradshaw WE, Holzapfel CM (2011) Genetic shift in photoperiodic response correlated with global warming. Proc Natl Acad Sci U S A 98:14509–14511

Carlton RR, West JS, Smith P, Fitt BDL (2012) A comparison of GHG emissions from UK field crop production under selected arable systems with reference to disease control. Eur J Plant Pathol 133:333–351. doi:org/10.1007/s10658-012-9961-0

Chakraborty S, Dutta S (2003) How will plant pathogens adapt to host plant resistance at elevated CO2underachangingclimate.NewPhytol159:733–742.doi:10.1046/j.1469-8137.2003.00842.x

Clements DR, DiTommaso A (2011) Climate change and weed adaptation: can evolution of invasive plants lead to greater range expansion than forecasted? Weed Res 51:227–240. doi:10.1111/j.1365-3180.2011.00850.x

Czajkowski R, Pérombelon MCM, van Veen JA, van der Wolf JM (2011) Control of blackleg and tuber soft rot of potato caused by *Pectobacterium* and *Dickeya* species: a review. Plant Pathol 60:999–1013. doi:10.1111/j.1365-3059.2011.02470.x

Daane KM, Caltagirone LE (1989) Biological control of black scale in olives. Calif Agric 43(1):9–11

Desneux N, Wajnberg E, Wyckhuys KAG, Burgio G, Arpaia S, Narvaez-Vasquez CA, Gonzalez-Cabrera J, Catalan Ruescas D, Tabone E, Frandon J, Pizzol J, Poncet C, Cabello T, Urbaneja A (2010) Biological invasion of European tomato crops by *Tuta absoluta*: ecology, geographic expansion and prospects for biological control. J Pest Sci 83:197–215. doi:10.1007/s10340-010-0321-6

Dukes JS, Mooney HA (2000) Does global change increase the success of biological invaders? Trends Ecol Evol 14:135–139. doi:org/10.1016/S0169-5347(98)01554-7

European Commission (2009a) White Paper from the European commission to the Council, the European Parliament, the European Economic and Social Committee and the Committee of the Regions – adapting to climate change: towards a European framework for action (COM(2009) 147/4)

European Commission (2009b) Regulation (EC) No 1107/2009 of the European Parliament and of the Council of 21 October 2009 concerning the placing of plant protection products on the market and repealing Council Directives 79/117/EEC and 91/414/EEC

European Commission (2009c) Directive 2009/128/EC of the European Parliament and of the Council of 21 October 2009 establishing a framework for Community action to achieve the sustainable use of pesticides

Fenton B, Woodford JAT, Malloch G (1998) Analysis of clonal diversity of the peach-potato aphid, *Myzus persicae* (Sulzer), in Scotland, UK and evidence for the existence of a predominant clone. Mol Ecol 7:1475–1487. doi:10.1046/j.1365-294x

Fernandes FR, de Albuquerque LC, Giordano LDB, Boiteux LS, de Avila AC, Inoue-Nagata AK (2008) Diversity and prevalence of Brazilian bipartite begomovirus species associated to tomatoes. Virus Gene 36:251–258. doi:10.1007/s11262-007-0184-y

Grunwald NJ, Garbelotto M, Goss EM, Heungens K, Prospero S (2012) Emergence of the sudden oak death pathogen *Phytophthora ramorum*. Trends Microbiol 20:131–138

Gutierrez AP, Ponti L, Cossu QA (2009) Effects of climate warming on olive and olive fly (*Bactrocera oleae* (Gmelin)) in California and Italy. Clim Change 95:195–217. doi:10.1007/s10584-008-9528-4

Hatfield JL, Boote KJ, Kimball BA et al (2011) Climate impacts on agriculture: implications for crop production. Agron J 103:351–370. doi:10.2134/agronj2010.0303

Hovmøller MS, Yahyaoui AH, Milus EA, Justesen AF (2008) Rapid global spread of two aggressivestrainsofawheatrustfungus.MolEcol17:3818–3826.doi:10.1111/j.1365-294X.2008.03886.x

Hovmøller MS, Walter S, Justesen AF (2010) Escalating threat of wheat rusts. Science 329:369. doi:10.1126/science.1194925

Hovmøller MS, Sørensen CK, Walter S, Justesen AF (2011) Diversity of *Puccinia striiformis* on cereals and grasses. Annu Rev Phytopathol 49:197–217. doi:10.1146/annurev-phyto-072910-095230

Huang Y-J, Pirie EJ, Evans N et al (2009) Quantitative resistance to symptomless growth of *Leptosphaeria maculans* (phoma stem canker) in *Brassica napus* (oilseed rape). Plant Pathol 58:314–323. doi:10.1111/j.1365-3059.2008.01957.x

Hutchinson GE (1957) Concluding remarks. Cold Spring Harbor Symp Quantum Biol 22:415–427. doi:10.1101/SQB.1957.022.01.039
Hyvönen T, Luoto M, Uotila P (2012) Assessment of weed establishment risk in a changing European climate. Agric Food Sci 21:348–360
IPCC (2007) Climate change, 73 pp. doi:http://www.ipcc.ch/ipccreports/ar4-syr.htm
Klerks MM, van Gent-Pelzer M, Franz E et al (2007) Physiological and molecular responses of *Lactuca sativa* to colonization by *Salmonella enterica* Serovar Dublin [down-pointing small open triangle]. Appl Environ Microbiol 73:4905–4914. doi:10.1128/AEM.02522-06
Licciardi S, Assogba-Komlan F, Sidick I et al (2008) A temporary tunnel screen as an eco-friendly method for small-scale growers to protect cabbage crop in Benin. Int J Trop Insect Sci 27:152–158. doi:org/10.1017/S1742758407883184
Liu D, Trumble JT (2007) Comparative fitness of invasive and native populations of the potato psyllid(*Bactericeracockerelli*).EntomolExpAppl123:35–42.doi:10.1111/j.1570-7458.2007.00521.x
Madgwick JW, West JS, White RP et al (2011) Impacts of climate change on wheat anthesis and Fusarium ear blight in the UK. J Plant Pathol 129:117–131. doi:10.1007/s10658-010-9739-1
Malcolm JR, Markham A, Neilson RP, Garaci M (2002) Estimated migration rates under scenarios of global climate change. J Biogeogr 29:835–849. doi:10.1046/j.1365-2699.2002.00702.x
Maron JL, Vilà M, Bommarco R, Elmendorf S, Beardsley P (2004) Rapid evolution of an invasive plant. Ecol Monogr 74:261–280. doi:org/10.1890/03-4027
McDonald A, Riha S, Ditommaso A, Degaetano A (2009) Climate change and the geography of weed damage: analysis of US maize systems suggests the potential for significant range transformations. Agric Ecosyst Environ 130:131–140. doi:org/10.1016/j.agee.2008.12.007
Milus EA, Kristensen K, Hovmøller MS (2009) Evidence for increased aggressiveness in a recent widespread strain of *Puccinia striiformis* f. sp. *tritici* causing stripe rust of wheat. Phytopathology 99:89–94. doi:10.1094/PHYTO-99-1-0089
Oerke EC (2006) Crop losses to pests. J Agric Sci 144:31–43. doi:10.1017/S0021859605005708
Parmesan C (2006) Ecological and evolutionary response to recent climate change. Annu Rev Ecol Evol Syst 37:637–669. doi:10.1146/annurev.ecolsys.37.091305.110100
Radcliffe EB, Ragsdale DW (2002) Aphid-transmitted potato viruses: the importance of understanding vector biology. Am J Potato Res 79:353–386. doi:10.1007/BF02870173
Rejmanek M (1996) A theory of seed plant invasiveness: the first sketch. Biol Conserv 78:171–181. doi:org/10.1016/0006-3207(96)00026-2
Samietz S, Stoeckli S, Hirschi M, Spirig C, Höhn H, Calanca P, Rotach MW (2015) Modelling the impact of climate change on sustainable management of the codling moth (*Cydia pomonella* L.) as key pest in apple. Acta Hortic 1068:35–42
Sarfraz M, Keddie AB, Dosdall LM (2005) Biological control of the diamondback moth, *Plutella xylostella*: a review. Biocontrol Sci Technol 15:763–789. doi:10.1080/09583150500136956
Scherm H (2004) Climate change: can we predict the impacts on plant pathology and pest management? Can J Plant Pathol 26:267–273. doi:10.1080/07060660409507143
Shaw MW, Osborne TM (2011) Geographic distribution of plant pathogens in response to climate change. Plant Pathol 60:31–43. doi:10.1111/j.1365-3059.2010.02407.x
Slayton RB, Turabelidze G, Bennett SD, Schwensohn CA, Yaffee AQ et al (2013) Outbreak of Shiga toxin-producing *Escherichia coli* (STEC) O157:H7 associated with Romaine Lettuce Consumption, 2011. PLoS One 8(2):e55300
Stephens AEA, Kriticos DJ, Leriche A (2007) The current and future potential geographical distribution of the oriental fruit fly, Bactrocera dorsalis (Diptera: *Tephritidae*). Bull Entomol Res 97:369–378. doi:10.1007/978-3-642-27281-3_30
Stoeckli S, Hirschi M, Spirig C, Calanca P, Rotach MW (2012) Impact of climate change on voltinism and prospective diapause induction of a global pest insect – Cydia pomonella (L.). PLoS One 7:e35723. doi:10.1371/journal.pone.0035723
Stratonovitch P, Storkey J, Semenov MA (2012) A process-based approach to modelling impacts of climate change on the damage niche of an agricultural weed. Glob Chang Biol 18:2071–2080. doi:10.1111/j.1365-2486.2012.02650.x

Tahiri A, Sekkat A, Bennani A, Granier M, Delvare G, Peterschmitt M (2006) Distribution of tomato-infecting begomoviruses and *Bemisia tabaci* biotypes in Morocco. Ann Appl Biol 149:175–186. doi:10.1111/j.1744-7348.2006.00083.x

Thuiller W, Albert C, Araújo MB, Berry PM, Cabeza M, Guisan A, Hickler T, Midgley GF, Paterson J, Schurr FM, Sykes MT, Zimmermann NE (2008) Predicting global change impacts on plant species' distributions: future challenges. Perspect Plant Ecol Evol Syst 9:137–152. doi:org/10.1016/j.ppees.2007.09.004

Trumble JT, Butler CD (2009) Climate change will exacerbate California's insect pest problems. Calif Agric 63:73–78

Van Overbeek LS, Van Doorn J, Wichers JH, Van Amerongen A, Van Roermond HJW, Willemsen PTJ (2014) The arable ecosystem as battleground for emergence of new human pathogens. Front Microbiol 5:1–17. doi:10.3389/fmicb.2014.00104

Vayssieres JF, Sinzogan A, Korie S, Ouagoussounon I, Thomas-Odjo A (2009) Effectiveness of spinosad bait sprays (GF-120) in controlling mango-infesting fruit flies (Diptera: *Tephritidae*) in Benin. J Econ Entomol 102:515–521. doi:org/10.1603/029.102.0208

Walther GR, Post E, Convey P, Menzel A, Parmesan C, Beebee TJC, Fromentin JM, Hoegh-Guldberg O, Bairlein F (2002) Ecological responses to recent climate change. Nature 416:389–395. doi:10.1038/416389a

West JS, Townsend JA, Stevens M, Fitt BDL (2012) Comparative biology of different plant pathogens to estimate effects of climate change of crop diseases in Europe. Eur J Plant Pathol 133:315–331. doi:10.1007/s10658-011-9932-x

Willmer P, Stone G, Johnston I (2004) Environmental physiology of animals, 2nd edn. Wiley-Blackwell, Oxford, 768 pp

Woodford JAT (1998) The impact of cropping policy on methods to control potato leafroll virus. Aspect Appl Biol 17:163–171

Zappalà L, Biondi A, Alma A, Al-Jboory IJ, Arnò J, Bayram A, Chailleux A, El-Arnaouty A, Gerling D, Guenaoui Y, Shaltiel-Harpaz L, Siscaro G, Stavrinides M, Tavella L, Vercher Aznar R, Urbaneja A, Desneux N (2013) Natural enemies of the South American moth, *Tuta absoluta*, in Europe, North Africa and Middle-East, and their potential use in pest control strategies. J Pest Sci 86:635–647. doi:10.1007/s10340-013-0531-9

Zhou XL, Harrington R, Woiwod IP, Perry JN, Bale JS, Clark SJ (1995) Effects of temperature on aphid phenology. Glob Chang Biol 1:303–313. doi:10.1111/j.1365-2486.1995.tb00029.x

Ziska LH (2010) Elevated carbon dioxide alters chemical management of Canada thistle in no-till soybean. Field Crop Res 199:299–303. doi:org/10.1016/j.fcr.2010.07.018

Ziska LH (2011) Global climate change and carbon dioxide: assessing weed biology and management. In: Hillel D (ed) Handbook on climate change and agroecosystems: impacts, adaptation, and mitigation. Imperial College Press, London, pp 191–208

Chapter 2
Phosphorus Dynamics and Management in Forage Systems with Cow-Calf Operation

Gilbert C. Sigua

Abstract Phosphorus fertilization is a vital component of productive farming. Phosphorus is an essential macronutrient that is required to meet global food requirements and make crop and livestock production profitable. While adequate levels of phosphorus in the soil are essential to grow crops, phosphorus has the potential to induce eutrophication in our water systems. Controlling phosphorus inputs is the thus considered the key to reducing eutrophication and managing ecological integrity. Forage-based cow-calf operations may have detrimental impacts on the chemical status of groundwater and streams and consequently on the ecological and environmental status of surrounding ecosystems. Relatively, little information exists regarding possible magnitudes of phosphorus losses from grazed pastures. Whether or not phosphorus losses from grazed pastures are significantly greater than background losses and how these losses are affected by soil, forage management, or stocking density are not well understood. The goal of this paper is to demonstrate the various effects of differing pasture fertility, animal behavior, and grazing management systems on the levels and changes of soil P in subtropical beef cattle pastures that will improve our understanding of P dynamics, cycling, and management in the agroecosystem. From our Florida experience perspectives, the following critical results are worthwhile mentioning: (1) environmentally, soil phosphorus in Florida pastures are declining; (2) soil phosphorus in pasture fields with no phosphorus fertilization were consistently lower than those of the fertilized fields by about 49.1 % to 40.9 % from 1988 to 2000, respectively; (3) soil phosphorus concentrations in 1988 of about 94.1 mg kg^{-1} and in 2000 of about 69.2 mg kg^{-1} were not high enough to be of environmental concern, so annual additions of phosphorus-fertilizer would be still practical to sustain plant and animal productivity in subtropical beef cattle pastures; (4) congregation zones in pastures with beef cattle operations in three regions of Florida are not phosphorus-rich, therefore may not contribute more phosphorus to surface and groundwater supply; and (5) slope aspect and slope position could be of relative importance in controlling spatial distribution of soil phosphorus. Effective use and cycling of phosphorus therefore is critical for pasture productivity and environmental stability in subtropics. This will

G.C. Sigua (✉)
USDA-Agricultural Research Service, Coastal Plains Soil, Water and Plant Research Center, Florence, SC 29501, USA
e-mail: gilbert.sigua@ars.usda.gov

E. Lichtfouse (ed.), *Sustainable Agriculture Reviews*, Sustainable Agriculture Reviews 17, DOI 10.1007/978-3-319-16742-8_2

help to renew the focus on improving inorganic fertilizer efficiency in subtropical beef cattle systems, and maintaining a balance of phosphorus removed to phosphorus added to ensure healthy forage growth and minimize phosphorus runoff. Additionally, if the overall goal is to reduce phosphorus losses from animal-based agriculture then there is a crucial need to balance off-farm phosphorus inputs in feed and fertilizer with outputs to the environment. Consequently, this paper has provided fundamental information on the source and transport control strategies that can provide the basis to increase phosphorus efficiency in agroecosystem with cow-calf operation.

Keywords Phosphorus dynamics • Phosphorus management • Cow-calf operation • Forage-based pasture • Subtropics • Phosphorus cycling • Eutrophication

2.1 Introduction

Agroecosystems, including crop, pasture, and rangelands constitute over 70 % of the continental United States and thereby play dominant roles in the management of the nation's watersheds and water resources. Pasturelands in the eastern USA are diverse as consequence of climatic conditions, previous land use, landscape position, dominant soil types, and management inputs. Grassland pastures are often consigned to locations in the landscape where more profitable horticultural and field crops cannot be grown. Unfortunately, a large portion of pastures in the eastern USA are relatively unmanaged or poorly managed to achieve full production potential (Franzluebbers et al. 2012).

South and central Florida encompasses a number of nationally and internationally renowned aquatic and semi-aquatic sensitive ecosystems, including Lake Okeechobee and the Everglades. The majority of Florida's cow herd is located in the Kissimmee/Lake Okeechobee watershed, where the public is becoming more concerned about high levels of phosphorus and nitrogen entering the lake and subsequently flowing out into the Everglades. An additional newer concern is the effect of phosphorus being transported from the interior by man-made waterways that lead directly to coastal areas where nutrients promote red tide and impact off-shore reefs. Because the land north of the lake is flat and poorly drained, most of the runoff occurs when the water table is close to the surface during the rainy season. The perception is that runoff, and therefore phosphorus discharge, is high during this time because rainfall is in excess of evapotranspiration and the sandy soils have low phosphorus retention capacity. Society relies on adequate freshwater resources to support populations, agriculture, industry, wildlife habitat, aquatic ecosystems, and a healthy environment.

Forage-based cow-calf operation shown in Fig. 2.1 has been suggested as one of the major sources of non-point source of nutrients, especially phosphorus and nitrogen, which are contributing to the degradation of water quality in lakes, reservoirs, rivers, and aquifers in south Florida. A report published by Shigaki et al. (2006)

Fig. 2.1 Typical forage-based cow-calf operation. Forage-based livestock systems have been blamed as one of the major sources of non-point source of phosphorus and nitrogen pollution that are contributing to the degradation of water quality in lakes, reservoirs, rivers, and ground water aquifers in Florida

claimed that beef cattle and stored cattle manure are responsible for 6 % of U.S. methane emissions is quite intriguing. There is a need to establish baseline information concerning the impact of pasture-based cattle production systems in this unique subtropical region on environmental factors associated with movement of nutrients, the ability of such systems to accumulate and store carbon, and the potential of such systems to incorporate urban and municipal waste materials such as lake sediments from dredging into pastures. Developing and implementing practices that can reduce the potential climate impact of beef cattle production is therefore equally important. Forage-beef cattle research programs must adopt an integrated approach that will lead to the development of appropriate sustainable pasture technologies that optimize beef cattle ranching sustainability and profitability. Livestock producers in the southeastern USA rely heavily on forages and pastures to reduce production costs and remain competitive. Thus, both actual and perceived environmental problems associated with beef cattle production systems need to be addressed when new management systems are being developed.

Integration of environmental protection and sustainability requirements into sectoral policies is a key element for successful socio-economic development as well as for improvement and implementation of environmental policy. It is important to find a balance between the use of land and natural resources for agricultural production and society's needs and values relating to the protection of the environment. Sustainable agriculture calls for natural resources to be managed in a way that ensures benefits are also available in the future. It takes into account the preservation of the overall balance and value of the natural capital stock and the need for agriculture to be viable and competitive. The nature of land management and land use are important issues within sustainable agriculture. A crucial key issue in order to minimize environmental impacts and enhance sustainability of beef cattle operations is to evaluate how different livestock management practices would affect the environment, including water quality, flora and fauna specie richness, and soil and landscape integrity.

Another equally important issue is concerning the balance of fertility management for forage-livestock agro-ecosystem that may result in increased nutrient use efficiency and, with less likelihood of nutrient loss to the environment due to leaching and/or runoff. The development and application of an integrated research approach based on soil, hydrological, geomorphic, and ecological principles to provide a better, watershed-scale understanding of water availability, plant-animal genetic resources, and other issues within sustainable agriculture have been identified as significant research needs. A further research effort on optimizing forage-based cow-calf operations to improve pasture sustainability and water quality protection is warranted and economically indispensable. Protecting landscapes, habitats and biodiversity, including genetic resources for food and agriculture, is an essential consideration.

2.1.1 Grass-Based Pasture System, Environmental Quality and Natural Resources

Beef cattle graze more than five million hectares of pasturelands in the State of Florida. Such large herds naturally generate large quantities of manure and other waste. Because of this, forage-based livestock systems have been blamed as one of the major sources of non-point source of phosphorus and nitrogen pollution that are contributing to the degradation of water quality in lakes, reservoirs, rivers, and ground water aquifers in Florida (Allen et al. 1976, 1982; Bogges et al. 1995; Edwards et al. 2000) and other cattle-producing states. Despite widespread concern, limited data has been available to measure nutrient losses to adjacent bodies of water from pastures for grazing and hay production. Cattle manure contains appreciable amounts of nitrogen and phosphorus (0.6 and 0.2 %, respectively), and portions of these components can be transported into receiving waters during severe rainstorms (Khaleel et al. 1980). Work in other regions of the country has shown that when grazing animals become concentrated near water bodies, or when they have unrestricted long-term access to streams for watering, sediment and nutrient loading can be high (Thurow 1991; Brooks et al. 1997). Additionally, there is a heightened likelihood of phosphorus losses from over-fertilized pastures through surface water runoff or percolation past the root zone (Schmidt and Sturgul 1989; Gburek and Sharpley 1998; Stout et al. 2000). Reduction of phosphorus transport to receiving water bodies has been the primary focus of several studies because phosphorus has been found to be the limiting nutrient for eutrophication in many Florida aquatic systems (Botcher et al. 1999; Sigua et al. 2000, 2006a; Sigua and Steward 2000; Sigua and Tweedale 2003).

Relatively little information exists regarding possible magnitudes of nutrient losses from grazed pastures in subtropical Florida. Recently, Sigua et al. (2004, 2005b, 2006b) found that the levels of soil phosphorus varied widely with different pasture management. The soil phosphorus levels of phosphorus-fertilized rhizoma peanut (*Arachis glabrata*) pastures that were grazed in spring and hayed in early fall were higher than bahiagrass (*Paspalum notatum*) pastures without phosphorus fertilizer that were grazed all year long. The soil test value of phosphorus' in grazed bahiagrass was about 23 % higher than that of grazed and hayed bahiagrass, suggesting that grazing followed by haying could have lowered the level of soil P. There was no spatial or temporal build up of soil P or other crop nutrients despite the annual application of fertilizers and daily in-field loading of animal waste. In fact, soil fertility levels showed a declining trend for crop nutrient levels, especially soil phosphorus ($y = 146.57 - 8.14 * year$; $r^2 = 0.75$) even though the fields had a history of phosphorus fertilization and the cattle were rotated into the legume fields. Their results further demonstrated that when nutrients are not applied in excess, cow-calf systems are slight exporters of phosphorus, potassium, calcium, and magnesium; removal of cut hay could increase export of nutrients. Water quality in lakes associated with cattle production was "good" (30–46 TSI) based upon the Florida Water Quality Standard. These findings indicate that properly managed livestock operations may not be major contributors to excess loads of nutrients (especially phosphorus) in surface water (Sigua et al. 2006a). In another study in south Florida, Arthington et al. (2003) reported that the presence of beef cattle at three stocking rates (1.5, 2.6, and 3.5 ha per cow) had no impact on nutrient loads (phosphorus and nitrogen) in surface runoff water compared with pastures containing no cattle.

Whether or not phosphorus losses from grazed pastures are significantly greater than background losses and how these losses are affected by soil, forage management, or stocking density are not well defined (Edwards et al. 2000; Sigua et al. 2004). Concern for losses of soil phosphorus by overland flow were noted when soil phosphorus exceeded 150 mg kg^{-1} in the upper 20-cm of soil (Johnson and Eckert 1995; Sharpley et al. 1996). This concentration of soil phosphorus should not be considered an absolute maximum number for soil phosphorus to have been harmful to water quality and the environment, but rather a good indicator of phosphorus accumulation in the soil. Sharpley (1997) noted that all soils do not contribute equally to phosphorus export from watersheds or have the same potential to transport phosphorus to runoff. In their studies, Coale and Olear (1996) observed that soil test phosphorus levels did not accurately predict total dissolved phosphorus. However, they noted that in all cases studied, soil test phosphorus levels were significantly related to an increase in total dissolved phosphorus.

In summary, a better understanding of soil phosphorus dynamics and other crop nutrient changes resulting from different management systems should allow us to better predict potential impact on adjacent surface waters. These issues are critical and of increasing importance among environmentalists, ranchers, and public officials in the state (Sigua et al. 2006a). Soil testing is currently the best management tool available to ensure that soils do not become overloaded with phosphorus, which increases the likelihood of their contribution to pollution in surface waters downstream (Norfleet et al. 1996; Sigua et al. 2006a).

2.1.2 Impact of Cattle Grazing Behavior on the Quality of Surface and Shallow Groundwater

For decades, understanding how nutrient resources vary across landscapes has been the focal point of much ecological, forest, and agricultural research (Watt 1947; Odum 1969; Foster et al. 1997; Canham et al. 2001; Sigua and Coleman 2010; Sigua et al. 2010, 2011b). Despite substantial measurements using both laboratory and field techniques, little is known about the spatial and temporal variability of phosphorus dynamics across the entire landscape, especially in agricultural landscapes with cow-calf operations (Fig. 2.2). Understanding cattle movement in pasture situations is critical to assess their impact on agro-ecosystems. Movement of free-ranging cattle varies due to spatial arrangement of forage resources (Senft et al. 1985), water (Holechek 1988; Ganskopp 2001), mineral feeders (Martin and Ward 1973), and shade (Sigua et al. 2011a). The breed of the animal also affects livestock distribution pattern.

Hammond and Olson (1994) and Bowers et al. (1995) reported that temperate British breeds (Angus and Hereford) of *Bos taurus* cows grazed less during the day than tropically adapted cows of the Senepol breeds, but compensated for reduced grazing activity during the hotter parts of the day by increasing time spent grazing at night. Grazing animals congregate close to the shade and watering areas during the warmer periods of the day (Mathews et al. 1994, 1999). White et al. (2001) claimed that there was a correlation between time spent in a particular area and the number of excretions and this behavior could lead to an increase in the concentration of soil nutrients close to shade and water. Sigua et al. (2005b) demonstrated that concentrations of total inorganic nitrogen, total phosphorus, and the degree of soil compaction varied significantly among different animal congregation sites. The highest concentrations of total inorganic nitrogen and total phosphorus were found at the shade and mineral feeder sites, respectively. Although the levels of total inorganic nitrogen and total phosphorus were high near the center of the congregation sites, their levels did not increase with soil depth and their concentrations decreased almost linearly away from the center of the congregation sites. This study suggests that congregation sites in beef cattle operations in Florida are not as nutrient rich as suspected, and because phosphorus levels at the center of sites were below 150 mg kg^{-1}, those sites may not

Fig. 2.2 Shallow ground water well is being sampled once a week to assess the impact of cattle grazing on the concentrations of nutrients in the ground water (USDA-ARS Research Station, Brooksville, FL)

contribute more nutrients to surface and groundwater supply under Florida conditions (Sigua 2010). Again, this concentration of soil phosphorus should not be considered an absolute maximum number for soil phosphorus to become harmful to water quality and the environment, but rather a good indicator of current phosphorus accumulation in the soil. Furthermore, since there was no evidence of a vertical build up or horizontal movement of total phosphorus in the landscape, Sigua et al. (2005a) surmised that cattle congregation sites may not be considered a substantial source of nutrients at the watershed level.

In summary, grazing animals can impact the movement and cycling of nutrients through the soil and plant system, and thus on fertility of pasture soils (Haynes 1981; Haynes and Williams 1993). Grazing can accelerate and alter the timing of nutrient transfers, and increase amounts of nutrients cycled from plants to soils (Klemmendson and Tiedemann 1995). Areas in pastures where animals congregate such as mineral feeders, water troughs, and/or shaded/trees can be important point sources of nutrient pollution and are often perceived to have higher levels of soil nutrients compared with less frequented parts of the landscape (Sigua and Coleman 2007, 2010; Sigua et al. 2010). Spatial distribution and movement patterns of cattle are particularly valuable in allocating and assessing impacts of utilization on a given landscape. The arrangement of food, water, and shelter and their concurrent interactions with topographic features obviously influences the distribution of animals and their simultaneous use of pasture resources (Ganskopp 1987).

2.2 Cow-Calf Operation and Pasture Management in Florida

Eleven million ha of grazingland in the subtropical (23.5–30° N Lat) United States supports about 30 % of the U.S. beef cow herd. Florida's beef production ranks 10th among beef producing states in the United States and 4th nationally among states in number of herds with more than 500 brood cows. Florida's beef cattle had sales of more than $443 million in 2004. The majority of Florida's cow herd is located in the Kissimmee/Lake Okeechobee watershed, a place where the public is becoming more concerned about high levels of phosphorus entering the lake and subsequently flowing out into waterways and the Everglades.

Florida is a large state with a considerable variability in soils and climate. In north Florida, there are some clay-loam soils with good moisture-holding capacity that are quite productive (Chambliss 1999). Coming down the peninsula, soils are dominated by sandy ridges and flatwoods. In general, the flatwood soils with their higher moisture-holding capacity are more productive than the upland deep, droughty sands. The warm growing season is longer in south Florida than in north Florida while winter temperatures are usually lower in north Florida than in south Florida. These differences in climate, soils and length of growing season affect the types of forage that can be grown. Nevertheless, Florida's relatively mild climate, together with 127 plus cm of annual rainfall, affords a better opportunity for nearly 12 months of grazing than in any other state except Hawaii (Chambliss 1999).

Bahiagrass (*Paspalum notatum*) is a common pasture used for beef cattle across Florida. Fertility and management practices have been based on University of Florida's recommendations as described by Chambliss (1999). Pastures are being grazed during spring of the year. After the start of summer rainy season, pastures that are to be hayed are being dropped out of the grazing cycle (usually starting in July). Pasture fields with bahiagrass are normally fertilized in the spring with 90 kg N ha^{-1} and 45 kg K_2O ha^{-1}. Grazing cattle at the United States Department of Agriculture-Subtropical Agricultural Research Station in Brookville, FL and other Florida ranches are being rotated among pastures on a 3-day grazing interval with 24 days of rest between pastures. The average number of grazing cattle was about 2.9 animal units per hectare and grazing days of about 5.5 on monthly basis as shown in Table 2.1. In addition, the number of days grazed each month, average number of animals per hectare and estimated total feces excreted along with the estimated total phosphorus in feces and from urine are also shown in Table 2.1.

Table 2.1 Monthly summary of grazing activity and excreted phosphorus in pastures at Brooksville, FL

Months	Average days grazed per pasture	Average number of animals per ha	Animal unit per month[a]	Monthly manure P excreted[b]
January	13.8	2.6	1.0	1.73
February	9.4	2.5	0.8	1.27
March	13.5	2.1	0.9	1.43
April	12.0	2.0	0.7	1.18
May	12.7	2.1	0.7	1.24
June	12.0	2.4	0.9	1.42
July	6.9	3.3	0.7	1.19
August	9.1	3.6	0.8	1.39
September	8.2	4.8	1.1	1.80
October	6.7	5.3	1.1	1.85
November	6.6	3.6	0.8	1.34
December	9.4	3.7	1.2	1.97
Average	**10.0**	**3.2**		**1.48**
Total	–	–	**10.8**	**17.81**

[a]Animal Units per Month, AUM (450 kg cow/calf unit)
[b]Total Manure Excreted (kg as excreted) = [(Number of AUM* Total Annual Animal Manure Excretion/12)] Total Manure Excretion (as excreted) per animal per year = 10.4 metric tons (Kellogg et al. 2000)

2.3 Eutrophication Associated with Animal-Based Agriculture: Overview

Eutrophication is frequently a result of nutrient pollution, such as the release of sewage effluent, urban stormwater run-off, and run-off carrying excess fertilizers into natural waters. Nutrients such as phosphorus or nitrogen and other pollutants may enter from a number of sources (Fig. 2.3). However, it may also occur naturally in situations where nutrients accumulate such as depositional environments or where they flow into systems on an ephemeral basis. Eutrophication generally promotes excessive plant algal growth and decay, favors certain weedy species over others, and is likely to cause severe reductions in water quality (Schindler 1974).

In aquatic environments, enhanced growth of choking aquatic vegetation or phytoplankton such as algal bloom disrupts normal functioning of the ecosystem, causing a variety of problems such as a lack of oxygen in the water, needed for fish and shellfish to survive (Fig. 2.3). The water then becomes cloudy, colored a shade of green, yellow, brown, or red. Human society is impacted as well; eutrophication decreases the resource value of rivers, lakes, and estuaries such that recreation, fishing, hunting, and aesthetic enjoyment are hindered. Health-related problems can occur where eutrophic conditions interfere with drinking water. Many drinking

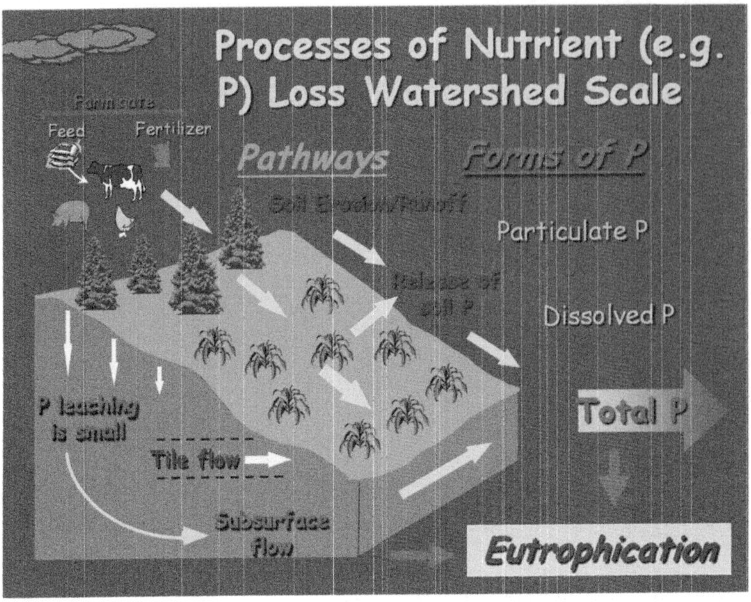

Fig. 2.3 Different processes describing nutrient losses from farm gate at watershed scale causing eutrophication (Source: Sigua 2010: 631–648)

water supplies throughout the world may experience periodic massive surface blooms of *cyanobacteria* (Kotak et al. 1993). These blooms contribute to a wide range of water-related problems including summer fish kills, and unpalatability of drinking water (Palmstrom et al. 1988).

2.4 Case Studies: Effects of Pasture and Grazing Management on Soil Phosphorus Dynamics and Management (Florida Experiences)

2.4.1 Case Study #1: Levels and Changes of Soil Phosphorus in Subtropical Beef Cattle Pastures with Bahiagrass and Rhizoma Peanuts

2.4.1.1 Study Sites and Experimental Methods

The Subtropical Agricultural Research Station (**STARS**) is a cooperative research unit of the United States Department of Agriculture-Agricultural Research Service (USDA-ARS) and the University of Florida and is located seven miles north of Brooksville, Florida. The station has three major pasture units with combined total area of about 1,538 ha with 1,295 ha in permanent pastures. Cattle used for nutritional, reproductive, and genetic research on the station include about 500 head of breeding females with a total inventory of about 1,000 head of cows, calves, and bulls. Most of the soils at STARS can be described as well-drained, loamy, siliceous hyperthermic family of the Grossarenic Paleudults. Forage production potential of the soils in the station is generally low to medium; the main limitation being droughtiness.

Soil analyses were conducted in 1988, 1997, and 2000. Soil samples were collected using a steel bucket type hand auger to a depth of 25 cm. Soil samples were air-dried and passed through a 2-mm mesh sieve prior to chemical extraction of soil phosphorus and other soil chemical properties. Soil chemical analyses were conducted at the University of Florida-Institute of Food and Agricultural Sciences Soil Testing Laboratory, Gainesville, Florida. Soil phosphorus was analyzed following the procedures outlined in Mehlich 1 (0.05 $\underline{N}$ HCl in 0.025 $\underline{N}$ H_2SO_4) method.

The objective of this study was to investigate the long term effect of pasture management (grazing alone vs. grazing + haying) on soil phosphorus dynamics in subtropical beef cattle pastures with bahiagrass and rhizoma peanuts (*Arachis glabrata*) with phosphorus or without phosphorus fertilization in Brooksville, Florida from 1988 to 2000.

2.4.1.2 Research Highlights

The levels of soil phosphorus varied widely and significantly ($p \leq 0.001$) among the different pasture management combinations in Main Station and Turnley pasture units (Fig. 2.4). The soil phosphorus levels of pastures in Main Station (averaged across years) with rhizoma peanuts that were grazed in spring and hayed (116.2 ± 42.1 mg kg^{-1}) in early fall were higher than those bahiagrass pastures (63.3 ± 17.7 mg kg^{-1}) that were grazed all year long. The soil test value of phosphorus in grazed bahiagrass pasture was about 23 % higher than that of bahiagrass pasture with grazing and haying, suggesting that grazing followed by haying of bahiagrass could have lowered the level of soil phosphorus in Turnley unit (Fig. 2.4). However, the average soil phosphorus levels did not vary significantly across location. The average levels of soil phosphorus from in Main Station and Turnley unit with grazed bahiagrass pasture were 63.3 ± 17.7 mg kg^{-1} and 71.7 ± 14.5 mg kg^{-1}, respectively. The average soil phosphorus levels in Main Station with grazed and hayed rhizoma peanut pastures was 116.2 ± 42.1 mg kg^{-1} compared with soil phosphorus levels in Turnley unit with grazed and hayed rhizoma peanut pasture of 123.3 ± 26.7 mg kg^{-1}.

The levels of soil phosphorus in Subtropical Agricultural Research Station beef cattle pastures with or without phosphorus fertilization from 1988 to 2000 are shown in Table 2.2. The levels of soil phosphorus between the fertilized and the unfertilized pastures were statistically different from each other in 1988 ($p \leq 0.001$), 1997 ($p \leq 0.0001$), and in ± 40.9, 121.8 ± 33.9, and 95.2 ± 19.9 mg kg^{-1} compared with soil phosphorus levels from pasture fields with no phosphorus fertilization of 71.2 ± 15.5, 60.3 ± 11.6, and 56.2 ± 19.1 in 1988, 1997, and 2000,

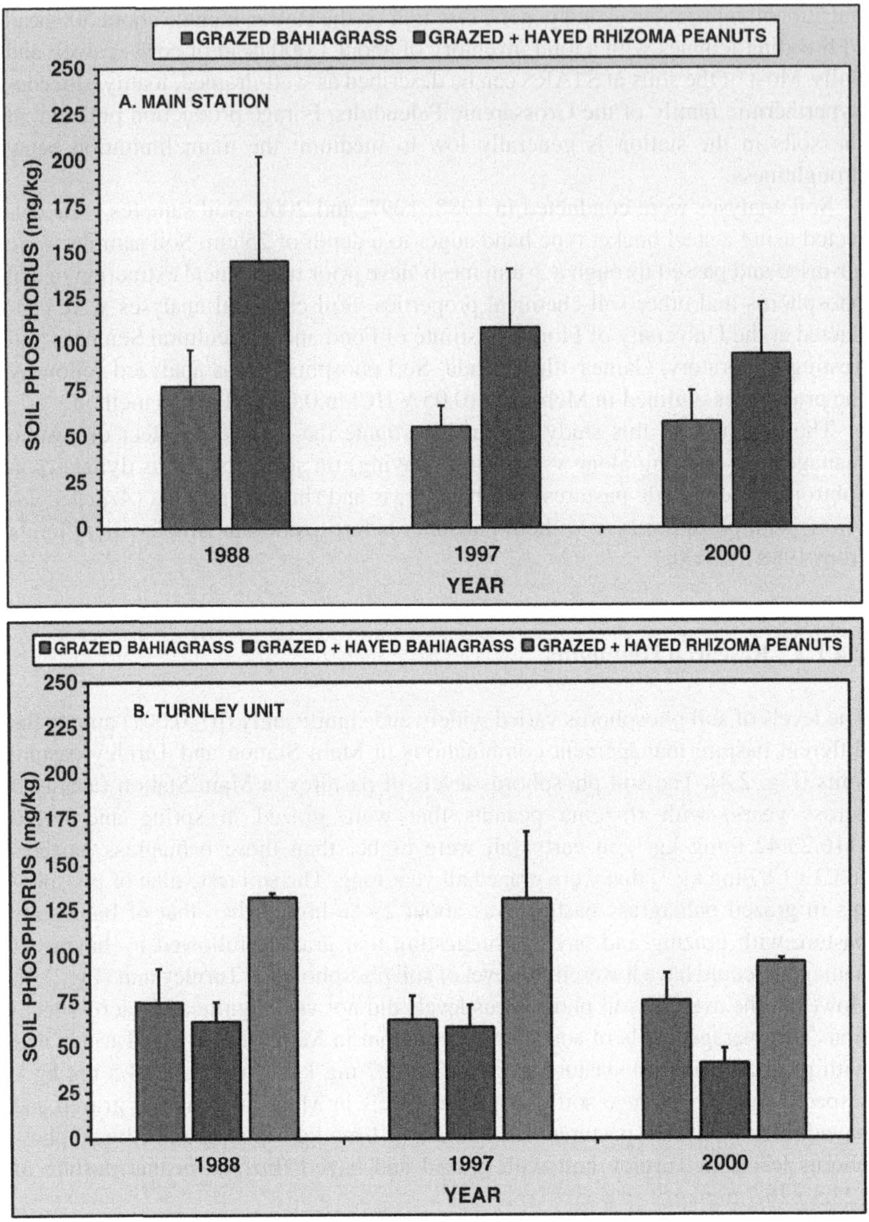

Fig. 2.4 Comparative levels of soil phosphorus between pasture locations under different pasture management such as grazed bahiagrass, grazed + hayed bahiagrass and grazed + hayed rhizoma peanuts in Brooksville, Florida (Source: Sigua et al. 2004: 975–990)

Table 2.2 Levels of soil phosphorus in STARS beef cattle pastures with or without phosphorus fertilization from 1988 to 2000

Treatment	1988	1997	2000	Mean
With P fertilizer	140.0 ± 40.9 a*	121.8 ± 33.9 a	95.2 ± 19.9a	119.0
No P fertilizer	71.2 ± 15.5 b	60.3 ± 11.6 b	56.2 ± 19.1b	62.6
Mean	105.6	91.1	75.7	

*Means on each column followed by common letter are not significantly different from each other at $p \leq 0.05$

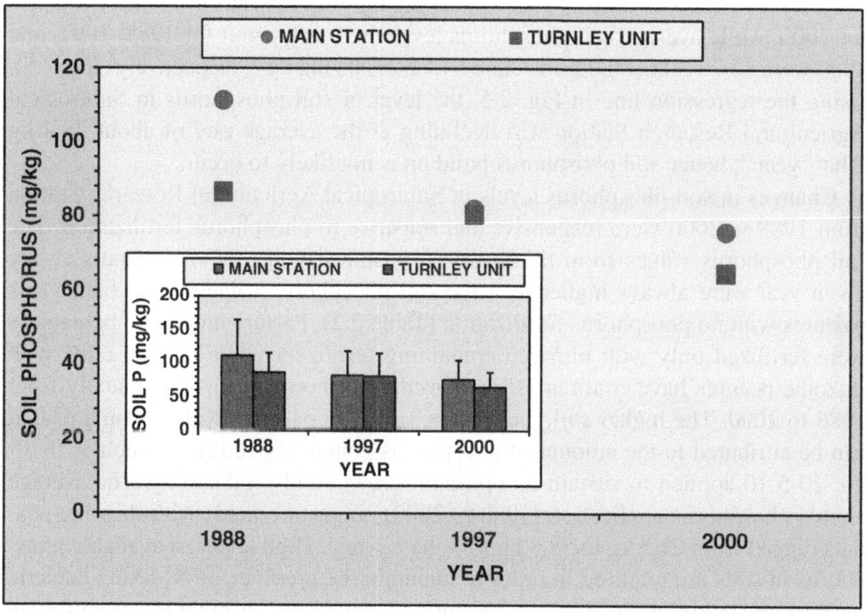

Fig. 2.5 Changes in soil phosphorus levels from 1988 to 2000 from two pasture locations (Main Station and Turnley unit) under different pasture management (Source: Sigua et al. 2006b: 115–134)

respectively. Levels of soil phosphorus in pasture fields with no phosphorus fertilization were consistently lower than those of the fertilized fields by about 49.1 %, 50.5 %, and 40.9 % in 1988, 1997, and 2000, respectively (Table 2.2). The average of soil phosphorus across years from the fertilized fields of 119.0 ± 4.9 mg kg^{-1} was significantly higher than those pasture fields with no phosphorus fertilization (62.8 ± 7.8 mg kg^{-1}).

During the last 12 years (1988–2000), soil test values for phosphorus in Main Station and Turnley unit have declined by about 17.6 and 10.8 mg kg^{-1} $year^{-1}$, respectively (Fig. 2.5). The regression models that best describe the changes and/or depletion rate of soil phosphorus in Main Station (MS) and Turnley Unit (TY) are given in Eqs. 2.1 and 2.2,

$$\textbf{Soil P}_{\textbf{MS}} = \mathbf{-17.6t + 125.1} \qquad R^2 = \mathbf{0.87}^{**} \qquad \mathbf{p \le 0.001} \tag{2.1}$$

$$\textbf{Soil P}_{\textbf{TY}} = \mathbf{-10.8t + 98.9} \qquad R^2 = \mathbf{0.94}^{**} \qquad \mathbf{p \le 0.001} \tag{2.2}$$

in which **Soil PMS** and **Soil PTY** = soil P depletion (mg kg^{-1} $year^{-1}$) in Main Station (MS) and Turnley unit (TY) and **t** = time (year). The average level of soil P in MS for 1988 was 111.3 ± 53.3 mg kg^{-1}, 82.0 ± 36.5 mg kg^{-1} in 1997, and 76.0 ± 28.8 mg kg^{-1} for 2000, while average soil phosphorus levels in Turnley unit for 1988, 1997, and 2000 were 82.7 ± 30.6, 80.0 ± 38.3, and 64.7 ± 25.8.5 mg kg^{-1}, respectively (Fig. 2.5). Using the regression line in Fig. 2.5, the level of soil phosphorus in Subtropical Agricultural Research Station was declining at the average rate of about 28.4 kg P ha^{-1} $year^{-1}$, hence soil phosphorus build up is not likely to occur.

Changes in soil phosphorus levels in Subtropical Agricultural Research Station from 1988 to 2000 were responsive and sensitive to phosphorus fertilization. The soil phosphorus values from the fertilized pastures with rhizome peanuts at any given year were always higher than the soil phosphorus values in the bahiagrass pastures with no phosphorus fertilization (Table 2.2). Pasture units with bahiagrass were fertilized only with nitrogen containing fertilizers while pasture units with rhizome peanuts have continuously received phosphorus fertilizers annually from 1988 to 2000. The higher soil phosphorus values in pastures with rhizoma peanut can be attributed to the amount of phosphorus-containing fertilizers such as 0-10-20; 20-5-10 applied to sustain its optimum growth and productivity. The average annual phosphorus application (1988 to 2000) on pasture fields with rhizome peanuts ranged from 22.5 kg to 38.8 kg P_2O_5 ha^{-1} $year^{-1}$. High levels of available phosphorus in soils are required in order to maintain the presence of N_2-fixing bacteria in pasture fields with rhizoma peanuts. Conversely, bahiagrass pastures in Subtropical Agricultural Research Station with no phosphorus-fertilizer application received 225 kg ha^{-1} of NH_4NO_3 or about 79 kg N ha^{-1} annually. Adjie (1999) claimed that N is the most limiting nutrient to warm-season grass production in Florida. Blue (1970) reported that oven-dry forage yields of bahiagrass were 3–4 Mg ha^{-1} without applied nitrogen and 12 Mg ha^{-1} or more for 224 kg ha^{-1}.

2.4.1.3 Conclusion

Results of this study has brought up a renewed focus on substantially improving the fertilizer efficiency in subtropical beef pasture, so that we can maximize benefits from every unit of fertilizer applied in the soil. Soil testing program in the station should continue to measure the amount of soil phosphorus that is proportional to what is available in the field and also looking at alternative soil phosphorus test that are better predictor of the loss and/or build up of total and dissolved phosphorus to soil and water systems.

2.4.2 *Case Study #2: Spatial Distribution of Soil P and Herbage Mass in Beef Cattle Pastures: Effects of Slope Aspect and Slope Position*

2.4.2.1 Study Site and Experimental Methods

This study (2004–2006) was conducted at the Land Use Unit (28°37′22.8″–28°37′38.2″N; 82°20′07.7″–82°20′31.1″W) of the Subtropical Agricultural Research Station, United States Department of Agriculture-Agricultural Research Service located seven miles north of Brooksville, FL. The research station has three major pasture units with a combined total area of about 1,558 ha with 1,295 ha in permanent pastures. Cattle used for nutritional, reproductive, and genetic research on the station include about 500 heads of breeding females with a total inventory of about 1,000 head of cows, calves, and bulls. Cattle production at the station is forage-based with bahiagrass the predominant forage species (approx. 1,000 ha).

At the beginning of 1990s, bahiagrass pastures were fertilized annually in the early spring (March) with 77 kg N, 10 kg P and 37 kg K ha^{-1} based on the revised fertilizer recommendation suggested by Chambliss (1999). Historically, grazing cattle were rotated among pastures to allow rest periods of 2–4 weeks based on herbage mass. The timing of movement for rotationally grazed cattle was determined by the herd manager's perception of herbage mass based on plant height and not based on pasture measurement. Starting in 2000, cattle were rotated twice weekly (3- or 4-days grazing period). We anticipated 24 days of rest between pastures, but herd numbers required more frequent grazing periods. During this study, the average number of 3.2 cow-calf pair ha^{-1} grazed each pasture for about 10 day each month. This cattle grazing rotation yielded an average annual total manure excretion of about 9,347 kg ha^{-1} or about 17.8 kg ha^{-1} of total phosphorus from manure excretions (Table 2.1).

Soil samples were collected at 0–20 and 20–40 cm soil depth on contiguous south-, north-, east- and west-facing slopes across different landscape positions (top slope, 10–30 %; middle slope, 5–10 % and bottom slope, 0–5 %) of 100 ha bahiagrass-based pastures during four summer seasons of 2003–2006. A total of 72 soil samples (year = 4; slope position = 3; soil depth = 2; number of replication = 3) were collected and analyzed from each slope aspect (east, north, south, and west). Soil samples were air-dried and passed through a 2-mm mesh sieve prior to chemical extraction of soil total P. Soil total P was extracted with double acid (0.025 *N* H_2SO_4 + 0.05 *N* HCl) as described by Mehlich (1953) and analyzed using an inductively coupled plasma spectrophotometer (ICP). The degree of soil saturation with P (DPS) as described in Eq. 2.3 was computed using the P, iron, and aluminum contents (mg kg^{-1}) of the soil (Hooda et al. 2000).

$$\mathrm{DPS}(\%)=\left(\left[\mathrm{P}\right]\times 100\right)/\left[\mathrm{Fe}+\mathrm{Al}\right] \quad (2.3)$$

The reason for conducting this study was to determine the effects of different slope aspects (North-, South-, West- and East-facing slope) and slope positions

(Top, Middle and Bottom) typical on most forage-based pastures in subtropics on spatial distribution of soil P and herbage mass of forages.

2.4.2.2 Research Highlights

The concentrations of phosphorus in soils varied among slope aspect ($p \leq 0.0001$), interaction effect of year and slope aspect ($p \leq 0.001$), interaction of slope aspect and slope position ($p \leq 0.01$), and among years ($p \leq 0.0001$) (Fig. 2.6). Degree of phosphorus saturation in soils was affected by soil depth ($p \leq 0.0001$), slope aspect ($p \leq 0.001$) and year ($p \leq 0.0001$). Although results were not significantly different, the upper soil depth (0–20 cm) had the highest concentrations of soil phosphorus. The lowest amount of soil phosphorus was found in the lowest (20–40 cm) soil depth. The north and south started with much higher phosphorus and decline more rapidly than that of the east- and south-facing slopes. Average concentrations of soil phosphorus in top slope, middle slope and bottom slope were all statistically comparable. Soil phosphorus declined rapidly over the years (Fig. 2.7) and the decline was different among slope aspects ($p \leq 0.01$). The decline of soil phosphorus over time (2003–2006) could be associated with decreasing of precipitation in 2005 and 2006. The overall average and yearly average of phosphorus saturation in soils were consistently the same across slope positions, but soils collected between soil depth of 0 and 20 cm had significantly higher degree ($p \leq 0.0001$) of phosphorus

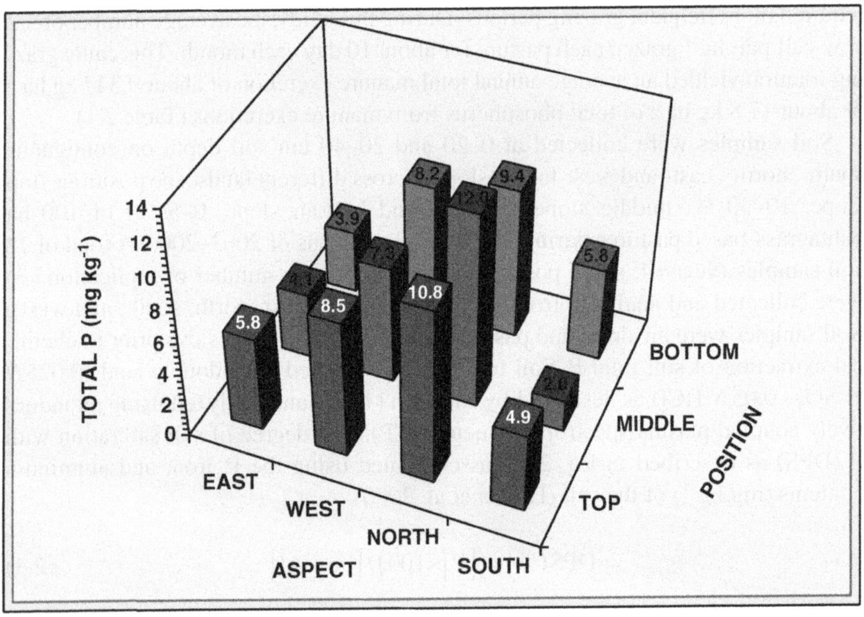

Fig. 2.6 Interaction effect of slope aspect and slope position on the levels of soil phosphorus in forage-based pasture with cow-calf operation (Source: Sigua and Coleman 2010: 240–247)

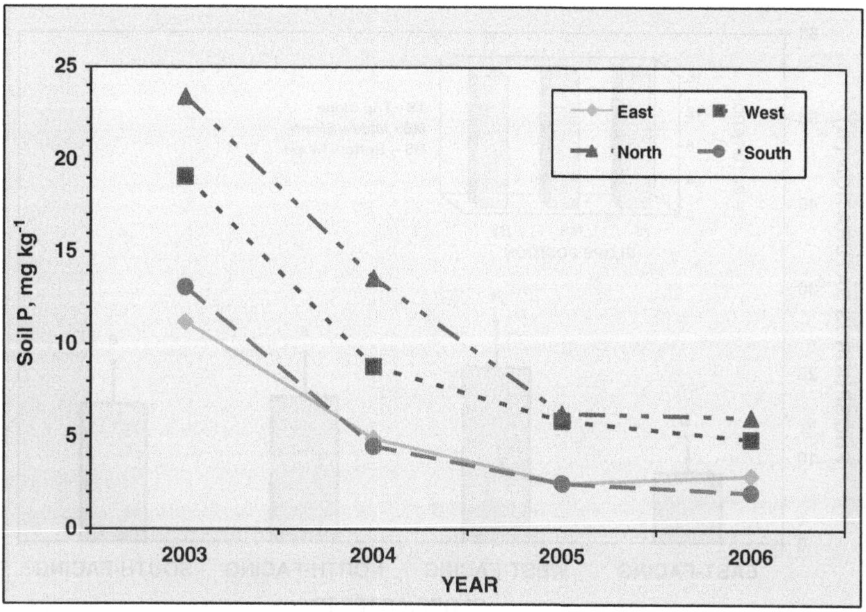

Fig. 2.7 Temporal (2003–2006) distribution of soil phosphorus in forage-based pastures at different slope aspects (Source: Sigua et al. 2011a: 59–70)

saturation than soils collected between 20 and 40 cm. Degree of phosphorus saturation (averaged across years) was similar in soils taken from the west-, north- and south-facing slopes, but their values were significantly higher than that of soils taken from the east-facing slope (Fig. 2.8). Unlike soil P, the degree of P saturation in soils was not affected by slope position.

The average level of soil phosphorus in the mineral feeders (averaged across distance) of 34.05 ± 0.44 mg kg^{-1} was not high enough to be of environmental concern. Losses of soil phosphorus by overland flow can become a big concern when the concentrations for soil phosphorus exceeded 150 mg kg^{-1} in the upper 20 cm of soils.

Overall, there was a very low net gain of soil phosphorus at any of the slope aspect (8.22 mg kg^{-1}) and slope position (8.20 mg kg^{-1}), but there had been no movement of soil phosphorus into the soil pedon since the average degree of phosphorus saturation in the upper 20 cm was 34 % while the average degree of P saturation is soils at 20–40 cm was about 11 %. Several studies (Heckranth et al. 1995; Hooda et al. 2000) have found that degree of phosphorus saturation in soils needs to exceed 45–60 % before dissolved reactive phosphorus becomes an environmental problem. Our results do not even approach this level of degree of phosphorus saturation, suggesting that phosphorus build-up and release is not a predicament at any slope aspect or slope position (Fig. 2.8).

Distribution and movement patterns of cattle are particularly valuable in allocating and assessing impacts of utilization on a given pasture. Our soil test values for

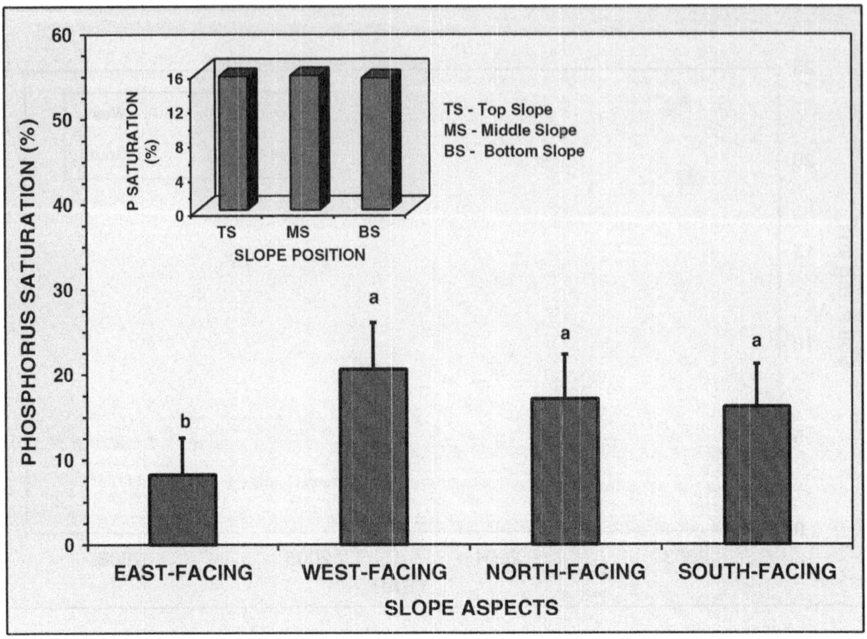

Fig. 2.8 Degree of soil phosphorus saturation at different slope aspects and slope positions of forage-based pastures with cow-calf operation (Source: Sigua et al., 2011a: 59–70)

phosphorus in east-, west-, north- and south-facing slope of 5.3, 9.7, 12.4 and 5.5 mg kg^{-1} respectively, were below the desired P levels for sustainable production of bahiagrass. Effective use and cycling of phosphorus is critical for pasture productivity and environmental stability. Periodic applications of additional phosphorus may be necessary to sustain agronomic needs and to offset the export of phosphorus due to animal production. Addition of organic amendments could represent an important strategy to protect pasture lands from excessive soil resources exploitation.

2.4.2.3 Conclusion

Both slope aspect and slope position had significant effects in controlling spatial distribution of soil P and herbage mass. Animals (cow-calf) tended to graze more at the bottom slope than in the middle slope or the top slope of the pastures and this non-uniform grazing distribution by livestock on landscapes can caused by many variables such as water location, minerals, herbage mass and raggedness of the terrain, which exist at a variety of scales. Since this study had produced contrasting results, additional field studies that were designed to determine the effects of different slope aspects and slope positions on most forage-based pastures on spatial distribution of soil phosphorus and other nutrients are still needed and warranted.

2.4.3 Case Study #3-Distribution of Soil Phosphorus: Effect of Animal Congregation Sites

2.4.3.1 Study Site and Experimental Methods

The study sites are located in the Turnley Unit (28.58–28.62°N; 82.26–82.29°W) of the USDA, ARS, STARS in Brooksville, Florida near Nobleton, Florida. Soils (Candler fine sand) at this location can be described as well-drained hyperthermic uncoated typic quartzipsamments (Hyde et al. 1977). The unit is about 870 ha divided into 47 pasture fields. A total of 206 ha are planted with rhizoma peanut (*Arachis glabrata*) and bahiagrass. The remainder of the pastures is primarily bahiagrass (BG; *Paspalum notatum*) except for 16 ha devoted to small-plot forage research trials. Forage production potential of the soils in this pasture unit is generally low to medium, droughtiness being the main limitation. Baseline soil samples around the congregations sites (mineral feeders, water troughs, and shades) in established (>10 year), grazed beef cattle pastures were collected in the fall and spring of 2003, 2004, and 2005, respectively. Soil samples were collected (0–20 cm) at different locations around the congregation sites following a radial (every 90°) sampling pattern at 0.9, 1.7, 3.3, 6.7, 13.3, 26.7, and 53.3 m from the approximate center of mineral feeders, water troughs, and shaded areas.

The objective of this study was to test whether cattle congregation sites typical on most Florida ranches, such as mineral feeders, water troughs and shaded areas are more nutrient-rich and may contribute more nutrients to surface and groundwater supply than in other pasture locations under Florida conditions.

2.4.3.2 Research Highlights

The levels of soil P (Table 2.3) varied significantly with the interaction of congregation sites and distance of sampling away from the center of the sites. Levels of soil phosphorus in mineral feeders (34.05 ± 0.44 mg kg^{-1}) and shaded sites (32.22 ± 0.40 mg kg^{-1}) were statistically comparable, but were significantly higher than in water troughs (15.98 ± 0.39 mg kg^{-1}) site. The two highest concentrations of phosphorus were from the mineral feeders at 1.7 m (49.97 ± 1.28 mg kg^{-1}) and 0.9 m (49.23 ± 1.44 mg kg^{-1}) followed by shades at 0.9 m and 1.7 m (Table 2.3). The higher soil phosphorus near and around the mineral feeders can be attributed to the presence of P in the supplemental feeds.

The average level of soil phosphorus in the mineral feeders (averaged across distance) of 34.05 ± 0.44 mg kg^{-1} was not high enough to be of environmental concern. Losses of soil phosphorus by overland flow can become a big concern when the concentrations for soil phosphorus exceeded 150 mg kg^{-1} in the upper 20 cm of soils. The concentrations of soil phosphorus decreased almost linearly with distance away from the center of the mineral feeders ($-5.24x + 55.10$; $R^2 = 0.92$) and the shades ($-6.25x + 57.21$; $R^2 = 0.85$). However, the level of phosphorus around the water troughs ($0.25x + 16.91$; $R^2 = 0.09$) does not appear to change significantly with

Table 2.3 Average (mean ± std error) distribution of phosphorus among the different congregation sites in beef cattle pastures at varying distance away from the center of congregation sites

Congregation sites	Distance from center (m)	Total P (mg kg^{-1})†
1. Mineral feeders	0.9	49.23 ± 1.44 ab
	1.7	49.97 ± 1.28 a
	3.3	37.82 ± 1.14 abcd
	6.7	30.82 ± 1.08 bcde
	13.3	26.68 ± 1.08 cde
	26.7	21.39 ± 0.88 de
	53.3	23.09 ± 1.08 de
2. Shades	0.9	45.83 ± 1.08 ab
	1.7	44.58 ± 1.07 abc
	3.3	43.89 ± 1.06 abc
	6.7	40.98 ± 1.15 abc
	13.3	19.79 ± 0.78 de
	26.7	15.54 ± 0.78 e
	53.3	14.91 ± 0.91 e
3. Water troughs	0.9	18.48 ± 1.18 e
	1.7	15.66 ± 1.01 e
	3.3	13.12 ± 0.96 e
	6.7	15.98 ± 0.94 e
	13.3	17.77 ± 0.99 e
	26.7	15.76 ± 1.01 e
	53.3	12.72 ± 0.91 e

†Means in columns followed by common letter (s) are not significantly different from each other at $p \leq 0.05$

distance, staying close to about 13–18 mg kg^{-1} (Fig. 2.9). The lowest soil phosphorus level at the water troughs of 13.12 ± 0.39 mg kg^{-1} was well within the background level.

2.4.3.3 Conclusion

Results of the study suggested that cattle congregation sites may not be as nutrient-rich as previously thought, therefore may not contribute more nutrients to surface and groundwater supply under Florida conditions. Since there is no apparent build-up with soil depth (0–100 cm) and horizontal movement of total phosphorus in the landscape, it could be surmise that cattle congregation sites may be considered not a potential source of nutrients at the watershed level.

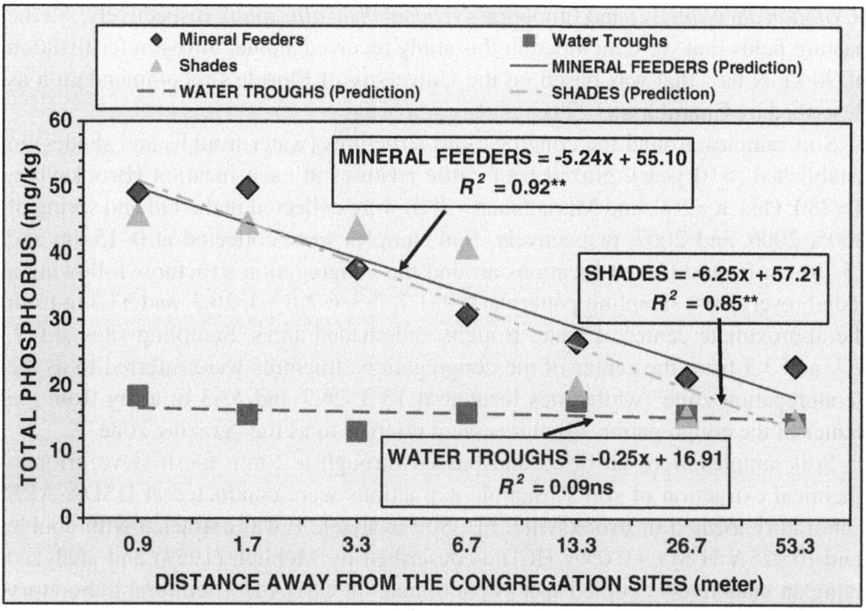

Fig. 2.9 Relationship of soil total phosphorus with distance away from the center of the congregation sites in forage-based pastures with cow-calf operations (Source: Sigua and Coleman 2007: 249–225)

2.4.4 *Case Study #4: Regional Distribution of Soil P across Congregation-Grazing Zones of Forage-Based Pastures with Cow-Calf Operations in Florida*

2.4.4.1 Study Site and Experimental Methods

The study sites were located in three (Brooksville, Ona and Marianna) Florida pastures with cow-calf operations. The site in Brooksville, FL (central region) was at Turnley Unit (82.29° W; 28.62° N) of the United States Department of Agriculture-Agricultural Research Service, Subtropical Agricultural Research Station. Soil (Candler fine sand) at this location can be described as well-drained hyperthermic uncoated typic quartzipsamments (Hyde et al. 1977). The study site in Ona, Florida (southern region) was located at the University of Florida Range Cattle Research and Education Center (82.92° W; 27.43° N) on a Pomona fine sandy soil (Sandy, siliceous, hyperthermic ultic alaquods). The study site in Marianna, FL (northern region) was located at the University of Florida North Florida Research and Education Center (85.18° W; 30.87° N) on a well drained acidic, sandy soil (fine loamy, kaolinitic, thermic kandiudults). Cattle production at these three pasture locations is forage-based with perennial tropical grass, bahiagrass (*Paspalum notatum*, Flugge), the predominant species. The other major forage species in Brooksville, Marianna and Ona are rhizoma peanuts (*Arachis glabrata*, Benth), bermudagrass

(*Cynodon dactylon*, L.) and limpograss (*Hemarthria altissima*), respectively. All the pasture fields that were included in this study received annual nitrogen fertilization of 90 kg N ha^{-1} that was based on the University of Florida's recommendation as described by Chambliss (1999).

Soil samples around the congregations structures (water troughs and shades) in established (>10 year), grazed beef cattle pastures at each location (Brooksville, n = 280; Ona, n = 260; and Marianna, n = 280) were collected in the fall and spring of 2005, 2006, and 2007, respectively. Soil samples were collected at 0–15 cm and 15–30 cm from different locations around the congregation structures following a radial (every 90°) sampling pattern at 0.9, 1.7, 3.3, 6.7, 13.3, 26.7, and 53.3 m from the approximate center of water troughs and shaded areas. Sampling sites at 0.9, 1.7, and 3.3 from the center of the congregation structures were referred to as the "congregation zone" while sites located at 13.3, 26.7 and 53.3 m away from the center of the congregation structures were referred to as the "grazing zone".

Soil samples were air-dried and passed through a 2-mm mesh sieve prior to chemical extraction of soil P. Sample extractions were conducted at USDA-ARS Laboratory located in Brooksville, FL. Soil available P was extracted with double acid (0.025 N H_2SO_4 + 0.05 N HCl) as described by Mehlich (1953) and analyzed using an inductively coupled spectrophotometer at USDA Horticultural Laboratory located in Fort Pierce, FL. The degree of soil saturation with P (DPS) was computed using the P, iron, and aluminum contents (mg kg^{-1}) of the soils (Hooda et al. 2000).

The objectives of this study were: (a) to determine whether cattle congregation sites typical on most Florida ranches, represented by water troughs and shaded areas, are more phosphorus-rich and may contribute more soluble phosphorus to surface water run-off and groundwater than other pasture locations; and (b) to assess the regional distribution of soil phosphorus across congregation-grazing zones of forage-based pastures with cow-calf operations in Florida.

2.4.4.2 Research Highlights

Soil phosphorus concentration varied among the pasture locations ($p \leq 0.0001$) and pasture zone ($p \leq 0.001$). There was an interaction between pasture location and pasture zone ($p \leq 0.001$) and interaction effects among year, pasture location, and pasture zone ($p \leq 0.0001$). Soil phosphorus concentration also varied with radial distance away from the center of the congregation structures ($p \leq 0.001$). Averaged across years and pasture zones, pasture located in Brooksville, FL (46.6 ± 5.3 mg kg^{-1}) had the greatest available soil phosphorus while pasture at Marianna, FL had the lowest concentrations of soil P (26.7 ± 1.4 mg kg^{-1}). The amount of soil phosphorus in pasture located in Ona, FL was about 45.5 ± 4.2 mg kg^{-1} (Table 2.4). The average concentrations of soil phosphorus at all three regions were still not high enough to be of environmental concern. Average concentration of phosphorus at all three pasture locations did not exceed the crop requirement threshold of 50 mg kg^{-1} identified by Beegle (2002) nor exceeded the water quality protection threshold of 150 mg kg^{-1} proposed by Sims et al. (2002). Losses of soil phosphorus by overland flow can become a big concern when the concentrations for soil phosphorus exceeded 150 mg kg^{-1}.

Table 2.4 Comparative concentrations of Mehlich-1 extractable P and percent P saturation in soils (0–20 cm) among the different pasture zones of three Florida pastures (Brooksville, Ona and Marianna)

Pasture location	Pasture zone	Sample number (n)	Soil P (mg kg^{-1})	P saturation (%)
1. Brooksville, FL	Congregation	211	39.7±4.3b†	13.2
	Grazing	187	51.6±5.1a	11.0
	Transition	70	49.2±6.4a	11.5
Mean			**46.6±5.3**	**11.9**
2. Ona, FL	Congregation	244	45.1±4.5b	53.8
	Grazing	224	39.2±3.4b	47.9
	Transition	82	52.2±4.7a	70.2
Mean			**45.5±4.2**	**57.3**
3. Marianna, FL	Congregation	240	26.5±1.3a	22.1
	Grazing	188	29.0±1.4a	21.1
	Transition	78	24.7±1.5a	16.8
Mean			**26.7±1.4**	**20.0**
Sources of variations			**F-value**	**F-value**
Year (Y)			140.9***‡	103.7***
Pasture location (PL)			114.2***	38.2**
Pasture zone (PZ)			61.7**	7.4**
PL×PZ			110.9***	1.6**ns**
Y×PL×PZ			11.1**	0.4**ns**

Source: Sigua et al. (2011b)

†Means in column within each subheading followed by common letter(s) are not significantly different from each other at $p \leq 0.05$

‡*** – ($p \leq 0.0001$) ** – ($p \leq 0.001$) * – ($p \leq 0.01$) *ns* not significant

Spatial trends of soil phosphorus in our study may be a function of feces and urine deposition where animals clustered. Where animals congregate may tend to develop some hot spots in the pasture. Our results did not support the idea that hot spots were likely had the highest concentration of soil phosphorus. Soil phosphorus concentrations in the congregation zones were comparable ($p \leq 0.05$) with the concentrations of soil phosphorus in the grazing zones at all the three regions, except for Brooksville site (Table 2.4). The grazing zone of pastures located in Brooksville, FL had the highest concentrations of soil phosphorus (51.6±5.1 mg kg^{-1}) while congregation zone of pastures in Marianna, FL had the lowest levels of soil phosphorus (26±1.3 mg kg^{-1}).

The concentrations of soil phosphorus decreased linearly with distance away from center of the congregation structures at all three Florida regions. Figure 2.10 shows the relationships between extractable soil phosphorus and distance away for the center of the shaded areas in Brooksville, Ona and Marianna, FL. The degree of phosphorus saturation in soils was significantly ($p \leq 0.0001$) affected by year, pasture location and pasture zone (Table 2.4). Averaged across years and pasture zones, the pastures at Ona, FL had the highest estimated degree of soil phosphorus saturation at 57.3 % followed by pastures located at Marianna, FL (20.0 %). Pastures at Brooksville had the lowest estimated degree of phosphorus saturation (12 %).

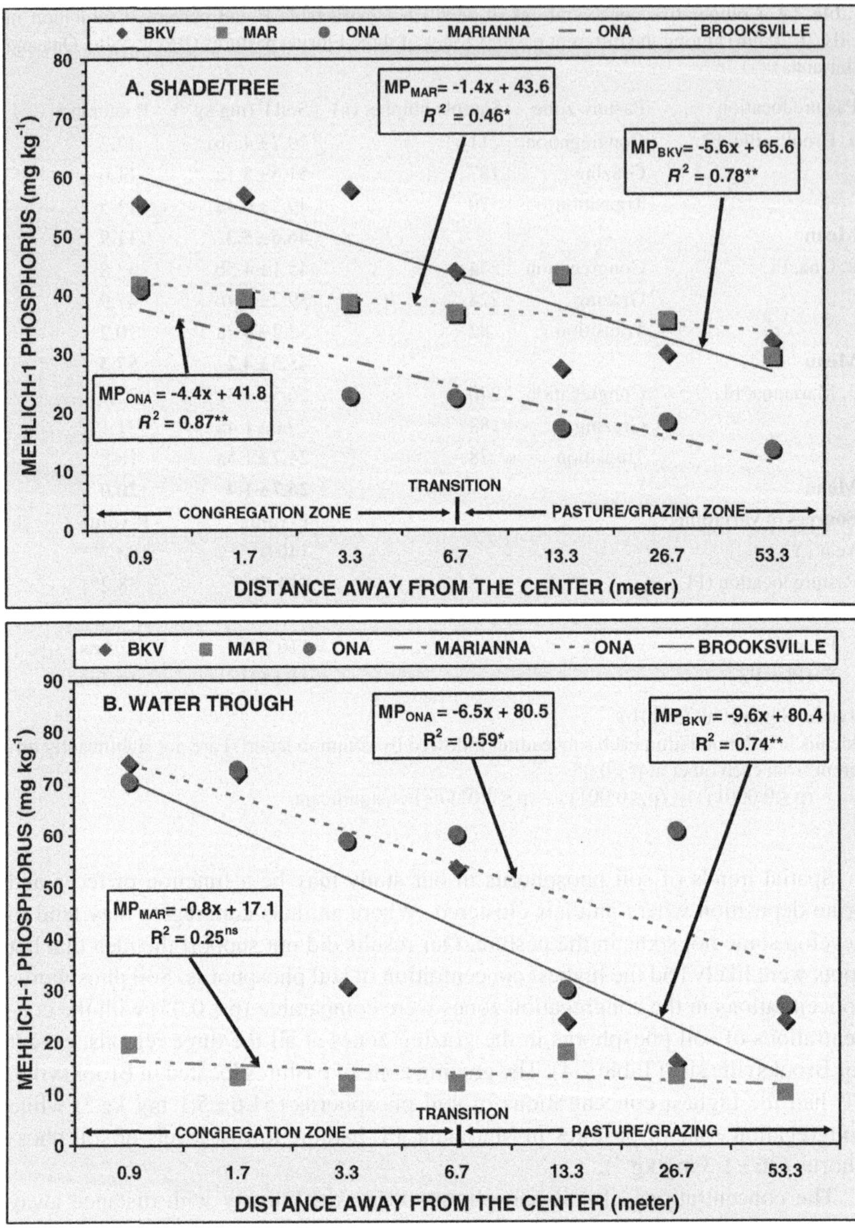

Fig. 2.10 Relationships between Mehlich extractable soil phosphorus (MP) and distance away from the center of the shaded areas in Brooksville (BKV), Ona (ONA) and Marianna (MAR), Florida (Source: Sigua et al. 2011b)

P. Soil P concentrations in the congregation zones were comparable ($p \leq 0.05$) with the at Brooksville, FL had the lowest estimated degree of soil phosphorus saturation (11.9 %). The degree of soil phosphorus saturation in the congregation zones (13.2 %) of pastures located in Brooksville, FL was comparable to the levels of phosphorus saturation in the grazing zones (11.0 %). Similarly, the degree of soil P saturation did not vary between the congregation zones (22.1 %) and grazing zones (21.1 %) of pastures at Marianna, FL. The degree of soil P saturation in congregation zones and grazing zones of pastures located in Ona, FL were 53.6 % and 47.9 %, respectively. The degree of soil P saturation in the three pastures were below the environmental threshold of P saturation (DPS ≥ 60 %), suggesting that P buildup and/or release is not a predicament anywhere in the pasture, including the congregation zones. These results may have significant implications for the transport of P to surface waters and our ability to predict and model losses of P from congregation zone or grazing zone of pastures with cow-calf operations.

2.4.4.3 Conclusion

The degree of soil phosphorus saturation in the three pastures were below the environmental threshold of phosphorus saturation (DPS ≥ 60 %), suggesting that phosphorus buildup and/or release is not a predicament anywhere in the pasture, including the congregation zones. These results may have significant implications for the transport of phosphorus to surface waters and our ability to predict and model losses of phosphorus from congregation zone or grazing zone of pastures with cow-calf operations.

Results from this study suggest that congregation zones in pastures with beef cattle operations in three regions of Florida are not phosphorus-rich, therefore may not contribute more phosphorus to surface and groundwater supply under Florida conditions. Averaged across years, phosphorus concentrations and soil phosphorus saturation of the congregation zones were comparable with soil phosphorus and soil phosphorus saturation in the grazing zones of all the three regions. Average phosphorus in all three pasture locations did not exceed the crop requirement threshold of 50 mg kg^{-1} and the water quality protection threshold of 150 mg kg^{-1}.

2.5 Balancing and Managing Soil Phosphorus Across Paddocks: Environmental Implications

The role of nutrient management in livestock systems takes on new meaning as producers and the public together consider economic and noneconomic issues. The intensification of livestock production with its associated increased demand for fodder has encouraged farmers to rely more heavily on chemical fertilizers and imported feeds, and very often the waste is considered as a disposal problem rather than useful source of plant nutrients (Hooda et al. 2000). It should be noted that for a farm to be sustainable, its phosphorus or nitrogen budget should balance, at least after soil reserves are brought up to desired levels for sustainable production. Nutrient balance in the ecosystem involves profitability of the agricultural enterprise and commitments to resource management to maintain quality of air, water and land resources. If there is a net loss of phosphorus, the farm's soils will eventually become depleted and if there is an excess, the likelihood of pollution is greater (Van Horn et al. 1996). Effective use and cycling of phosphorus is critical for pasture productivity and environmental stability. Phosphorus cycling in pastures is complex and interrelated and pasture management practices influence the interactions and transformations occurring within the phosphorus cycle (Fig. 2.11).

Increased loss of nutrients in agricultural runoff has potentially serious ecological and public health implications (Hooda et al. 2000). Nitrogen and phosphorus are particularly important as both are implicated in aquatic eutrophication (Levine and Schindler 1989). Any approach that controls the future contained excellent discussion on the source and transport management of nutrients in the watershed. Source management attempts to minimize the build-up in the soil above levels sufficient for optimum crop growth while transport management refers to efforts to control the movement of nutrients from soils to sensitive locations such as bodies of fresh water. As shown in Table 2.5, there are several measures available to minimize the potential for nutrient losses in agricultural runoff, which address sources and transport of phosphorus and/or nitrogen. Important measures to be considered are those that attempt to decrease the surplus of nutrients such as phosphorus or nitrogen in localized areas. Shigaki et al. (2006) suggested the following measures: (1) dietary phosphorus reduction; (2) feed additives nitrogen and/or phosphorus losses from agriculture to water must begin with the long-term objective of increasing

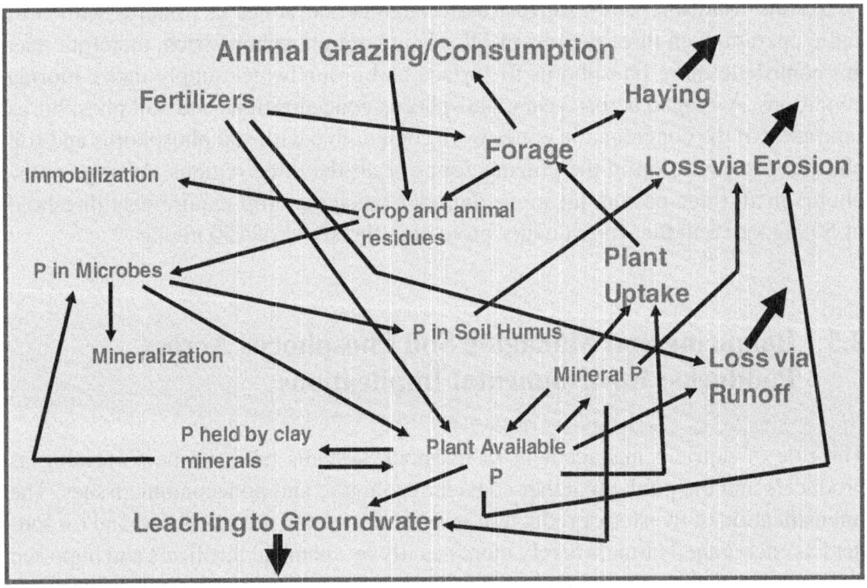

Fig. 2.11 Generalized schematic showing the different phosphorus compartments and phosphorus in forage-based pasture system with cow-calf operation (Source: Sigua 2010: 631–648)

Table 2.5 Modified best management practices for the control of diffuse sources of agricultural nutrients

Source best management practice – practices that minimize nutrients loss at the origin
Attempts to match animal requirements for nitrogen or phosphorus with feeds
Enzyme added to feeds to increase nutrient utilization by animals
Test soils and manures to optimize nitrogen and phosphorus management
Chemically treat manure to reduce phosphorus solubility such as alum, fly ash, etc.
Biologically treat manure such as microbial enhancement
Calibrate fertilizer and manure application equipment
Apply proper amount of fertilizer and
Careful timing of fertilizer application to avoid imminent heavy rainfalls
Transport best management practice – practices that minimize transport of nutrients
Minimize erosion, runoff and leaching of nutrients
Use cover crops to protect soil surface from erosion
Install filter strips and other conservation buffers to trap eroded phosphorus
Manage riparian zones, grassed waterways and wetlands to trap eroded nutrients
Stream bank fencing to exclude animal from water course
Install and maintain impoundments or small ponds to trap sediments and nutrients
Source and transport best management practice – systems approach to minimize nutrient losses
Retain crop residues and reduce tillage to minimize erosion and runoff
Proper grazing management to minimize erosion and runoff
Install and maintain manure handling systems
Implement nutrient management plan for the farms

Source: Shigaki et al. (2006: 194–209)

nitrogen and/or phosphorus efficiency by attempting to balance nitrogen or phosphorus inputs with nitrogen or phosphorus outputs within a watershed. Reducing nitrogen or phosphorus loss in agricultural runoff may be brought about by Best Management Practices (BMPs) that control the source and transport of nitrogen and/or phosphorus (Table 2.5).

Nutrient removal is accomplished only by removing forage as hay crop and transporting the nutrient away from the application site. Haying of rhizoma peanut-based pastures removes biomass-containing nutrients, whereas grazing largely recycles nutrients (Sigua et al. 2006b). Nutrients may enter the pasture system from a number of sources such as fertilizers, manures, urine, crop residues, atmospheric and could be lost via erosion, runoff, leaching to groundwater, and haying livestock is critical for productive crop growth and water quality protection. If more nutrients are added that can be used for productive forage growth, nutrients will build up in soil, creating high risk for runoff and water contamination. Effective use and cycling of nutrients is critical for pasture productivity and environmental stability. However, phosphorus cycling in pastures is complex and interrelated (Fig. 2.11). Each cycle has its complex set of interactions and transformations as well as interactions with the other cycles. Pasture management practices influence the interactions and transformations occurring within nutrient cycles.

A review paper published by Shigaki et al. (2006) on animal-based agriculture options for the future contained excellent discussion on the source and transport management of nutrients in the watershed. Source management attempts to minimize the build-up in the soil above levels sufficient for optimum crop growth while transport management refers to efforts to control the movement of nutrients from soils to sensitive locations such as bodies of fresh water (Table 2.5).

In addition to speeding up phosphorus recycling from the grass, grazing animals also can increase phosphorus losses in the system by increasing leaching potential due to concentrating phosphorus into small volumes of soil under dung and urine patches, redistributing phosphorus around the landscape, and removal of nitrogen or phosphorus in the form of animal products (see Fig. 2.11). The overall goal efforts to reduce phosphorus losses from animal-based agriculture should be to balance off-farm phosphorus inputs in feed and fertilizer with outputs to the environment. Source and transport control strategies can provide the basis to increase phosphorus efficiency in agricultural systems.

Finally, soil testing is still the best management tool to monitor soil fertility. Routine soil tests can help identify nutrient deficiencies. Based on soil test results, cost-effective fertilization programs can be developed to meet forage nutrient requirements and minimize production costs. To effectively implement any best management practices (BMPs), it is necessary to recognize the potential impact of agricultural phosphorus on surface and ground water and understand that different landscape locations and varying hydrologic conditions in forage-based landscape could affect the spatial and temporal variations of phosphorus losses at the watershed scale. If we understand where and how phosphorus is getting into our waters, we can implement BMPs to reduce or even eliminate phosphorus as a potential pollutant from our water supply.

2.6 Summary

The overall observations from this review paper could be briefly summarized as follows:

- Environmentally, soil phosphorus levels in Florida pastures are declining. During the past 12 years, there was no phosphorus build up in Subtropical Agricultural Research Station despite of the annual application of phosphorus-containing fertilizers in addition to the daily in-field loading of animal waste bi-products such as fecals, urine, etc. The average soil test values for phosphorus in Subtropical Agricultural Research Station, Brooksville, Florida have declined by about 28.3 %;
- Levels and changes in soil phosphorus levels in Florida pastures from 1988 to 2000 were responsive and sensitive to phosphorus fertilization. Levels of soil phosphorus in pasture fields with no phosphorus fertilization were consistently lower than those of the fertilized fields by about 49.1 %, 50.5 %, and 40.9 % in 1988, 1997, and 2000, respectively;

- Differences among the pasture units for soil phosphorus may still not of particular concern environmentally, but are important from a fertility management point of view. The levels of soil phosphorus in 1988 of about 94.1 mg kg^{-1} and in 2000 of about 69.2 mg kg^{-1} were not high enough to be of environmental concern, so annual additions of phosphorus-fertilizer would be still practical to sustain plant and animal productivity in subtropical beef cattle pasture units. Losses of soil phosphorus by overland flow are becoming a big concern when the test values for soil phosphorus exceeded 330 kg ha^{-1} in the upper 20-cm of soil;
- The highest average level of soil phosphorus in the mineral feeders of 34.05 ± 0.44 mg kg^{-1} was not high enough to be of environmental concern. If the sites at Florida pastures can be assumed to mimic those of commercial producers, then they probably are not a source of nutrients to pollute surface and ground water supply;
- Congregation zones in pastures with beef cattle operations in three regions of Florida are not phosphorus-rich, therefore may not contribute more phosphorus to surface and groundwater supply under Florida conditions. Averaged across years, phosphorus concentrations and soil P saturation of the congregation zones were comparable with soil phosphorus and soil phosphorus saturation in the grazing zones of all the three regions. Average phosphorus in all three pasture locations did not exceed the crop requirement threshold of 50 mg kg^{-1} and the water quality protection threshold of 150 mg kg^{-1}; and
- Slope aspect and slope position could be of relative importance in controlling spatial distribution of soil phosphorus. Averaged across years, soils on the north-facing slope contained the greatest amount of soil phosphorus when compared with other slope aspects.

2.7 Conclusions

Contrary to early perception, forage-based animal production systems with grazing are not likely one of the major sources of non-point source phosphorus pollution that are contributing to the degradation of water quality in lakes, reservoirs, rivers, and ground water aquifers, but perennially grass-covered pastures are associated with a number of environmental benefits. Continuous grass cover leads to the accumulation of soil organic matter, sequestering carbon in the soil and thereby reducing the potential CO_2 accumulation in the atmosphere. The increase in soil organic matter is also related to soil quality, with improvements in soil structure, aeration and microbial activity.

Effective use and cycling of phosphorus is critical for pasture productivity and environmental stability. This will help to renew the focus on improving inorganic fertilizer efficiency in subtropical beef cattle systems, and maintaining a balance of phosphorus removed to phosphorus added to ensure healthy forage growth and minimize phosphorus runoff. New knowledge based on the whole-farm approach is desirable to identify pastureland at risk of degradation and to prescribe treatments or management practices needed to protect the natural resources while maintaining an economically and environmentally viable operation. Therefore, a better understanding of soil phosphorus dynamics, phosphorus use efficiency, and other crop nutrient changes in pastures with cow-calf management systems should allow us to better predict the least risk of phosphorus losses to adjacent surface water and ground water.

References

Adjie MB (1999) Fertilizer treatment effects on forage yield and quality of tropical pasture grasses. Soil Crop Sci Soc Fla Proc 59:32–37

Allen LH Jr, Stewart EH, Knisel WG Jr, Stack RA (1976) Seasonal variation in runoff and water quality from the Taylor Creek watershed, Okeechobee County, FL. Soil Crop Sci Soc Fla Proc 35:126–138

Allen LH Jr, Ruddell JM, Rutter GJ, Yates P (1982) Land effects of Taylor Creek water quality. In: Kruse EG et al (eds) Environmentally sound water and soil management. American Society of Civil Engineers, New York, pp 67–77

Arthington J, Bohlen P, Roka F (2003) Effect of stocking rate on measures of cow-calf productivity and nutrient load in surface water runoff. University of Florida IFAS Extension, Gainesville. 4 pp

Beegle DB (2002) Soil fertility management. In: Martz F (ed) The agronomy guide. Publications Distribution Center, College of Agricultural Sciences, The Pennsylvania State University, University Park, pp 19–42

Blue WG (1970) Fertilizer nitrogen uptake by Pensacola bahiagrass from Leon fine sand, a spodosol. In: Norman MJT (ed) Proceedings of the XI international grassland congress. Surfers Paradise, Queensland, Australia, pp 389–392

Bogges CF, Flaig EG, Fluck RC (1995) Phosphorous budget-basin relationships for Lake Okeechobee tributary basins. Ecol Eng 5:143–162

Botcher AB, Tremwel TK, Campbell KL (1999) Phosphorus management in flatwood (Spodosols) soils. In: Reddy KR et al (eds) P biogeochemistry in subtropical ecosystems. Lewis Publishers, Boca Raton, pp 405–423

Bowers EJ, Hammond AC, Chase CC Jr, Olson TA (1995) Effect of breed on indicators of heat tolerance and grazing activity in lactating Angus and Brahman cows in Florida. J Anim Sci 73(Suppl. 1):131

Brooks KN, Folliot PN, Gregersen HM, DeBano LF (1997) Hydrology and the management of watersheds, 2nd edn. Iowa State University Press, Ames

Canham CD, Papaik MJ, Latty EF (2001) Interspecific variation in susceptibility to windthrow as a function of tree size and storm severity for northern temperate tree species. Can J For Res 31:1–10

Chambliss CG (1999) Florida forage handbook. University of Florida Cooperative Extension Service, Gainesville, FL. SP253

Coale FJ, Olear JA (1996) The relationship between soil test P level and the concentration of dissolved and potentially transportable P in field drainage water. Technical report, Chesapeake Research Consortium, Inc., Edgewater

Edwards DR, Hutchens TK, Rhodes RW, Larson BT, Dunn L (2000) Quality of runoff from plots with simulated grazing. J Am Water Res Assoc 36:1063–1073

Foster DR, Aber JD, Melillo JM, Bowden RD, Bazzaz FA (1997) Forest response to disturbance and anthropogenic stress. Bioscience 47:437–445

Franzluebbers A, Owens LB, Sigua GC, Cambardella C, Richard HL (2012) Soil organic carbon under pasture management. In: Leibig M, Franzluebbers A, Follet R (eds) Managing agricultural greenhouse gases: coordinated agricultural research through GRACEnet to address our changing climate. Elsevier Publishing, Boston, MA, pp 93–111. ISBN: 978-0-12-386897-8

Ganskopp D (1987) Slope use by cattle, feral horses, deer and bighorn sheep. Northwest Sci 61(2):74–81

Ganskopp D (2001) Manipulating cattle distribution with salt and water in large arid-land pastures: a GPS/GIS assessment. Appl Anim Behav Sci 73:251–262

Gburek WJ, Sharpley AN (1998) Hydrologic controls on P loss from upland agricultural watersheds. J Environ Qual 27:267–277

Hammond AC, Olson TA (1994) Rectal temperature and grazing time in selected beef cattle breeds under tropical summer conditions in subtropical Florida. Trop Agric (Trinidad) 71:128–134

Haynes RJ (1981) Competitive aspects of the grass-legume association. Adv Agron 33:227–261

Haynes RJ, Williams PH (1993) Nutrient cycling and soil fertility in grazed pasture ecosystem. Adv Agron 49:119–199

Heckranth GP, Brookes C, Poulton PR, Goulding KWT (1995) P leaching from soils containing different P concentrations in the Broadbalk experiment. J Environ Qual 24:904–910

Holechek JL (1988) An approach for setting stocking rate. Rangeland 10:10–14

Hooda PS, Rendell AR, Edwards AC, Withers PJ, Aitken MN, Truesdale VW (2000) Relating soil P indices to potential P release to water. J Environ Qual 29:1166–1171

Hyde AG, Law L Jr, Weatherspoon RL, Cheney MD, Eckenrode JJ (1977) Soil survey of Hernando County, FL. USDA-NRCS/University of Florida, Washington, DC/Gainesville, p 152

Johnson J, Eckert D (1995) Best management practices: land application of animal manure. The Ohio State University Extension Agronomy Facts. AGF-208-95. Columbus, 10 p

Kellogg RL, Lander CH, Moffit DC, Gollehon N (2000) Manure nutrients relative to the capacity of cropland and pastureland to assimilate nutrients: spatial and temporal trends for the United States. USDA-NRCS, Washington, DC, p 93

Khaleel R, Reddy KR, Overcash MR (1980) Transport of potential pollutants in runoff water from land areas receiving animal wastes: a review. Water Res 14:421–426

Klemmendson JO, Tiedemann AR (1995) Effects of nutrient stress. In: Bedunah DJ, Sosebee R (eds) Wildland plants: physiological ecology and developmental morphology. Society of Range Management, Denver, pp 414–439

Kotak BG, Kenefick SL, Fritz DL, Rousseaux CG, Prepas EE, Hrudey SE (1993) Occurrences and toxicological evaluation of cyanobacterial toxins in Alberta lakes and farm dugouts. Water Res 27:495–506

Levine SL, Schindler DW (1989) Phosphorus, nitrogen and carbon dynamics of experimental lake 303 during recovery from eutrophication. Can J Fish Aquat Sci 46:2–10

Martin SC, Ward DE (1973) Salt and meal-salt help distribute cattle use on semi-desert range. J Range Manag 26:94–97

Mathews BW, Sollenberger LE, Nair VD, Staples CR (1994) Impact of grazing management on soil nitrogen, P, potassium, and sulfur distribution. J Environ Qual 23:1006–1013

Mathews BW, Tritschler JP, Carpenter JR, Sollenberger LE (1999) Soil macronutrients distribution in rotationally stocked kikuyugrass paddocks with short and long grazing periods. Commun Soil Sci Plant Anal 30:557–571

Mehlich A (1953) Determination of P, Ca, Mg, K, Na, and NH_4: North Carolina Soil Test Division. Mimeo, Raleigh

Norfleet ML, Wood CW, Mullins GL, Hajek BF (1996) P in agriculture. Tech. Pamph. no. 2. Soil Quality Institute, USDA Natural Resources Conservation Service, Auburn

Odum EP (1969) The strategy of ecosystem development. Science 164:262–270

Palmstrom NS, Carlson RE, Cooke GD (1988) Potential links between eutrophication and formation of carcinogens in drinking water. Lake Reserv Manag 4:1–15

Schindler DW (1974) Eutrophication and recovery in experimental lakes: implications for lake management. Science 184:260–262

Schmidt KP, Sturgul SJ (1989) Nutrient and pesticide best management practices from Wisconsin farms. WDAPTCP technical bulletin ARM-1. University of Wisconsin and Wisconsin Department of Agriculture, Madison

Senft RL, Rittenhouse LR, Woodmanse RG (1985) Factors influencing patterns of cattle grazing behavior on shortgrass steppe. J Range Manag 38:82–87

Sharpley AN (1997) Dispelling common myths about P in agriculture and the environment. Technical report, Watershed Science Institute, USDA Natural Resources Conservation Service, Aiken Center, University of Vermont, Burlington

Sharpley AN, Daniel TC, Sims JT, Pote DH (1996) Determining environmentally sound soil P levels. J Soil Water Conserv 51:160–166

Shigaki F, Sharpley A, Prochnow LI (2006) Animal-based agriculture, P management and water quality in Brazil: options for the future. Sci Agric 63:194–209

Sigua GC (2010) Sustainable cow-calf operations and water quality. Agron Sustain Dev J 30(3):631–648

Sigua GC, Coleman SW (2007) Sustainable management of nutrients in forage-based pasture soils: effect of animal congregation sites. J Soils Sediments 6:249–253

Sigua GC, Coleman SW (2010) Spatial distribution of soil carbon in pastures with cow-calf operations: effects of slope aspect and slope position. J Soils Sediments 10(2):240–247

Sigua GC, Steward JS (2000) Establishing pollutant load reduction targets for the Indian River Lagoon, Florida. J Am Water Res Assoc 36:123–132

Sigua GC, Tweedale WA (2003) Watershed scale assessment of nitrogen and P loadings in the Indian River Lagoon Basin, FL. J Environ Manag 67(4):361–370

Sigua GC, Steward JS, Tweedale WA (2000) Water quality monitoring and biological integrity assessment in the Indian River Lagoon, Florida I. Status and trends (1988–1994). J Environ Manag 25:199–209

Sigua GC, Williams MJ, Coleman SW (2004) Levels and changes of soil P in the subtropical beef cattle pastures. Commun Soil Sci Plant Anal 35(7&8):975–990

Sigua GC, Adjei MB, Rechcigl JE (2005a) Cumulative and residual effects of repeated sewage sludge applications: forage productivity and soil quality implications in South Florida, USA. Environ Sci Pollut Res J 12:80–88

Sigua GC, Coleman SW, Williams MJ (2005b) Cattle congregation sites do not appear to impact soil nutrients build-up. Fla Cattlem Livest J 69(6):76–79

Sigua GC, Williams MJ, Coleman SW, Starks R (2006a) Nitrogen and P status of soils and trophic state of lakes associated with forage-based beef cattle operations in Florida. J Environ Qual 35:240–252

Sigua GC, Williams MJ, Coleman SW (2006b) Long-term effects of grazing and haying on soil nutrient dynamics in forage-based beef cattle operations. J Sustain Agric 29(3):115–134

Sigua GC, Hubbard R, Coleman SW (2010) Quantifying P levels in soils, plants, surface water and shallow groundwater associated with bahiagrass-based pastures. Environ Sci Pollut Res J 17:210–219

Sigua GC, Coleman SW, Albano J, Williams MJ (2011a) Spatial distribution of soil P and herbage mass in beef cattle pastures: effects of slope aspect and slope position. Nutr Cycl Agroecosyst J 89:59–70

Sigua GC, Meyer RO, Coleman SW, Mackowiak C, Adjei M, Chase CC, Albano J (2011b) Regional distribution of soil P across congregation-grazing zones of forage-based pastures with cow-calf operations in Florida. J Environ Prot 2(3):408–417

Sims JT, Maguire RO, Leytem AB, Gartley KL, Pautler MC (2002) Evaluation of Mehlich-3 as an agri-environmental soil P test for the Mid-Atlantic United States of America. Soil Sci Soc Am J 66:2016–2032

Stout WL, Weaver SR, Gburek WJ, Folmar GJ, Schnabel RR (2000) Water quality implications of dairy slurry applied to cut pastures in northeast USA. Soil Use Manag 16:189–193

Thurow TL (1991) Hydrology and erosion. In: Heitschmidt RK, Stuth JW (eds) Grazing management: an ecological perspective. Timber Press, Portland, pp 141–160

Van Horn HH, Newton GL, Kunkle KE (1996) Ruminant nutrition from an environmental perspective: factors affecting whole-farm nutrient balance. J Anim Sci 74:3082–3102

Watt AS (1947) Pattern and process in the plant community. J Ecol 35:1–22

White SL, Sheffield RE, Washburn SP, King LD, Green JT Jr (2001) Spatial and time distribution of dairy cattle excreta in an intensive pasture systems. J Environ Qual 30:2180–2187

Chapter 3
Conservation Tillage for Soil Management and Crop Production

Surjeet S. Manhas, Ajab S. Sidhu, and Khuswinder S. Brar

Abstract Conservation tillage affects chemical, biological, and physical properties of soil. The oxidation of organic matter is slow in conservation tillage which causes increase in organic matter of soil. Reduced tillage increases soil acidity in top layer of soil due to an accumulation of organic acids in the superficial layer. Soil microbial biomass and microbial activity were highly stratified under reduced tillage. Conservation tillage increased soil aggregate stability and improved soil structure due to increased labile organic matter, microbial biomass and microbial activity. Under conservation tillage, soil porosity improved as compared to conventional tillage due to beneficial effects of soil organic matter, microbial activity and residue cover. The gravimetric water content and infiltration rate are maximum under no-tillage as compared to conventional tillage. High concentration of immobile nutrients such as phosphorus and potassium resulted near the soil surface and decreasing with depth has been routinely observed with conservation tillage systems. The water-use efficiency is higher in no tillage as compared to conventional tillage. Under a wide range of environment conditions crop yield, net return and benefit: cost ratio obtained by no-till, reduced tillage, and stubble retention systems were equivalent or even a higher than those recorded under conventional tillage. Conservation tillage directly reduces carbon emissions by reducing fuel use and indirectly by sequestration of atmospheric C in the soil and biomass.

Keywords Conservation tillage • Organic matter • Soil structure • Soil biology • Crop yield • Contents

3.1 Introduction

Tillage systems influence physical, chemical, and biological properties of soil and have a major impact on soil productivity and sustainability. The main objective in agriculture production, so far, focused mostly on the increase of yield and

---

S.S. Manhas (✉) • A.S. Sidhu • K.S. Brar
Punjab Agricultural University, Ludhiana, Punjab 141004, India
e-mail: manhas30@gmail.com

E. Lichtfouse (ed.), *Sustainable Agriculture Reviews*,
Sustainable Agriculture Reviews 17, DOI 10.1007/978-3-319-16742-8_3

production. Meanwhile, economic, production and sustainable agriculture are in getting attention and improvement in product quality, reduction in production inputs, conservation of natural resources and environmental awareness gain importance (Ulusoy 2001). In the current cropping systems, all the crop residues are removed or burnt before mouldboard ploughing. Using these traditional practices, farmers seek to produce good seedbeds, conserve water and reduce variability in crop yields. However, in the long term, this traditional tillage tends to increase soil bulk density, reduce macro-porosity and macro-aggregates, resulting in less water and nutrient availability and consequently, crop yields become unstable and decline, especially in dry years (Qin et al. 2004). Since land preparation for double-cropping systems requires timeliness, especially when a moldboard plow is used, reduced tillage, mainly no-tillage systems, are becoming widespread.

Beneficial soil management is essential to maintain long-term productivity, long-term environmental stability and food safety. Sustainable soil management can be practiced through conservation tillage, high crop residue return, and crop rotation (Lal 2009). Conservation tillage, by most definitions, embraces crop production systems involving the management of surface residues (Unger et al. 1988). According to the Conservation Technology Information Center in West Lafayette, Indiana, USA, conservation tillage is defined as: "any tillage or planting system in which at least 30 % of the soil surface is covered by plant residue after planting to reduce erosion by water; or where soil erosion by wind is the primary concern, with at least 1,120 kg ha^{-1} flat small grain residue on the surface during the critical wind erosion period." No tillage, minimum tillage, reduced tillage and mulch tillage are terms synonymous with conservation tillage. According to Greenland (1981) and Lal and Hahn (1973) the no-till system of cultivation with crop residue mulches forms a basis for conservation farming because it conserves water, prevents erosion, maintains organic matter content at a high level, and sustains economic productivity. In addition, there are savings in machinery investment and in the time required for seedbed preparation (Lal 1974). Antapa and Angen (1990) report that retaining crop residues on the soil surface with conservation tillage reduces evapotranspiration, increases infiltration rates, and suppresses weed growth.

3.2 Soil Properties

Conservation tillage leaves most or part of crop residues on the soil surface, thus effecting chemical, biological, and physical properties of soil. Soil temperature, water content, bulk density, porosity, penetration resistance and aggregate distribution are some of the physical properties affected by tillage systems. Changes in soil physical properties due to use of no- tillage depend on several factors including differences in soil properties, weather conditions, history of management, intensity and type of tillage (Fabrizzi et al. 2005; Osunbitan et al. 2005).

3.2.1 Organic Matter

Organic carbon is considered an important indicator of soil fertility. There exists a strong relationship between the agronomic production and the soil organic carbon (SOC) pool, especially under low-input agriculture. Soil organic carbon pool is an essential determinant of soil quality because of its positive impact on the soil structure and aggregation, water and nutrient retention, biotic activity including the microbial biomass, erosion control, nonpoint-source pollution abatement, C sequestration, increase in the use efficiency, and increase in biomass production. Soil organic matter tends to stabilize at a certain level for a specific tillage system used in fields with a particular soil texture. Moldboard plowing buries essentially all residues and increases oxidation of organic matter. Walling (1990) reported that over the last 40 years, the amount of organic matter being returned to the soil had declined, primarily as a consequence of more intensive soil cultivation, the removal of crop residues, the replacement of organic manures with inorganic fertilizer, and the loss of grass leys from rotations. In addition, organic matter is being eroded from arable land to rivers disproportionately to its availability.

Soil organic matter is conserved with no-till because residue is left on the soil surface where oxidation of organic matter is slow (Wilkins et al. 2002) which then causes organic matter in the upper few inches to increase after several years. The SOM increases resulting from conservation tillage are attributed to the greater straw input and reduced biological oxidation associated with less soil disturbance by tillage (Chan et al. 2002). The quantity of SOM in the whole topsoil varies due to the interacting influence of climate, topography, soil type and crop management history (fertilizer use, tillage, rotation and time) (Kay and VandenBygaart 2002). Varvel and Wilhelm (2010) studied the use of conservation tillage systems for corn and soybean production and found soil organic carbon levels were maintained or even increased in all tillage system with the greatest increase obtained in systems with the least amount of soil disturbance which strongly support the adoption and use of conservation tillage systems for soil sustainability. Hunt et al. (1996) and Angers and Giroux (1996) found no-tillage systems increased SOC content, compared with moldboard plow and chisel plow systems, in the top 5-cm layer of soils with a range of soil textures, including loamy sand, silt loam, and silty clay loam. Conventional tillage and fallow practices are not recommended as a sustainable system in a semi-arid condition. In semi-arid climate regions where the correct management of crop residues is essential to achieve sustainable yields (Du Preez et al. 2001).

High temperatures conditions, limit the accumulation of organic carbon at the soil surface. Therefore, the simple determination of total contents may not be the best indicator of improvements in soil quality resulting from conservation tillage. Under arid or semi-arid climates, it is more useful to determine the stratification ratio of soil organic carbon. The stratification ratio of the organic carbon of a soil is a more precise indicator of quality (Franzluebbers 2002a, b). The stratification ratio is defined as the quotient between the surface value of organic C (often at a depth of 0–5 cm) and its value at greater depth, for example, the lower limit of the arable layer. The choice of both depths: surface and greater depths, is an important aspect to take into account. Hernanz et al. (2002) have also shown in soils from the Spanish Centre that both depths, 0–5 and 5–10 cm, similarly reflect the benefits obtained with conservation tillage. In general, in any edapho-climatic context, a high stratification ratio is a good index of soil quality (Franzluebbers 2002a, b; Mrabet 2002). Ratios above 2 would not be common in degraded soils, so this value can be considered as an approximate threshold under which the soil would have a poorer quality (with the lowest ratios corresponding to the poorest qualities). After 12 years of experimentation under edapho-climatic condition, the stratification ratio is greater in conservation tillage than in traditional tillage, although the threshold of 2 reached only when relating surface layers (0–5 cm and 5–10 cm depths) to the 25–35 cm depth was considered (Murillo et al. 2004). Switching from conventional tillage to a conservation tillage system will generally bring about an increase in organic matter, most noticeably at the surface and within the top 2 in. of the soil. In conventional tillage, plant residues, fertilizers, and other amendments become homogenized and diluted as they are mixed deeply into the soil. But with conservation tillage, the residues and amendments reside in a layer near the soil surface, and plant roots tend to proliferate in the top 2 in. of soil where the nutrients are located. Reicosky et al. (2002) found that 30 years of fall moldboard plowing reduced the SOC whether the above ground corn biomass was removed for silage or whether the stover was returned and plowed into the soil. Tillage tended to decrease the SOC content, although only no till combined with stover return to the soil resulted in an increase in SOC in the surface layer compared with moldboard plowed (Hooker et al. 2005).

Qin et al. (2007) demonstrated that soil organic matter down to 10 cm depth was up to 10 % higher in no-tillage than traditional tillage after 4 years in semiarid Inner Mongolia. Under semiarid condition, the mean SOM in the 0–5 cm soil layer was 17.9 g/kg for the three conservation tillage treatments (no-tillage with straw cover, subsoiling with straw cover and rototilling with straw cover), which was significantly greater than the 14.3 g/kg observed on the traditional tillage. The SOM difference between conservation and traditional tillage declined in the deeper layers, but were still significant at 20 cm depth. Mean SOM of no-tillage with straw cover was greater in the 0–5 cm layer than subsoiling with straw cover and rototilling with straw cover treatments, but not below 5 cm depth (He et al. 2009). At Frick, Switzerland, organic carbon (C_{org}) under reduced tillage in the 0–10 cm soil layer was 19 % higher than the initial values while no significant differences found in the 10–20 cm soil layer (Gadermaier et al. 2011). The increase in C_{org} took place only in the 0–15 cm soil layer, and no differences were reported below 15 cm (Koch and Stockfisch 2006). Alvarez (2005) found no differences in C_{org} accumulation between no-tillage and reduced tillage. Emmerling (2007) reported a relative increase in C_{org} of 7–10 % in the surface layer with no differences below the tilled layer after 10 years of reduced tillage under organic farming conditions. Other studies found an increase in C_{org} in the superficial layer but a decrease in the untilled soil layers (Kay and VandenBygaart 2002). Baker et al. (2007) argue that C_{org} gains in most cases are based only on near-surface samples (0–30 cm) and changes in C_{org} disappear when deeper sampling (below 30 cm) was included. Ogle et al. (2005) found

C_{org} increased by 16 % in 0–30 cm depth after 20 years of no-tillage in a temperate wet climate. The amount of C_{org} integrated over 30 cm soil depth under no-tillage and reduced tillage was 14 % higher than under conventional tillage while total carbon more in the 0–5 cm layer but similar in the 5–20 cm layer under no tillage Six et al. (1999).

3.2.2 Soil PH

Soil pH were significantly lower under reduced tillage as compared to conventional tillage. The decrease of pH was highest under RT in 0–10 cm as compared to 10–20 cm (Gadermaier et al. 2011). According to Rasmussen (1999) soil acidity under RT increases in the long run by 0.2–0.3 units in topsoil, this may be due to an accumulation of organic acids in the superficial layer. On the other hand, conventional tillage prevents ions from leaching by turning the soil and thus retards acidification of the topsoil (Friedel et al. 1996).

3.2.3 Soil Life

Microbial biomass and activity are considered to be early indicators of changes in soil properties induced by tillage regimes (Kandeler et al. 1999). Soil microbes decompose organic matter. In doing so, they derive energy for themselves by breaking the long chains of carbon molecules that compose organic matter. As they do this, many plant nutrients attached to these chains are released. These nutrients are absorbed by the organisms themselves or cycled to plants. When the plants or microbes die, nutrients can be recycled once again. Thus, soil microbial biomass serves as both a source and a sink for nutrients. The soil biological community in conservation systems differs from that found in conventional tillage systems. In conventional tillage, plowing and other tillage operations bury plant residues deeply into the soil, where they decompose rather quickly, primarily because bacteria are the decomposers. But in conservation tillage, when the soil is left undisturbed and plant residues remain at the surface, the primary decomposers are fungi. Decomposition is much slower. Slower decomposition results in an increased accumulation of organic materials and a rise in populations of assorted secondary decomposer organisms, such as nematodes and arthropods (insects and spiders). Earthworms and other tunneling invertebrates, such as the larvae of some beetles, proliferate in this soil environment. This community of decomposer organisms helps to maintain soil structure and improve infiltration and aeration within the soil. Plant root growth proliferates in the tunnels and channels these organisms provide.

Friedel et al. (1996) found a high dependence of microbial biomass distribution in the soil on the amount of fresh decomposable organic matter. Soil microbial biomass C (C_{mic}) and N (N_{mic}) and microbial activity (dehydrogenase activity) were highly stratified under reduced tillage, whereas they were relatively homogeneously distributed throughout the soil profile under conventional tillage. Soil microbial biomass was greater under reduced tillage in the 0–10 cm soil layer, C_{mic} being 37 % and N_{mic} 35 % greater than under conventional tillage. The C_{mic} to C_{org} ratio, which is considered to be an indicator of biological soil fertility (Stockfisch et al. 1999) was 14 % greater under reduced tillage than under conventional tillage in the 0–10 cm soil layer. Microbial activity (dehydrogenase activity) was greater by 57 % under reduced tillage compared to conventional tillage in the 0–10 cm soil depth. In the 10–20 cm layer, dehydrogenase activity was greater under reduced tillage by 17 % as compared to conventional tillage (Gadermaier et al. 2011). Angers et al. (1993) found C_{mic}-to-C_{org} ratio three times greater under reduced tillage compared to conventional tillage in 0–16 cm after 11 years of silage maize rotation and low input of organic matter. An increase in the C_{mic} to C_{org} ratio of 16 % in the superficial layer was also reported by Emmerling (2007).

A strong differentiation of the microbial biomass between tilled and untilled layers under reduced tillage and more in tilled layer as compare to untilled under reduced tillage was found by authors (Emmerling 2007; Von Lutzow et al. 2002). Emmerling (2007) found an increase in fungi in the upper soil layer under no-tillage and reduced tillage due to green fallow and cereal straw mulched and remained on the field, leaving high amounts of lignin and cellulose as a favorable substrate for the fungal population. Emmerling (2007) and Von Lutzow et al. (2002) found soil respiration and alkaline phosphomonoesterase significantly higher in 0–15 cm under reduced tillage but no difference in the soil layer below. Some author found no difference in microbial biomass C in the untilled soil layer (Kandeler et al. 1999; Friedel et al. 1996) while others reported less microbial biomass in the untilled layer (Emmerling 2007 and Stockfisch et al. 1999). Alvarez and Alvarez (2000) reported more active microbial biomass in the 0–5 cm layer under no tillage and no difference in total soil microbial biomass in the 0–5 cm layer under no tillage. Chan (2001), Kladivko (2001) and Birkas et al. (2004) also reported more earthworms, nematodes under no tillage. The number of earthworms and their activity increase in conservation tillage compared with conventional tillage. Ploughing disrupts earthworm soil habitats, especially deep burrowing species and exposes earthworms to predation and desiccation (Holland 2004). In the same way, the increase of fresh organic matter in organic farming is an additional resource stimulating trophic and burrowing activity of earthworms (Glover et al. 2000; Shepherd et al. 2000).

3.2.4 Soil Structure

Tillage destroys aggregate stability and promotes compaction and crusting. Conservation tillage systems help to build and preserve the aggregate stability of soils. Good soil structure is important in allowing crop plants to yield well and resist erosion caused by the action of rainfall, melting snow and wind. Soil organic matter, concentrated near the soil surface with conservation tillage and especially labile

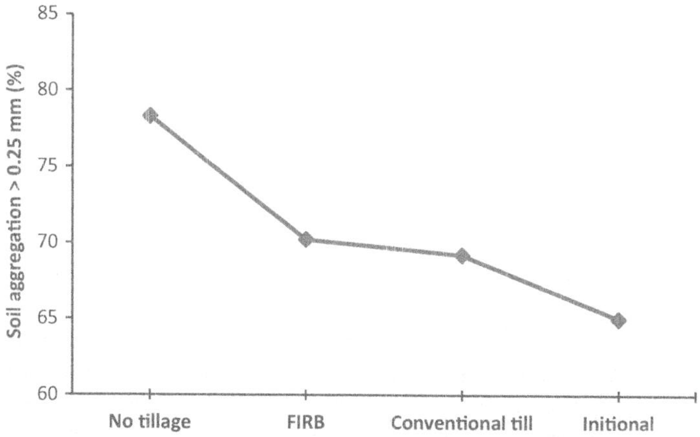

Fig. 3.1 Effect of tillage methods on soil aggregation in maize-wheat system (Jat et al. 2005)

organic matter (Ball et al. 2005), encourages microbial activity leading to increased soil aggregate stability and improved soil structure. Conservation tillage practices were associated with a greater percentage of macro-aggregates (>0.25 mm) than conventional (Fig. 3.1). Mean macro-aggregates in 0–30 cm soil depth at Daxing were 22.1 and 12.0 % greater under shallow tillage and no tillage than conventional tillage and the improvements at Changping were 18.9 % under shallow tillage and 9.5 % under no tillage (Zhang et al. 2009). These results were consistent with the increase in aggregation occurring as a result of greater biological activity in minimum tilled soils, demonstrated by Tisdall and Oades (1982) and with a reduction in the breakdown of surface soil aggregates as a result of residue cover of soil surface and the absence of tillage (Oyedele et al. 1999). Soils from conservation tillage contained more macro-aggregates (13–37 %) than those under traditional tillage throughout the soil profile while percentage of micro-aggregates was 25–59 % greater in traditional tilled soils (He et al. 2009). Improved aggregate stability under conservation tillage, particularly under no tillage management, was also a consequence of increased soil organic matter and reduced disturbance of the soil by tillage (Oyedele et al. 1999; Zhang et al. 2007). Fungal hyphae, more abundant in the surface layer in conservation tillage play an important role in aggregating and stabilizing soil structure.

3.2.5 Bulk Density

If bulk density becomes too high, it can limit plant root growth. For this reason, bulk density is frequently identified as indicator of soil quality (Logsdon and Karlen 2004). Osunbitan et al. (2005) observed greater bulk density in no-till system in the 5–10 cm soil depth. Jabro et al. (2009) in a 22 years study on a sandy loam soil found that the tillage practices (no-till, spring till, and fall and spring till) apparently had not significantly influenced the soil bulk density and only slight differences were observed in bulk density. These findings are in agreement with those of Anken et al. (2004) and Lampurlanes and Cantero-Martinez (2003). Based on 8 years studies, Zhang et al. (2003) reported that the mean soil bulk density was 0.8–1.5 % lower in sub-soiling with retention of all surface plant residues and no tillage treatments (consisted of zero tillage; planting was through the previous plant residues.) than in conventional tillage (consisted of manually removing all plant residues from the soil surface, followed by mouldboard ploughing). The crop residue retention has been reported to increase soil organic carbon and biotic activity (Karlen et al. 1994; Tiarks et al. 1974; Schjonning et al. 1994), thereby decreasing bulk density, particularly near the soil surface in the sub-soiling with retention of all surface plant residues and no tillage. Fabrizzi et al. (2005) found greater soil bulk density under conservation tillage than conventional tillage. The soil bulk density values found under no tillage were the lowest, passing from 1.26 to 1.18 Mgm^{-3} at 10–30 cm, as increasing soil depth (Sessiz et al. 2010). Fabrizzi et al. (2005) observed changes in bulk density in tillage system could also be related to the type of machinery used for harvest and the soil compaction at harvest. Osunbitan et al. (2005); Hammad and Davelbeit (2001) and Olaoye (2002) observed significantly higher bulk density under no tillage cultivation when compared to conventional tillage treatment. Li et al. (2007) obtained no-tillage with residues retained reduced mean bulk density by 0.06 Mgm^{-3} on silt loam soils.

3.2.6 *Porosity*

Porosity is a measure of the total pore space in the soil. Porosity characteristics differ among tillage systems (Benjamin 1993). Soil porosity characteristics are closely related to the soil physical behavior, root penetration and water movement (Pagliai and Vignozzi 2002; Sasal et al. 2006). The increased porosity is especially important for the crop development since it may have a direct effect on the soil aeration and enhances the root growth (Oliveira and Merwin 2001). The improved root growth would hence increase plant water as well as nutrient uptake. Zhang et al. (2003) compared the mean aeration porosity in the top 0–0.30 m between conservation tillage and conventionally tilled soil. The results illustrated an improvement in the soil porosity under conservation tillages (sub-soiling with retention of all surface plant residues; and zero tillage and planting was through the previous plant residues) was most probably related to the beneficial effects of soil organic matter caused by minimum tillage and residue cover. Within the conservation tillage treatments, sub-soiling with retention of all surface plant residues produced more aeration porosity than zero tillage, but the effect on capillary porosity appeared to be reversed in the 0–0.30 m soil layer. The highest value of porosity was found under zero tillage method (55.47 %) as compared to other method of tillage at 20–30 cm depth (Sessiz et al. 2010). Mean total porosity was 42 % on conservation tillage plots and 38 % on traditional tillage plots. The increased porosity was largely due to an increase in macroporosity and mesoporosity in the conservation tillage. In the 0–10 cm soil layer, macroporosity and mesoporosity on conservation tillage were

14 and 4.6 % greater, but microporosity was 5.9 % less than traditional tillage. In deeper soil layers, conservation tillage treatments also had significantly greater (75 %) macroporosity in the 10–20 cm soil layer, as well as a 17 % increase in mesoporosity in the 20–30 cm soil layer. Mean microporosity in the 10–30 cm soil layer was reduced by 19.5 % (He et al. 2009). Zhang et al. (2006) found an increase in mesoporosity in the 0–10 cm soil layer of only 1.6 % compared to ploughing during 3-years.

3.2.7 Soil Temperature

Soil temperature has an inverse relationship with the amount of residue cover (Radke 1982). The decrease in soil temperature is due to the influence of surface residue by reflecting solar radiation and insulating the soil surface (Shinners et al. 1993). Solar energy at the soil surface is partitioned into soil heat flux, sensible heat reflection, and latent heat for water evaporation. Heat flux in soils depends on the heat capacity and thermal conductivity of soils, which vary with soil composition, bulk density, and water content (Hillel 1998; Jury et al. 1991). Soil particles have a lower heat capacity and greater heat conductivity than water, thus dry soils potentially warm and cool faster than wet soils. Tillage processes alter rates of soil drying and heating because tillage disturbs the soil surface, it also increases air pockets in which evaporation occurs, and ultimately accelerates soil drying and heating. Hillel (1998) explains this phenomena or change in soil temperature with tillage due to the change in soil thermal conductivity, where tillage caused a lower soil thermal conductivity compared to that of the untilled soil. Soil disturbance due to tillage can shift or change the air to soil particles volume by creating additional air pockets that can be responsible for reducing the heat capacity of the tilled zone. The average maximum daily spring soil temperature under corn and soybean (Glycine max L. Merr.) residue was reduced by an average of 5.25 °C at a 5 cm soil depth (Kaspar et al. 1990). Therefore, early spring corn growth and development could significantly be reduced under no-tillage conditions. The increase in soil temperature in the surface layer is due to the influence of tillage on surface residue cover, where residue normally reduces surface temperature by reflecting solar radiation and insulating the soil surface (Shinners et al. 1993; Van Wijk et al. 1959). Residues from corn, wheat, and grass sod maintain cooler soil than residue from soybeans and other crops that produce less residue or residue that decomposes rapidly. Crop residues reduce the evaporation of water from soil by shading, causing a lower surface soil temperature (Klocke et al. 2009).

3.2.8 Infiltration Rate and Water Content

Infiltration is the process by which water on the ground surface enters into the soil. Infiltration is governed by two forces viz; gravity, and capillary action. Tillage disturbs the natural channels that have formed in a soil. The increase in porosity when soil is tilled may not result in an increase in the infiltration rate because of disruption of the vertical continuity of the pores (Kooistra et al. 1984). No-tillage systems are very effective in reducing evaporation from soil, to increase the water holding capacity and soil moisture and increase water infiltration. The use of soil covers reduces water evaporation and therefore water is available for crop production. No tillage systems increase soil water infiltration substantially compared to the infiltration of the moldboard-ploughed soil (Fig. 3.2). The covered surface of no-tillage fields acts as a protective skin for the soil. This soil skin reduces the impact of raindrops and buffers the soil from temperature extremes as well as reducing water evaporation. The plant roots are important in forming new channels (Parker and Jenny 1945). Antapa and Angen (1990) reported that retaining crop residues on the soil surface with conservation tillage would reduce evapo-transpiration and increase infiltration rate. Increased earthworm population under conservation tillage favoured water flow and infiltration (Hangen et al. 2002).

Jabro et al. (2009) in a long term study evaluated that zero tillage plots had greater gravimetric water content (0.141 g g^{-1}), followed by sub-soiling with retention of all surface plant residues having 0.139 g g^{-1}, and followed by intensive tillage with a mean of 0.135 g g^{-1}. Zhang et al. (2009) showed soil mean gravimetric water content values averaged across three tillage systems (zero tillage and planting was through the previous plant residues; sub-soiling with retention of all surface plant residues; and intensive tillage) were 0.144, 0.136, and 0.135 g g^{-1} at 0 to 5 cm, 5 to 10 cm, and 10 to 15 cm depths, respectively. The impact of no-tillage system on conserving soil moisture in the top 5 cm following 12 years of various tillage systems was documented in a study by Karlen et al. (1994) in which the gravimetric soil moisture of the no-tillage system had a gravimetric water content of 32.4 %, compared to 25.5 and 23.1 % for chisel plow and moldboard plow systems, respectively. Fortin (1993) found that bare row no-tillage and conventional tillage had a lower water content from planting to emergence than no-tillage with in-row residue cover, whereas the inter row water content of both no-tillage was higher than that for conventional tillage. Under

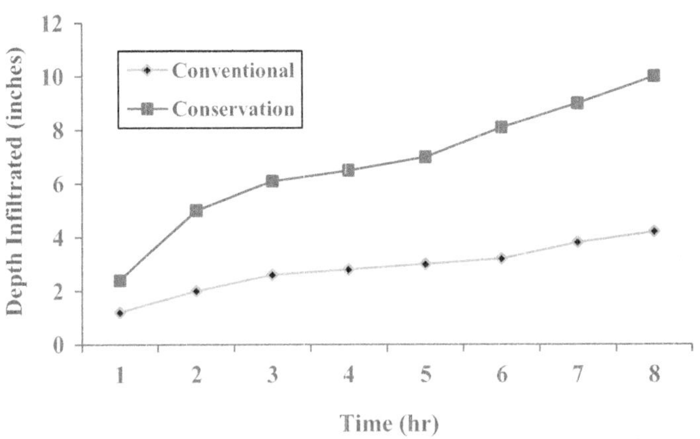

Fig. 3.2 A comparison of the average depth of water infiltrated in the conventional and conservation tillage treatments (Martin et al. 2004)

edapho-climatic conditions (rainfed condition, arid or semi-arid climates,) conservation tillage (reduced tillage) can lead to important improvements in the water storage in the soil profile (Moreno et al. 1997, 2001). Rong (2004) conducted that no-tillage increased soil water content by 4 % (0–20 cm) compared to traditional ploughing on silt loam soils. Under semiarid agriculture, mean soil water storage in the 0–30 cm layer was about 10 % greater on the conservation tillage plots (59 mm) than on traditional tillage (54 mm) (He et al. 2009). No tillage had the greatest soil water storage and conventional tillage the least, with reduced tillage had intermediate values in the 30-cm surface under rain fed condition in Northeast China (Shuang et al. 2013).

3.2.9 *Hydraulic Conductivity*

The hydraulic conductivity of a soil is a measure of the soil's ability to transmit water when submitted to a hydraulic gradient. Soil macropores and aggregations under no tillage formed by decayed roots can be preserved under no tillage whereas conventional tillage breaks up the continuity of these macropores. Macropores generally occupy a small fraction of the soil volume but their contribution to water flow in soil is high. Mahboubi et al. (1993) found that no-tillage resulted in higher saturated hydraulic conductivity compared with conventional tillage after 28 years of tillage on a silt loam soil in Ohio. Whereas, Chang and Lindwell (1989) did not observed any changes in the saturated hydraulic conductivity after 20 years of tillage in a clay loam soil in Alberta. Heard et al. (1988) reported that saturated hydraulic conductivity of silt clay loam soil was higher when subjected to 10 years of tillage than no-tillage in Indiana. Jabro et al. (2009) reported that the Soil saturated hydraulic conductivity (K_s) was slightly influenced by tillage and varied from 3.295 mm h^{-1} for intensive tillage to 5.297 mm h^{-1} for no tillage. Continuous tillage of 11 years had developed a compacted layer that impeded water movement at a depth of approximately 10–15 cm (Pikul and Aase 2003).

3.2.10 *Nutrients Pools and Availability in Soil*

In conservation tillage, especially no tillage, there is a greater pool of soil labile N from microbial activity in the surface layer. However, this pool has a slower turnover rate caused by the decrease of the decay rate of SOM (Balesdent et al. 2000; Kay and VandenBygaart 2002). Pekrun et al. (2003) indicated that net N immobilization can occur with slow SOM turn over during the transition period from conventional to conservation tillage. Conservation tillage can usually lead to greater accumulation of surface nutrients, compared to traditional tillage, with soil ploughing (Franzluebbers and Hons 1996; Holanda et al. 1998; Roldan et al. 2005; Li et al. 2007). Stratification of relatively immobile nutrients, such as phosphorus and potassium, with high concentrations near the soil surface and decreasing concentrations with depth has been routinely observed where no till and other conservation tillage systems have been used for at least 3–4 years. This stratification results from both the addition of fertilizer to the soil surface and from the "cycling" of nutrients, in which roots take up nutrients from well below the soil surface; some of these nutrients are then deposited on the soil surface in the form of crop residue A stratification of soluble P_{CO2} and K_{CO2}, was found after 6 years under RT. P_{CO2} in the 0–10 cm soil layer was greater by 72 % in reduced tillage than in conventional tillage plots, while the exchangeable $P_{Aac-EDTA}$ was only greater by 27 %. K_{CO2} was greater by 40 % in reduced tillage than in conventional tillage in the 0–10 cm layer and $K_{Aac-EDTA}$ was greater by 23 %. There were no significantly tillage effects in the 10–20 cm soil layer (Gadermaier et al. 2011).

Rasmussen (1999) reported a significant increase in plant available P in 0–5 cm soil depth under reduced tillage while available P in 10–20 cm remained stable or even decreased. Vu et al. (2009) found a concentration of plant available P in 0–10 cm under no tillage. A high accumulation of C_{org} was closely related to organic P dynamics, as organic P accumulates only when C availability is high (Bunemann et al. 2006). Plant available K in the top layer was greater under reduced tillage, whereas there were no differences between reduced tillage and ploughed soil in 10–20 cm (Rasmussen 1999). Diaz-Zorita and Grove (2002) reported, surface enrichment of P under non tillage systems (following a similar distribution pattern to that of organic C) was observed in soils to which phosphate fertilizers have not been applied. The quantity of Olsen's P was 34.5 % greater under conservation tillage than under traditional tillage in the 0–5 cm layer. Below 5 cm, this pattern was reversed with traditional tillage containing 8.0–24 % more Olsen's P than conservation tillage. A similar increase was found for total N on the conservation tillage plots, but differences were only significant for the 0–5 cm layer. According to Sisti et al. (2004) the combined action of conservation tillage and the input of fresh organic matter as leguminous residues increased the soil C and, in the long term, improved the mineral N supply to crops. Conservation tillage usually improves the availability of surface phosphorus by converting it into organic phosphorus. In standard tillage systems this phosphorus would be remixed into the soil profile, whereas in conservation tillage it accumulates at the surface (Robbins and Voss 1991; Zibilske et al. 2002). Motta et al. (2002) showed, soils under no tillage usually have higher surface Ca contents compared to traditional tillage, often attributed to the greater exchange capacity of the soil.

3.3 Agronomic Parameters

3.3.1 Water Use Efficiency

The benefits of conservation tillage become even more obvious for water use efficiency. Water use efficiency ranged from 3.8 to 5.4 kg/ha/mm for conservation tillage and from 3.6 to 4.5 kg/ha/mm for traditional tillage (He et al. 2009). The application of no-tillage practices can have positive consequences on the water use in crop production. The water-use efficiency was higher in no tillage with mulching and subsoil tillage with mulching as compared to conventional tillage (Ziyou et al. 2007). Precipitation use efficiency was higher in subsoiling with mulch and no tillage with mulching as compared to conventional tillage and reduced tillage (Ke Jin et al. 2007). Retention of crop residues on soil surface, along with fertilization with organic manure and involvement of legumes in crop rotation coupled with minimum/no-tillage practices play an important role to improving water use efficiency (Sainju et al. 2008; Mohammad et al. 2003; Dalal and Chan 2001; Lampurlanes et al. 2001). A number of studies from both irrigated and rain-fed regions around the United States where no-tillage is used have reported annual irrigation savings of as much as 4 to 5 in. (10–13 cm) Klocke et al. (2009). In Nebraska, switching from conventional tillage to no-tillage under center-pivot irrigation has been shown to save 3 to 5 in. (8–13 cm) of water annually Pryor (2006).

3.3.2 Penetration Resistance

Soil porosity, structure, and strength are impacted by excessive soil compaction and are often differentiated by penetration resistance (Croissant et al. 1991; Voorhees 1983). Penetration resistance is a common measure of soil strength, where increased penetration resistance restricts root growth (Singh et al. 1992; Taylor and Ratliff 1969; Voorhees et al. 1975). It was determined that penetration resistance of no-tillage was slightly higher compared with that of chisel plow in the top 10 cm of the soil (Erbach et al. 1992). Under a wheat–sorghum–fallow crop rotation, no-tillage had a greater surface penetration resistance than a minimum tillage system (Unger and Jones 1998). Fabrizzi et al. (2005), Bayhan et al. (2006) and Sessiz et al. (2010) had shown increases of penetration resistance values under no tillage compared with that in conventional tillage. Soil compaction causes low porosity, reduced infiltration, increased penetration resistance and limited root growth.

3.3.3 Weed Pressures

Tillage influence weed populations by the combined effects of mechanical destruction of weed seedlings and by changing the vertical distribution of weed seeds in the soil. Tillage also acts indirectly on weed populations, through the changes in soil conditions, influence weed dormancy, germination and growth. Weed seeds are more uniformly distributed in the topsoil with conventional tillage, but are mainly located in the first few centimeters of soil under conservation tillage (Moonen and Barberi 2004; Mohler et al. 2006). Perennial and annual grasses are more highly represented in conservation tillage than in conventional (Moonen and Barberi 2004). In conservation tillage, the seed–soil contact, modified by interference with crop residues, could be less advantageous for germination and emergence of small-seeded weeds (Bond and Grundy 2001). Nevertheless, a greater proportion of the seed bank germinates in conservation tillage (Kouwenhoven et al. 2002) favouring the emergence of grass weeds and other species with a large rate of seed production (El Titi 2003). For dicotyledonous weeds, the impact of tillage systems depends on the species (Moonen and Barberi 2004). For instance, conventional tillage tends to increase some annual dicotyledons, such as *Chenopodium* sp. and *Papaver rhoeas*, when their persistent seeds are brought back to the surface by ploughing (Locke et al. 2002). With conservation tillage, the germination of older and deeper located persistent weed seeds was slowed down (Moonen and Barberi 2004). Weeds with creeping roots or rhizomes are favoured by the absence of tillage (Torresen et al. 2003). However, conservation tillage with tines or discs can also assist their development by disrupting and dispersing their rhizomes, especially *Agropyrum repens*. Kouwenhoven et al. (2002) suggested that shallow ploughing (12–20 cm) was the best reduced tillage in shrink/swell soils for controlling weeds, especially perennials. In conservation tillage, seed predation is increased (El Titi 2003) and soil disturbance responsible for weed seed germination is decreased, both leading to less weed seed return to the soil and in the long term a depleted weed seed bank.

Wheat sown after puddle rice recorded higher population of weeds compared to sown after unpuddled rice. Conventionally tilled wheat resulted the highest mean population of weeds followed by zero-tilled wheat. This was due to the fact that intensive tillage operation in conventional tillage treatment brought out the weed seeds from sub surface to favourable moist upper soil layer for good germination. Contrary to this, in zero-tillage treatment weed seeds remained in sub surface due to puddling carried out during paddy transplanting and failed to germinate because of unfavorable condition (Singh et al. 2008). In rice crop, highest mean population and dry weight of weeds (*Cyperus rotundus*, *Fimbristylis miliaceae* and *Alternenthera sessilis*) was recorded with direct seeding by zero-till-drill without tillage and lowest being with transplanting methods (Singh et al. 2008). Controlling weeds is essential for profitable production with any tillage system. With less tillage, weed control becomes more dependent on herbicides. However, effective herbicides are available for controlling most weeds in conservation tillage systems. Herbicide selection and application rate, accuracy, and timing become more important.

3.3.4 Yield and Quality

Crop germination, emergence, and growth are largely regulated by soil temperature, aeration, and moisture content, by nutrient availability to roots, and by mechanical impedance to root growth. All of these factors are affected by tillage. Samarajeewa et al. (2006) pointed out that conservation tillage systems could be more productive than conventional tillage systems as a result of improved soil quality and water use efficiency of plants. Under a wide range of environment conditions crop yield obtained by no-till, reduced tillage, and stubble retention systems were equivalent or even a higher than those recorded under conventional tillage (Barzegar et al. 2003).

Yields of forage crops were greater under reduced tillage as compared to conventional tillage (Gadermaier et al. 2011). Zhao et al. (2007) compared no-tillage with full residue cover and ploughing with all residues removed. Their results indicate that no-tillage improved mean yields by up to 7 % in wheat and potato crop. On average, the yields on conservation tillage treatments were greater than those on the traditional tillage, with significant differences after 6 years of continuous conservation tillage. The mean yield advantage of conservation tillage was relatively small (6 %) in the first 4 years of the experiment, but this increased to a mean value of 13 % in the subsequent 6 years (He et al. 2009). De Vita et al. (2007) studied effects of no tillage and conventional tillage on wheat yield. They found that greater yield is obtained with no tillage than as with conventional tillage. Protein, oil and ash content of corn were not affected statistically by tillage method (Sessiz et al. 2010). According to De Vita et al. (2007) no significant differences in grain protein content by tillage method. Yusuf et al. (1999) reported that soybean grain oil and protein content were not affected by soil tillage system, but significantly affected by years. Lueschen et al. 1991, in a corn-soybean rotation in Minnesota, found an increase of 6.30 Mg ha^{-1} in yield of the no tillage system above the mouldboard ploughing system in a dry year. Kapusta et al. (1996), studied the effects of tillage systems for 20 years and found equal corn yield for no-till, reduced till, and conventional tillage systems despite the lower plant population in no-till. Average maize and soybean yields over 11 years were 33 and 31 % higher, respectively, under no-tillage than under conventional tillage (Thiagalingam et al. 1996).

Rice wheat is a dominant cropping system in India and happy seed drill is used for direct sowing of wheat (Figs. 3.3 and 3.4). These two crops together contribute 74 % of the total food grain production in the country. Wheat crop establishment

Fig. 3.3 Zero till sowing of wheat by happy seed drill in farmer field in (Punjab, India)

Fig. 3.4 Field view of zero till sown wheat by happy seed drill in farmer field (Punjab, India)

after rice is also an important factor in improving the productivity of rice-wheat system. Plant stand of wheat is influenced by soil type, moisture content, tillage and time of seeding. Poor plant stand and delayed wheat sowing after rice limits the productivity of wheat. Delayed sowing of wheat beyond November reduces grain yield by 30–50 kg/ha/day (Randhawa et al. 1976). The average winter wheat yields over 6 years on tillage with mulching and subsoil tillage with mulching plots were significantly higher than that in conventional tillage or reduced tillage plots (Ziyou et al. 2007). Wheat crop grown by zero-tillage techniques was able to compete with crop grown by conventional tillage method due to favourable effect of early sowing, early emergence of seedlings and availability of higher moisture and nutrients and produced grain and straw yield and harvest index at par with conventional tillage method (Gupta et al. 2007; Dash and Varma 2003). Zero tillage recorded significantly higher number of spikes, grains/spike and wheat yield compared with the conventional tillage possibly due to the cumulative effect of zero tillage and one week advance sowing (Tripathi and Chauhan 2000). Aslam et al. (1999) found that no-tillage produced 10 % more wheat grain compared to farmers practices of minimum three ploughings with planking in rice-based cropping system in Pakistan. Hunt et al. (1997) reported that no yield loss was found when no-till system was used in winter wheat agriculture after cotton. Gürsoy et al. (2010) stated that permanent raised beds proved to be an excellent option for wheat and offered potential benefits in terms of higher crop yield. Gwenzi et al. (2009) reported that the effect of tillage methods on crop yields was inconsistent in an irrigated wheat-cotton rotation throughout the

6-year period and the lower wheat yield under minimum and no-tillage resulted from the higher weed infestation and poor crop stand, also reported by Li et al. (2008).

3.3.5 Nutrient Uptake

Nutrients usually are stratified in conservation tillage systems because of the lack of substantial mechanical soil mixing. Nutrient levels tend to be higher near the soil surface where the nutrients are applied and where crop residues decay. The maximum uptake of nutrients by wheat was recorded in rice plots seeded by zero-till-drill without tillage after spray of glyphosate followed by dry seeding in conventional tillage. Wheat sown in reduced tillage depleted significantly higher amount of nutrients followed by zero tilled wheat owing to higher grain yield Parihar (2004). Uptake of N, P or K by wheat crop was usually greater under zero tillage than under conventional tillage, and there were no interactions between tillage and crop residue (Singh et al. 2008). Nitrogen, phosphorus, potassium, sulfur and zinc uptake in maize and soybean grain were higher under no-tillage as compared to conventional tillage (Thiagalingam et al. 1996). Total grain N content is often found to be greater under no-tillage system due to greater N use efficiency in no-tillage crops than crops grown in conventional tillage systems (Angle et al. 1993). Several studies have found that no-tillage increased total grain N uptake slightly compared with conventional tillage and generally equaled that of conservation tillage (Sainju and Singh 2001). On the other hand, some studies found N deficiencies are more common in no-tillage than conventional tillage systems (Mehdi et al. 1999) translating into less grain N uptake.

3.4 Economics

Conservation tillage is becoming increasingly attractive to farmers because clearly reduces production cost relative to conventional tillage (De Vita et al. 2007). The zero tillage sowing was more remunerative than the conventional method of sowing in wheat crop (Gupta et al. 2007). In relation to CT (conventional tillage), the economic benefit of reduced tillage, no tillage with mulching and subsoil tillage with mulching increased 62, 1,754, and 1,467 Yuan ha^{-1}, respectively, and the output/input ratio of conservation tillage was higher than that of conventional tillage (Ziyou et al. 2007). Wheat sown in zero-tilled soil accrued the highest net return and benefit: cost ratio as compared to conventional tilled soil (Singh et al. 2008). Reduced tillage was a good compromise for both soybean and corn crops, as it were associated with higher economic returns for farmers in Northeast China (Shuang et al. 2013). Gursoy et al. (2010) stated that permanent raised beds proved to be an excellent option for wheat and offered potential benefits in terms of lower production costs and higher crop yield. Zero tillage gave maximum benefit cost ratio, which was at par with raised bed system but significantly higher than conventional tillage (Kumar et al. 2004).

3.5 Energy Use-Efficiency

Effective energy use in agriculture is one of the conditions for sustainable agricultural production, since it provides financial savings, fossil resources preservation and air pollution reduction (Uhlin 1998). Energy and environmental security are major problems facing our global economy (Barbir and Ulgiati 2008). Tillage is a main operation in a crop production system that affected energy input (Claus and Villy 2005). In the Sacramento Valley, an average savings of 50 % for fuel and 72 % for time have been reported with one-pass tillage equipment compared with the standard tillage program of disking and land planing (Upadhyaya et al. 2001). Conventional tillage method had highest fuel consumption (33.48 L ha^{-1}) and lowest field efficiency (0.29 ha h^{-1}) as compared to the other tillage method. Direct seeding method, no tillage had the lowest fuel consumption (6.6 L ha^{-1}) with maximum field efficiency (1.87 ha h^{-1}). The conventional method requires five times more fuel comparing the no tillage method. Beside this, no tillage methods had six times more field efficiency comparing the conventional method (Sessiz et al. 2010). Energy use-efficiency and energy productivity were significantly more under zero tillage than conventional tillage. Each mega joule of input energy produced significantly maximum wheat yields under zero tillage as compared to conventional tillage. Zero tillage registered 8.30 and 8.13 % more energy productivity than conventional tillage during 2003–2004 and 2004–2005 respectively. This could be attributed to lesser energy (operation time, manual labour and fuel) requirement under zero tillage than conventional tillage (Srivastava et al. 2000; Gupta et al. 2007).

3.6 Environment

3.6.1 Impacts on Greenhouse Gas Emissions

Agricultural eco-systems represent an estimated 11 % of the earth's land surface and include some of the most productive and carbon-rich soils. As a result, they play a significant role in the storage and release of C within the terrestrial carbon cycle (Lal 1995). During 2005, agriculture accounted for 10–12 % of the total global human caused emissions of greenhouse gases, according the Inter-governmental Panel on Climate Change (IPCC 2007). Conservation tillage can directly reduce carbon emissions of a farming system by reducing fuel use. The reduction in fuel consumption for tillage depends on the amount of subsoil tillage required and/or the reduction in the number of trips across the field needed to prepare the land for planting. Also, crop residues maintained on the soil surface can enhance soil carbon storage. Improved carbon sequestration under conservation tillage depends on the climate, management history, and soils of the system (Baker et al. 2007; Manley et al. 2005). The sequestration of atmospheric C in the soil and biomass would not only reduce greenhouse effect, but also helps to maintain or restore the capacity of a soil to perform its production and environmental functions on a sustainable basis. Thus, sequestration of atmospheric C into the soils, maintain or renew soil fertility and mitigating carbon dioxide emissions to the atmosphere. In the northern Great Plains of United States, traditional farming systems, such as conventional tillage with wheat -fallow, have resulted in a decline in soil organic C (SOC) by 30–50 % of their original levels in the last 50–100 years (Peterson et al. 1998). Halvorson et al. (2002a) observed that no-till with continuous cropping increased C sequestration in the drylands of the northern Great Plains by 233 kg ha^{-1} $year^{-1}$ compared to a loss of 141 kg ha^{-1} $year^{-1}$ in conventional tillage. The use of no-till has allowed producers to increase cropping intensity (Peterson et al. 2001) because no-till conserves surface residues and retains water in the soil profile more than the conventional tillage Farhani et al. (1998). Thus reduced tillage and increased cropping intensity could conserve C and N in a soil (Sainju et al. 2007). An extensive review of conservation tillage impacts on soil organic carbon in the Southeast United States showed that a change from conventional to conservation tillage would sequester an additional 400 ± 35 lbs C/acre annually (Franzluebbers 2010).

3.6.2 Soil Erosion

Conservative tillage involving soil management practices that resulted protection against water and wind erosion (Holland 2004; Lopez et al. 2003). Conservation tillage has greater impacts on erosion rates than on runoff and infiltration (Leys et al. 2010). Maintaining crop residue on the soil surface can reduce the severity of erosion (Dickey et al. 1985). Residue absorbs the impact of raindrops, thereby reducing the amount of soil dislodged. It also intercepts water as it moves down the slope, which allows soil particles to settle. The residue left on the surface by conservation tillage systems slows the wind near the soil surface, thereby reducing the movement of soil particles into the air. Conservation tillage leaves organic mulch at the soil surface, which reduces run-off, increases the surface soil organic matter, promoting greater aggregate stability which restricts soil erosion (Franzluebbers 2002a).

3.7 Conclusion

Conservation tillage increased soil acidity, soil aggregate stability, soil porosity and improved soil structure due to increased labile organic matter, microbial biomass and microbial activity. The gravimetric water content and infiltration rate are maximum under no-tillage as compared to conventional tillage. High concentration of immobile nutrients such as phosphorus and potassium stratified near the soil surface in conservation tillage systems. The water-use efficiency is higher in no tillage as compared to conventional tillage. Under a wide range of environment conditions crop yield, net return and benefit: cost ratio obtained by no-till, reduced tillage, and stubble retention systems were equivalent or even a higher than those recorded under conventional tillage. Conservation tillage directly reduces carbon emissions by reducing fuel use and indirectly by sequestration of atmospheric C in the soil and biomass.

Acknowledgements We thank all the scientists whose reviews are used to compile the manuscript.

References

Alvarez R (2005) A review of nitrogen fertilizer and conservation tillage effects on soil organic carbon storage. Soil Use Manag 21:38–52

Alvarez CR, Alvarez R (2000) Short-term effects of tillage systems on active soil microbial biomass. Biol Fertil Soils 31:157–161

Angers DA, Giroux M (1996) Recently deposited organic matter in soil water-stable aggregates. Soil Sci Soc Am J 60(5):1547–1551

Angers DA, N'dayegamiye A, Cote D (1993) Tillage- induced differences in organic matter of particle-size fractions and microbial biomass. Soil Sci Soc Am J 57:512–516

Angle JS, Gross CM, Hill RL, McIntosh MS (1993) Soil nitrate concentrations under corn as affected by tillage, manure, and fertilizer applications. J Environ Qual 22:141–147

Anken T, Weisskopf P, Zihlmann U, Forrer H, Jansa J, Perhacova H (2004) Long term tillage system effects under moist cool conditions in Switzerland. Soil Tillage Res 78:171–183

Antapa PL, Angen TV (1990) Tillage practices and residue management in Tanzania. In: Organic-matter management and tillage in humid and sub-humid Africa. 4957 IBSRAM proceedings 10. Bangkok: IBSRAM

Aslam M, Majid A, Gill MA (1999) Zero tillage wheat production technology: prospects and threats. Sci Technol Dev 18:17–23

Baker JM, Ochsner TE, Venterea RT, Griffis TJ (2007) Tillage and soil carbon sequestration – what do we really know? Agric Ecosyst Environ 118:1–5

Balesdent J, Chenu C, Balabane M (2000) Relationship of soil organic matter dynamics to physical protection and tillage. Soil Tillage Res 53:215–230

Ball BC, Bingham I, Rees RM, Watson CA, Litterick A (2005) The role of crop rotations in determining soil structure and crop growth conditions. Can J Soil Sci 85:557–577

Barbir F, Ulgiati S (2008) Benefits, strategies and environmental costing: proceedings of the NATO advanced research workshop on sustainable energy production and consumption and environmental costing, Naples, Italy, 4–7 July 2007

Barzegar AR, Asoodar MA, Khadish A, Hashemi AM, Herbert SJ (2003) Soil physical characteristic and chickpea yield response to tillage treatments. Soil Tillage Res 71:49–57

Bayhan YB, Kayisoglu E, Gonulol H, Yalcin H, Sungur N (2006) Possibilities of direct drilling and reduced tillage in second crop silage corn. Soil Tillage Res 88:1–7

Benjamin J (1993) Tillage effects on near-surface soil hydraulic properties. Soil Tillage Res 26:277–288

Birkas M, Jolankai M, Gyuricza C, Percze A (2004) Tillage effects on compaction, earthworms and other soil quality indicators in Hungary. Soil Tillage Res 78:185–196

Bond W, Grundy AC (2001) Non-chemical weed management in organic farming systems. Weed Res 41:383–405

Bunemann EK, Heenan DP, Marschner P, McNeill AM (2006) Long-term effects of crop rotation, stubble management and tillage on soil phosphorus dynamics. Aust J Soil Res 44:611–618

Chan KY (2001) An overview of some tillage impacts on earthworm population abundance and diversity – implications for functioning in soils. Soil Tillage Res 57:179–191
Chan KY, Heenan DP, Oates A (2002) Soil carbon fractions and relationship to soil quality under different tillage and stubble management. Soil Tillage Res 63:133–139
Chang C, Lindwall CW (1989) Effect of long term minimum tillage practices on some physical properties of Chernozemic clay loam. Can J Soil Sci 69:433–449
Claus GS, Villy N (2005) Operational analyses and model comparison of machinery systems for reduced tillage. Biosyst Eng 92(2):143–155
Croissant RL, Schwartz HF, Ayers PD (1991) Soil compaction and tillage effects on dry bean yields. J Prod Agric 4:461–464
Dalal RC, Chan KY (2001) Soil organic matter in rainfed cropping systems of the Australian cereal belt. Aust J Soil Res 39:435–464
Dash R, Varma SC (2003) Management of weeds nitrogen and tillage operations in wheat sown after puddle rice. Indian J Agric Sci 73(5):286–288
De Vita P, Di Paolo E, Fecondo G, Di Fonzo N, Pisante M (2007) No-tillage and conventional tillage effects on durum wheat yield, grain quality and soil moisture content in southern Italy. Soil Tillage Res 92:69–78
Diaz-Zorita M, Grove JH (2002) Duration of tillage management affects carbon and phosphorus stratification in phosphatic Paleudalfs. Soil Tillage Res 66:165–174
Dickey EC, Shelton DP, Jasa PJ, Peterson TR (1985) Soil erosion from tillage systems used in soybean and corn residues. Trans Am Soc Agric Eng 28(4):1124–1129
Du Preez CC, Steyn JT, Kotze E (2001) Long-term effects of wheat residue management on some fertility indicators of a semi-arid plinthosol. Soil Tillage Res 63:25–33
El Titi A (2003) Implications of soil tillage for weed communities. In: El Titi A (ed) Soil tillage in agroecosystems. CRC Press, Boca Raton, pp 147–185
Emmerling C (2007) Reduced and conversation tillage effects on soil ecological properties in an organic farming system. Biol Agric Hortic 24:363–377
Erbach DC, Benjamin JG, Cruse RM, Elamin MA, Mukhtur S, Choi CH (1992) Soil and corn response to tillage with paraplow. Trans ASAE 35:1347–1354
Fabrizzi KP, Gorcia FO, Costa JL, Picone LI (2005) Soil water dynamics, physical properties and corn and wheat responses to minimum and no-tillage systems in southern Pampas of Argentina. Soil Tillage Res 81:57–69
Farhani HJ, Peterson GA, Westfall DG (1998) Dryland cropping intensification: a fundamental solution to efficient use of precipitation. Adv Agron 64:197–223
Fortin MC (1993) Soil temperature, soil water, and no-till corn development following in-row residue removal. Agron J 85:571–576
Franzluebbers AJ (2002a) Soil organic matter stratification ratio as an indicator of soil quality. Soil Tillage Res 66:95–106
Franzluebbers AJ (2002b) Water infiltration and soil structure related to organic matter and its stratification with depth. Soil Tillage Res 66:197–205
Franzluebbers AJ (2010) Achieving soil organic carbon sequestration with conservation agricultural systems in the southeastern United States. Soil Sci Soc Am J 74:347–357
Franzluebbers AJ, Hons FM (1996) Soil-profile distribution of primary and secondary plant-available nutrients under conventional and no tillage. Soil Tillage Res 39:229–239
Friedel JK, Munch JC, Fischer WR (1996) Soil microbial properties and the assessment of available soil organic matter in a haplic Luvisol after several years of different cultivation and crop rotation. Soil Biol Biochem 28:479–488
Gadermaier F, Berner A, Fliebach A, Friedel JK, Mader P (2011) Impact of reduced tillage on soil organic carbon and nutrient budgets under organic farming. Renew Agric Food Syst. doi:10.1017/S1742170510000554, pp 1–13
Glover JD, Reganold JP, Andrews PK (2000) Systematic method for rating soil quality of conventional, organic, and integrated apple orchards in Washington State. Agric Ecosyst Environ 80:29–45
Greenland DJ (1981) Soil management and soil degradation. J Soil Sci 32:301–322

Gupta M, Bali AS, Sharma S, Dixit AK (2007) Potential role and influence of zero tillage technology on energy saving in rice (Oryza sativa)-wheat (Triticurm aestivum) system. Indian J Agric Sci 77(10):657–659

Gursoy S, Sessiz A, Malhi SS (2010) Short-term effects of tillage and residue management following cotton on grain yield and quality of wheat. Field Crop Res 119(2–3):260–268

Gwenzi W, Gotosa J, Chakanetsa S, Mutema Z (2009) Effects of tillage systems on soil organic carbon dynamics, structural stability and crop yields in irrigated wheat (Triticum aestivum L.)-cotton (Gossypium hirsutum L.) rotation in semi-arid Zimbabwe. Nutr Cycl Agroecosyst 83:211–221

Halvorson AD, Peterson GA, Reule CA (2002) Tillage system and crop rotation effects on dryland crop yields and soil carbon in the central Great Plains. Agron J 94:1429–1436

Hammad EA, Dawelbeit MI (2001) Effect of tillage and field condition on soil physical properties, cane and sugar yields in vertisols of kenana sugar estate, Sudan. Soil Tillage Res 62:101–109

Hangen E, Buczko U, Bens O, Brunotte J, Huttl RF (2002) Infiltration patterns into two under conventional and conservation tillage: influence of the spatial distribution of plant room structures and soil animal activity. Soil Tillage Res 63:181–186

He J, Kuhn NJ, Zhang XM, Zhang XR, Li HW (2009) Effects of 10 years of conservation tillage on soil properties and productivity in the farming–pastoral ecotone of Inner Mongolia, China. Soil Use Manag 25:201–209

Heard JR, Kladivko EJ, Mannering JV (1988) Soil macroporosity, hydraulic conductivity and air permeability of silty soils under long-term conservation tillage in Indiana. Soil Tillage Res 11:1–18

Hernanz JL, Lopez R, Navarrete L, Sanchez-Gironv V (2002) Long-term effects of tillage systems and rotations on soil structural stability and organic carbon stratification in semiarid central Spain. Soil Tillage Res 66:129–141

Hillel D (1998) Environmental soil physics. Academic, San Diego

Holanda FSR, Mengel DB, Paula MB, Carvaho JG, Bertoni JC (1998) Influence of crops rotations and tillage systems on phosphorus and potassium stratification and root distribution in the soil profile. Commun Soil Sci Plant Anal 29:2383–2394

Holland JM (2004) The environmental consequences of adopting conservation tillage in Europe: reviewing the evidence. Agr Ecosyst Environ 103:1–25

Hooker BA, Morris TF, Peters R, Cardon ZG (2005) Long-term effects of tillage and cornstalk return on soil carbon dynamics. Soil Sci Soc Am J 69:188–196

Hunt PG, Karlen DL, Matheny TA, Quisenberry VL (1996) Changes in carbon content of a Norfolk loamy sand after 14 years of conservation or conventional tillage. J Soil Water Conserv 51(3):255–258

Hunt PG, Bauer PJ, Matheny TA (1997) Crop production in a wheat-cotton double crop rotation with conservation tillage. J Prod Agric 10(3):462–465

IPCC (2007) In: Solomon S, Qin D, Manning M, Chen Z, Marquis M, Tignor KBM, Miller HL (eds) Climate change 2007: the physical science basis. Contribution of Working Group I to the Fourth Assessment Report of the Intergovernmental Panel on Climate Change. Cambridge University Press, Cambridge/New York, p 996

Jabro JD, Stevens WB, Evans RG, Iversen WM (2009) Tillage effects on physical properties in two soils of the Northern Great Plains. Appl Eng Agric 25:377–382

Jat ML, Srivastava A, Shanna SK, Gupta RK, Zaidi PH, Rai HK, Srinivasan G (2005) Evaluation of Maize-wheat cropping system under double-no-till practice in Indo-Gangetic Plains of India. Proceedings of the ninth Asian Regional Maize workshop, 5–9 Sept, Beijing, China

Jin K, Cornelis WM, Schiettecatte W, Lu J, Yao Y, Wu H, Gabriels D, De Neve S, Cai D, Jin J, Hartmann R (2007) Effects of different management practices on the soil–water balance and crop yield for improved dryland farming in the Chinese Loess Plateau. Soil Tillage Res 96(1–2):131–144

Jury WA, Gardner WR, Gardner WH (1991) Soil physics, 5th edn. Wiley, Toronto

Kandeler E, Tscherko D, Spiegel H (1999) Long-term monitoring of microbial biomass, N mineralisation and enzyme activities of a chernozem under different tillage management. Biol Fertil Soils 28:343–351

Kapusta G, Krausz RF, Matthews JL (1996) Corn yield is equal in conventional, reduced, and no tillage after 20 years. Agron J 88(5):812–817

Karlen DL, Wollenhaupt NC, Erbach DC, Berry EC, Swan JB, Eash NS, Jordahl JL (1994) Long-term tillage effects on soil quality. Soil Tillage Res 32:313–327

Kaspar TC, Erbach DC, Cruse RM (1990) Corn response to seed-row residue removal. Soil Sci Soc Am J 54:1112–1117

Kay BD, VandenBygaart AJ (2002) Conservation tillage and depth stratification of porosity and soil organic matter. Soil Tillage Res 66:107–118

Kladivko EJ (2001) Tillage systems and soil ecology. Soil Tillage Res 61:61–76

Klocke NL, Currie RS, Aiken RM (2009) Soil water evaporation and crop residues. Trans ASABE 52(1):103–110

Koch HJ, Stockfisch N (2006) Loss of soil organic matter upon ploughing under a loess soil after several years of conservation tillage. Soil Tillage Res 86:73–83

Kooistra MJ, Bouma J, Boersma OH, Jager A (1984) Physical and morphological characterization of undisturbed and disturbed ploughpans in a sandy loam soil. Soil Tillage Res 4:405–417

Kouwenhoven JK, Perdok UD, Boer J, Oomen GJM (2002) Soil management by shallow mouldboard ploughing in The Netherlands. Soil Tillage Res 65:125–139

Kumar S, Pandey DS, Rana NS (2004) Effect of tillage, rice residue and nitrogen management practices on yield of wheat and chemical properties of soil under rice-wheat system. Indian J Agron 49(4):223–225

Lal R (1974) No-tillage effects on soil properties and maize (Zea mays L.) production in Western Nigeria. Plant Soil 40:321–331

Lal R (1995) Tillage and mulching effects on maize yield for seventeen consecutive seasons on a tropical alfisol. J Sustain Agric 5:79–93

Lal R (2009) Challenges and opportunities in soil organic matter research. Eur J Soil Sci 60:158–169

Lal R, Hahn SK (1973) Effects of methods of seedbed preparation, mulching and time of planting on yam in Western Nigeria. Symposium, Proceedings of the international Society for Tropical Root Crops. Institute of Tropical Agriculture, Ibadan, Nigeria, 2–12 December 1973

Lampurlanes J, Cantero-Martinez C (2003) Soil bulk density and penetration resistance under different tillage and crop management systems and their relationship with barley root growth. Agron J 95:526–536

Lampurlanes J, Angas P, Cantero-Martinez C (2001) Root growth, soil water content and yield of barley under different tillage systems on two soils in semiarid conditions. Field Crop Res 69:27–40

Leys A, Govers G, Gillijns K, Berckmoes E, Takken I (2010) Scale effects on runoff and erosion losses from arable land under conservation and conventional tillage: the role of residue cover. J Hydrol 390(3–4):143–154

Li FB, Niu YZ, Gao WL, Liu JG, Bian XM (2008) Effects of tillage styles and straw return on soil properties and crop yields in direct seeding of rice. Chin J Soil Sci 39:549–552

Li HW, Gao HW, Wu HD, Li WY, Wang XY, He J (2007) Effects of 15 years of conservation tillage on soil structure and productivity of wheat cultivation in northern China. Aust J Soil Res 45:344–350

Locke MA, Reddy KN, Zablotowicz RM (2002) Weed management in conservation crop production systems. Weed Biol Manag 2:123–132

Logsdon SD, Karlen DL (2004) Bulk density as a soil quality indicator during conversion to no-tillage. Soil Tillage Res 78:143–149

Lopez MV, Moret D, Gracia R, Arrue JL (2003) Tillage effects on barley residue cover during follow in semiarid Aragon. Soil Tillage Res 72:53–64

Lueschen WE, Evans SD, Ford JH (1991) Soybean production as affected by tillage in a corn and soybean management system: I. Cultivar response. J Prod Agric 4(4):571–579

Mahboubi AA, Lal R, Fausey NR (1993) Twenty-eight years of tillage effect on two soils in Ohio. Soil Sci Soc Am J 57:506–512

Manley J, Van Kooten G, Moeltner K, Johnson D (2005) Creating carbon offsets in agriculture through no- till cultivation: a meta-analysis of costs and carbon benefits. Clim Change 68(1):41–65

Martin EC, Adu-Tutu KO, Mc Closkey WB, Husman SH, Clay P, Ottman M (2004) Conservation tillage effects on infiltration and irrigation advance times in Arizona cotton. Arizona Cotton Report, p 138. WWW.ABSTRACT.LIB-EBOOK.COM

Mehdi BB, Madramootoo CA, Mehuys GR (1999) Yield and nitrogen content of corn under different tillage practices. Agron J 91:631–636

Mohammad W, Shah Z, Shah SM, Iqbal MM (2003) Rotational benefits of legumes to subsequent rainfed wheat in a low N soil. Pak J Soil Sci 22:19–27

Mohler CL, Frisch JC, McCulloch CE (2006) Vertical movement of weed seed surrogates by tillage implements and natural processes. Soil Tillage Res 86:110–122

Moonen AC, Barberi P (2004) Size and composition of the weed seedbank after 7 years of different cover-crop maize management systems. Weed Res 44:163–177

Moreno F, Pelegrín F, Fernández JE, Murillo JM (1997) Soil physical properties, water depletion and crop development under traditional and conservation tillage in southern Spain. Soil Tillage Res 41:25–42

Moreno F, Murillo JM, Pelegrín F, Fernan-Dez JE (2001) Conservation and traditional tillage in years with lower and higher precipitation than the average (south-west Spain). In: García-Torres L, Benites J, Martínez- Vilela A (eds) Conservation agriculture, a world- wide challenge. ECAF, FAO, Cordoba, pp 591–595

Motta ACV, Reeves DW, Touchton JT (2002) Tillage intensity effects on chemical indicators of soil quality in two coastal plain soils. Commun Soil Sci Plant Anal 33:913–932

Mrabet R (2002) Stratification of soil aggregation and organic matter under conservation tillage systems in Africa. Soil Tillage Res 66:119–128

Murillo JM, Moreno F, Giron IF, Oblitas MI (2004) Conservation tillage: long term effect on soil and crops under rainfed conditions in south-west Spain (Western Andalusia). Spanish J Agric Res 2(1):35–43

Ogle SM, Breidt FJ, Paustian K (2005) Agricultural management impacts on soil organic carbon storage under moist and dry climatic conditions of temperate and tropical regions. Biogeochemistry 72:87–121

Olaoye JO (2002) Influence of tillage on crop residue cover, soil properties and yield components of cowpea in derived savannah ectones of Nigeria. Soil Tillage Res 64:179–187

Oliveira MT, Merwin IA (2001) Soil physical conditions in a New York orchard after eight years under different groundcover management systems. Plant Soil 234:233–237

Osunbitan JA, Oyedele DJ, Adekalu KO (2005) Tillage effects on bulk density, hydraulic conductivity and strength of a loamy sand soil in southwestern Nigeria. Soil Tillage Res 82:57–64

Oyedele DJ, Schjonning P, Sibbesen E, Debosz K (1999) Aggregation and organic matter fractions of three Nigerian soils as affected by soil disturbance and incorporation of plant material. Soil Tillage Res 50:105–114

Pagliai M, Vignozzi N (2002) Soil pore system as an indicator of soil quality. Adv Geoecol 35:69–80

Parihar SS (2004) Effect of crop establishment tillage, irrigation and nitrogen potential of rice-wheat cropping system. Indian J Agron 49(1):1–5

Parker ER, Jenny H (1945) Water infiltration and related soil properties as affected by cultivation and organic fertilization. Soil Sci 60:353–376

Pekrun C, Kaul HP, Claupein W (2003) Soil tillage for sustainable nutrient management. In: El Titi A (ed) Soil tillage in agroecosystems. CRC Press, Boca Raton, pp 83–113

Peterson GA, Halvorson AD, Havlin OR, Lyon DG, Tanaka DL (1998) Reduced tillage and increasing cropping intensity in the Great Plains conserve soil carbon. Soil Tillage Res 47:207–218

Peterson GA, Westfall DG, Peairs FB, Sherrod L, Poss D, Gangloff W, Larson K, Thompson DL, Ahuja LR, Koch MD, Walker CB (2001) Sustainable dryland agroecosystem management. Tech Bull TB012 Agric Exp Stn, Colorado State Univ, Fort Collins

Pikul JL Jr, Aase JK (2003) Water infiltration and storage affected by subsoiling and subsequent tillage. Soil Sci Soc Am J 67:859–866

Pryor R (2006) Switching to no-till can save irrigation water. Univ Nebraska-Lincoln Ext Pub EC196-3. www.ianrpubs.unl.edu

Qin R, Stamp P, Richner W (2004) Impact of tillage on root systems of winter wheat. Agron J 96:1523–1530

Qin HL, Gao WS, Ma YC, Yang SQ, Zhao PY (2007) Effects of no-tillage on soil properties affecting wind erosion during fallow in Ecotone of north China. Ecol Sinica 27(9):3778–3784

Radke JK (1982) Managing early season soil temperatures in the northern corn belt using configured soil surfaces and mulches. Soil Sci Soc Am J 46:1067–1071

Randhawa AS, Gill KS, Dhillon SS (1976) Time of sowing and productivity of wheat varieties. Prog Farm 13:210–222

Rasmussen KJ (1999) Impact of ploughless soil tillage on yield and soil quality: a Scandinavian review. Soil Tillage Res 53:3–14

Reicosky DC, Evans SD, Cambardella CA, Allmaras RR, Wilts AR, Huggins DR (2002) Continuous corn with moldboard tillage: residue and fertility effects on soil carbon. J Soil Water Conserv 57(5):277–284

Robbins SG, Voss RD (1991) Phosphorus and potassium stratification in conservation tillage systems. J Soil Water Conserv 46:300

Roldan A, Salinas-Garcia JR, Alguacil MM, Diaz E, Caravaca F (2005) Soil enzyme activities suggest advantages of conservation tillage practices in sorghum cultivation under subtropical conditions. Geoderma 129:178–185

Rong JF (2004) Development and application of moveable erosion wind tunnel. Dissertation, China Agricultural University, Beijing, China

Sainju UM, Singh BP (2001) Tillage, cover crop, and kill- planting date effects on corn yield and soil nitrogen. Agron J 93:878–886

Sainju UM, Lenssen A, Caesar-Tonthat T, Waddell J (2007) Dryland plant biomass and soil carbon and nitrogen fractions on transient land as influenced by tillage and crop rotation. Soil Tillage Res 93:452–461

Sainju UM, Senwo ZN, Nyakatawa EZ, Tazisong IA, Reddy KC (2008) Tillage, cropping systems, and nitrogen fertilizer source effects on soil carbon sequestration and fractions. J Environ Qual 37:880–888

Samarajeewa KBDP, Horiuchi T, Oba S (2006) Finger millet (*Eleucine corocana* L. Garetn) as a cover crop on weed control, growth and yield of soybean under different tillage systems. Soil Tillage Res 90:93–99

Sasal MC, Andriulo AE, Taboada MA (2006) Soil porosity characteristics and water movement under zero tillage in silty soils in Argentinian Pampas. Soil Tillage Res 87:9–18

Schjonning P, Christensen BT, Carstensen B (1994) Physical and chemical properties of a sandy loam receiving animal manure, mineral fertilizer or no fertilizer for 90 years. Eur J Soil Sci 45:257–268

Sessiz A, Alp A, Gursoy S (2010) Conservation and conventional tillage methods on selected soil physical properties and corn (*Zea mays* l.) yield and quality under cropping system in Turkey. Bulgarian J Agric Sci 16(5):597–608

Shepherd M, Harrison R, Cuttle S, Johnson B, Shannon D, Gosling P, Rayns F (2000) Understanding soil fertility in organically farmed soils. In: UK Ministry of Agriculture, Fisheries and Food report. (OF0164). Report ADAS Consulting Ltd

Shinners KJ, Nelson WS, Wang R (1993) Effects of residue-free band width on soil temperature and water content. Trans ASAE 37:39–49

Shuang L, Xing-Y Z, Jingyi Y, Drury CF (2013) Effect of conservation and conventional tillage on soil water storage, water use efficiency and productivity of corn and soybean in Northeast China. Acta Agric Scandinavica Sect B-Soil Plant Sci 63(5):383–394

Singh KK, Colvin TS, Erbach DC, Mughal AQ (1992) Tilth index: an approach to quantifying soil tilth. Trans ASAE 35:1777–1785

Singh G, Singh OP, Kumar V, Kumar T (2008) Effect of methods of establishment and tillage practices on productivity of rice (*Oryza sativa*)-wheat (*Triticum aestivum*) cropping system in lowlands. Indian J Agric Sci 78(2):163–166

Sisti CPJ, Dos Santos HP, Kohhan R, Alves BJR, Urquiaga S, Boddey RM (2004) Change in carbon and nitrogen stocks in soil under 13 years of conventional or zero tillage in southern Brazil. Soil Tillage Res 76:39–58

Six J, Elliott ET, Paustian K (1999) Aggregate and soil organic matter dynamics under conventional and no-tillage systems. Soil Sci So Am J 63:1350–1358

Srivastava AP, Panwar JS, Garg RN (2000) Influence of tillage on soil properties and wheat productivity in rice-wheat cropping system. Indian J Agric Sci 70:207–210

Stockfisch N, Forstreuter T, Ehlers W (1999) Ploughing effects on soil organic matter after twenty years of conservation tillage in Lower Saxony, Germany. Soil Tillage Res 52:91–101

Taylor HM, Ratliff LF (1969) Root elongation rates of cotton and peanuts as a function of soil strength and soil water content. Soil Sci 108:113–119

Thiagalingam K, Dalgliesh NP, Gould NS, Mc Cown RL, Cogle AL, Chapman AL (1996) Comparison of no-tillage and conventional tillage in the development of sustainable farming systems in the semi-arid tropics. Aust J Exp Agric 36:995–1002

Tiarks AE, Mazurak AP, Chesnin L (1974) Physical and chemical properties of soil associated with heavy applications of manure from cattle feedlots. Soil Sci Soc Am Proc 38:826–830

Tisdall JM, Oades JM (1982) Organic matter and water stable aggregates in soil. Eur J Soil Sci 33:141–163

Torresen KS, Skuterud R, Tandsaether HJ, Bredesen Hagemo M (2003) Long-term experiments with reduced tillage in spring cereals. I. Effects on weed flora, weed seedbank and grain yield. Crop Prot 22:185–200

Tripathi SC, Chauhan DS (2000) Evaluation of fertilizer and seed rate in wheat under different tillage conditions after transplanted rice. Indian J Agric Sci 70:574–576

Uhlin H (1998) Why energy productivity is increasing: an analysis of Swedish agriculture. Agric Syst 56(4):443–465

Ulusoy E (2001) Objectives of agricultural techniques in changing conditions and conceptions in 2000 years. In: 20th National Agricultural Mechanization Congress, Sanliuifa, Turkey

Unger PW, Jones OR (1998) Long-term tillage and cropping systems affect bulk density and penetration resistance of soil cropped to dryland wheat and grain sorghum. Soil Tillage Res 45:39–57

Unger PW, Langdale GW, Papendick RI (1988) Role of crop residues - improving water conservation and use. In: Hargrove WL (ed) Cropping strategies for efficient use of water and nitrogen, American Society of Agronomy special publication no.51. American Society of Agronomy, Madison, pp 69–100

Upadhyaya SK, Lancas KP, Santos-Filho AG, Raghuwanshi NS (2001) One-pass tillage equipment outstrips conventional tillage method. Calif Agric 55(5):44–47

Van Wijk WR, Larson WE, Burrows WC (1959) Soil temperature and the early growth of corn from mulched and unmulched soil. Soil Sci Soc Am Proc 23:428–434

Varvel GE, Wilhelm WW (2010) Long-term soil organic carbon as affected by tillage and cropping systems. Soil Sci Soc Am J 74(3):915–921

Von Lutzow M, Leifeld J, Kainz M, Kogel-Knabner I, Munch JC (2002) Indications for soil organic matter quality in soils under different management. Geoderma 105:243–258

Voorhees WB (1983) Relative effectiveness of tillage and natural forces in alleviating wheel-induced soil compaction. Soil Sci Soc Am J 47:129–133

Voorhees WB, Farrell DA, Larson WE (1975) Soil strength and aeration effects on root elongation. Soil Sci Soc Am J 39:948–953

Vu DT, Tang C, Armstrong RD (2009) Tillage system affects phosphorus form and depth distribution in three contrasting Victorian soils. Aust J Soil Res 47:33–45

Walling DE (1990) Linking the field to the river: sediment delivery from agricultural land. In: Boardman J, Foster IDL, Dearing JA (eds) Soil erosion on agricultural land. Wiley, Chichester, pp 129–152

Wilkins DE, Simens MC, Albrecht SL (2002) Changes in soil physical characteristics during transition from intensive tillage to direct seeding. Trans ASAE 45(4):877–880

Yusuf R, John I, Siemens C, Bullock G (1999) Growth analysis of soybean under no- tillage and conventional tillage systems. Agron J 91

Zang Y, Gao HW, Zhou XX (2003) Experimental study on soil erosion by wind under conservation tillage. Trans Am Soc Agric Eng 19:56–60

Zhang W, Hou LB, Zhang B, Wen J, Wang G, Jiang WC, Jia Y (2006) Effects of different cultivation ways on soil physical capability in western semi-arid area of Liaoning Province. J Arid Land Resour Environ 3:149–153

Zhang GS, Chan KY, Oates A, Heenan DP, Huang GB (2007) Relationship between soil structure and runoff/soil loss after 24 years of conservation tillage. Soil Tillage Res 92:122–128

Zhang X, Li H, He J, Wang Q, Golabi MH (2009) Influence of conservation tillage practices on soil properties and crop yields for maize and wheat cultivation in Beijing, China. Aust J Soil Res 47:362–371

Zhao MQ, Hao JG, She DQ, Zhao SJ, Fan XQ, Liu HT (2007) Conservation tillage in the interlock region of the farming area and pastoral area. J Agric Mech Res 2:122–125

Zibilske LM, Bradford JM, Smart JR (2002) Conservation tillage induced changes in organic carbon, total nitrogen and available phosphorous in a semi-arid alkaline subtropical soil. Soil Tillage Res 66:153–163

Ziyou S, Zhang J, Wenliang W, Dianxiong CJV, Jiang G, Huang JG, Hartmann R, Gabriels D (2007) Effects of conservation tillage practices on winter wheat water-use efficiency and crop yield on the Loess Plateau, China. Agric Water Manag 87(3):307–314

Chapter 4
Salinity and Crop Productivity

Alfred Oghode M. Okorogbona, Lavhelesani R. Managa, Patrick O. Adebola, Hanyeleni M. Ngobeni, and Tsunduka B. Khosa

Abstract Salinity stress decreases crop production by 50 % in irrigated farming systems of the arid and semi-arid regions, worldwide. At application rate of 720 t beef feedlot ha^{-1} per year, large increases in exchangeable Na and K decreases crop yields on agricultural land treated with high levels of manure and this was attributed to salt and ammonium toxicity in the soil. Increases in salt level of irrigation water from electrical conductivity (EC) of 0.5 dS m^{-1} to EC of 2.8 dS m^{-1}, which had a sea salt level of 0.2 % decreased fruit vegetable biomass by 4.5 mg $plant^{-1}$ whilst an increase in the EC of irrigation water from 0.5 dS m^{-1} to EC of 5.4 dS m^{-1}, which had sea salt level of 0.4 % decreased fruit vegetable biomass by 9 mg $plant^{-1}$. Apart from the decrease in crop yield caused by salinity, this phenomenon also decreases the productivity level of ruminant animals, particularly animals which their feeding predominantly depends on the grazing of grass and forage terra firma.

Salinity has been reported as one of the major environmental factors playing a key role in soil degradation. Studies from various parts of the world indicated that the process of degradation is related to the dynamics of ground water, which is dependent on soil type, cropping systems and the condition of irrigation networks. Presently, research focus on crop productivity is propelled towards providing for today and future generations. Hence, the need to preserve arable land for agriculture becomes imperative. To this end, this paper reviews the effect of salinity on germination, seedling emergence; crop growth and crop yield. The paper also reviews the relationship between soil salinity and crop breeding.

We reviewed that the response of crops' seedlings to salt level is crop and soil (growth medium) specific. While tomato, onion and cabbage had 85 %, 99.6 % and 86.5 % seedling survival rate, respectively, in a healthy soil with very low EC-0.08 mmh cm^{-1} 25 °C, 7 days after transplanting, the survival rate of these vegetables were found to be 3.0 % for tomato; 0 % for onion and 15 % for cabbage. Our review indicates that the salt level in terms of EC, to which most leafy vegetables can

A.O.M. Okorogbona (✉) • L.R. Managa • P.O. Adebola • H.M. Ngobeni
Plant Breeding Division, Agricultural Research Council (ARC) - Roodeplaat, Vegetable and Ornamental Plant Institute (VOPI), P Bag X293, Pretoria 0001, South Africa
e-mail: alfredoghode@yahoo.com

T.B. Khosa
Department of Water and Sanitation, Private Bag X313, Pretoria, 185 Francis Baard Street, Pretoria 0001, South Africa

E. Lichtfouse (ed.), *Sustainable Agriculture Reviews*,
Sustainable Agriculture Reviews 17, DOI 10.1007/978-3-319-16742-8_4

tolerate in goat manure, used as nutrient source, for the cultivation of these crops, is between 480 and 650 mS m^{-1}. The decrease in the growth of grain crops as a result of increasing salt level in irrigation water was also reviewed. In regard to this, our review indicates a decrease in the growth of rice in terms of the crop height, when the salt level of irrigation water was increased from 0 to 12 dS m^{-1} at 35 days after planting; 65 days after planting and at maturity stage. It was also reviewed that at 10 dS m^{-1} ECe salinity level, only crops with high salt tolerant ability have no reduction in yield while sensitive plants have their yield drastically declined due to the excessiveness of salt in the soil. It was revealed in this review, that in soil irrigated with non-saline water, the interaction effects between salinity stresses and genotypes on plant height were slightly different, among the varieties of a particular crop species. However, in the case of soil irrigated with saline water, there were significant differences in the ability of the various varieties of the crop species to tolerate high salt level.

In this review, a few strategies were pointed out as means of overcoming the decrease in crop productivity associated with soil degradation induced by salinity stress. Among these strategies is the use of organic mulches or gypsum or a combination of both to soil irrigated with saline water. The exploitation of the genetic variability of available germplasm for the identification of tolerant genotype, which may sustain a considerable yield under salt affected soils was also indicated as a strategy. The growing of halophytic plants was also pointed to be among the strategies. The use of halophytes was recommended due to their capacity to sequester sodium chloride (NaCl) in vacuoles and to produce compatible osmotic characteristic in the cytoplasm. Also, due to the ability of the plants having a diversity of secondary mechanisms to handle excess salt. Apart from the consideration always given to soil analysis before crop cultivation, the need to also consider the analysis of electrical conductivity of the irrigation water that has been considered for use in crop husbandry is imperative to avoid the reduction in crop yield caused by salinity.

Keywords Salt stress • Germination • Seedling emergence • Crop growth • Crop yield • Crop breeding

4.1 Introduction

In Africa, peasant farming, gardening and large scale commercial crop production are becoming tangible activities providing a huge part of the rural population remarkable prospects, as means of livelihood. This is because many people who engage in the agricultural sector of the economy, realises income to provide for themselves and their households. Growing crops in Africa, particularly in the sub-Saharan regions of the continent, is interrupted by an increasing soil productivity crisis, which is inclined with soil degradation. Soil degradation in Africa has been attributed partly to low nutrient availability, soil alkalinity and soil acidity problems. Ali et al. (2004) identified soil salinity stress as the most serious environmental problem that decreases crop growth rate, crop development and crop yield in arid and semi-arid regions of many parts of the continent. According to El-Kharbotly et al. (2003), salinity does not only directly decrease crop production but also indirectly affects the availability of animal feeds in arid and semi-arid regions of the world, as it seriously inflicts environmental threat that decreases the growth rate or production of grassland cover grazed and browsed on by ruminant animals.

Efficient incorporation of water and nutrients to low fertility soil provides a stable base to support roots for good plant growth and development. This restores the soil nutrient balance in sustainable farming systems. Two forms of salinity that decreases crop productivity include water and soil salinity.

4.1.1 Saline Water

Owing to the method employed in the supply of water to soil for plant use, the cultivation of crop seriously affects the physico-chemical properties of arable land, as it favours high capillary rise of irrigation water, which causes high salt concentration in the upper most part of the soil profile (Jamin and Doucet 1994). According to Jamin and Doucet (1994), the permanent presence of irrigation water in the channels required by the irrigation method (channels permanently filled with water) and the use of minimum water through hand sprinkling of water on soil surface, are both responsible for high capillarity rise and the concentration of salts in the upper 20th (twentieth) centimetre of the soil layer. Another opinion pointed out by PSI (1999) indicated that the process of soil degradation is related to the dynamics of ground water, which is dependent on soil type, cropping systems and the state of irrigation networks. Irrigation source influences the productivity of soils to a large extent, considering the composition of the water supplied to a farm land. For example, some water could have high alkalinity or acidity level. Some could be clean and free of any form of pollution. Others could have high salt level such as sodium chloride and others-with high nitrate level, due to their proximity to waste water treatment plants. Due to the scarcity of water for household and industrial use, the water processed for drinking and other domestic purposes are not channeled to irrigation reservoirs, and as such, partially treated water become the only available water for irrigation purpose in farmlands. An example of partially treated water is the high nitrate water shown in Fig. 4.1, where water from Zeekoegat waste water treatment plant falls directly to a stream leading to Roodeplaat dam in which its water is channeled to different irrigation dams in the surrounding environments.

The water from the Zeekoegat waste treatment plant joins the in stream water to which water falling from its dam wall gets to various irrigation dams and or irrigation reservoirs, supplying water used by several farmlands in the area. The source of water used for irrigation in the Roodeplaat farmlands is presented in Fig. 4.2.

Fig. 4.1 High nitrate content water from Zeekoegat waste water treatment plant. The water is directly from the waste treatment plant, where solid waste from human are separated from sewage water. This water flows directly to the Roodeplaat Dam, where the water used for irrigation activities are sourced. Zeekoegat is a farming environment situated at the Northern part of Pretoria metropolis, South Africa. One of the Pretoria Municipality sewage water treatment plants is situated at Zeekoegat. This water is considered highly unsafe for drinking

The Roodeplaat dam is sometimes referred to as Pienaars river. Two sewage treatment works feed treated effluent to the dam resulting to high eutrophic conditions. The hazard potential of the dam has been ranked high.

4.1.2 Saline Soil

The continuous use of arable land to meet up with the world's food demand disregards the need for balance through rudimentary agricultural practices including crop rotation and shifting cultivation. This necessitates the need to replenish the fertility of nutrient depleted soil with fertilisers. Farmers use both organic and inorganic fertilisers for high crop yield. The use of organic fertilisers including animal manure has positive and negative effects in the context of nutrient supply, crop growth and environmental quality. The main positive effect of applying nutrient sources to soil is to improve crop growth and yield as a result of plant uptake of the nutrients contained in the fertilisers (Maerere et al. 2001; Gosh et al. 2004; Kihanda et al. 2004). The addition of fertilisers at excessive rate disrupts soil natural balances, which in essence limit crop productivity and agricultural development

Fig. 4.2 Roodeplaat dam wall water spillage. This is the dam wall where water passes to various irrigation dams used for the supply of water for crop cultivation in the surrounding farmlands. Here, the water from the sewage water treatment plant at Zeekoegat that usually undergoes partial treatment mixes with another effluent water and water from other channels, which all together flows to the stream until it branches to the various irrigation dam in Roodeplaat farmlands. The dam used to be mainly for irrigation purpose until recently when it also serves as resort for recreation activities

(Mathers and Stewart 1974; Pratt et al. 1976, 1978; Horton et al. 1981; Sutton et al. 1984, 1986; Chang et al. 1991; Rhoades et al. 1999; Turner et al. 2010). The negative crop response to the addition of fertilisers particularly animal manure at high rates has been attributed to soil salinity (Shortall and Liebhardt 1975). Plant growth and development are negatively affected by salinity–a major environmental stress that limits agricultural production worldwide.

4.1.3 Salinity in Arid and Semi-arid Region

According to Shannon et al. (1994), an increasing proportion of the world's food supply is produced with irrigation in semi-arid regions. In arid and semi-arid regions of the world, agricultural productivity is very low as soil salinity decreases crop production to a great extent in these places (Shannon 1997; Munns 2002; Tas and Basar 2009). As pointed out by McWilliams (1986); Evengelou (1994) together with Turhan and Ayaz (2004), reductions in the production of agricultural crops in these areas sometimes reach up to 50 %, due to problems associated with either soil or water salinity. For example, at application rate of 720 t beef feedlot ha^{-1} per year, Murphy et al. (1972) found large increases in exchangeable Na and K, which resulted in the corn silage yields being decreased in areas treated with high levels of manure. This was attributed to salt and ammonium toxicity in the soil. In an experiment carried out on soil amended with cattle feed lot manure, applied at the rate of 112 t ha^{-1}, Mathers and Stewart (1974) indicated that a saturated paste extract of electrical conductivity greater than 3 mmho cm^{-1} was observed in the seed zone at seeding time. This resulted in the plants being stressed early in the season. The percentage of exchangeable Na and K was also increased but not enough to seriously retard plant growth. Evidence of problem associated with soil salinity is demonstrated in Fig. 4.3, where Amaranth seedling growth rate was reduced due to excessive application of manure from goat, as compared to Fig. 4.4, where seeds planted the same day with the plants in Fig. 4.3 germinated, developed into seedling and increased adequately in size at lower application rates of the goat manure.

Fig. 4.3 Amaranth seedlings' growth retarded due to high concentration of salt in the goat manure incorporated to a low fertility soil. Indications in the pots show that the organic nutrient source was applied in excess of the crop seed absorbable threshold. The seeds in these pots were incorporated into the soils the same day with those of the seedlings growing in Fig. 4.4. The *black crystals* formed on the surface of the soils indicate the excessiveness of the concentration of salts in the growth medium. The occurrence in this figure indicates that at the application rate of 150, 180 and 210 t ha^{-1}, the concentration of salt in the goat manure applied to the soil exceeded the maximum limit to which germination and seedling emergence could take place

Fig. 4.4 Amaranth plants observed to have germinated adequately and exhibited considerable seedling emergence. Amaranth plants showing increase in growth where the optimum salt concentration in the of goat manure could be determined between 30 and 90 t ha^{-1}. Indication from the figure establishes that the effect of salinity stress has commenced at the application rate of 90 t goat manure ha^{-1}. Note: the observation from the figure was identified at the early stage of seedling growth

The fact that salinity appears to be a serious threat to increase in crop yield is well established in literature. Several studies already ascertained the verity that this phenomenon negatively affects crops, right from planting to harvesting stage and even at storage. Most reviews on the effect of saline water and or saline soils on crop growth, development and yield, keeps on awarding limited information to a canopy of researches. These studies comprise how salinity affects crops at germination and seedling emergence stages; effects of salinity on the growth and development of different crops. Also, among these, are salinity tolerance ability of different crops; the effect of salinity on crop breeding and the effect of salinity on crop productivity. Instead, each study on salinity has one way or the other focused intently and or selectively on either one or just two of the aforementioned twirl of phrases that all surrounds how salinity causes draw-back on crop production in sustainable agriculture. Despite the fact that the primary sector of African agriculture is marked by low agricultural productivity, with little application of science and technology. A factor associated with the fact that rural farmers or low marginal input farmers, constituting majority of the population of African farming community, have little technical know-how of the scenery that encompasses neither the negative effect of saline water nor saline soil, on crop productivity. Research reports on how salinity variously affects sustainable crop production remain shroud to most African Agrarian setting. A situation, which is inclined with the assumption that, if seeds refuse to germinate or seedlings growth is retarded, it invariably means the seeds incorporated into the soil lack good viability. If plant growth or development and yield are poor, the problem could be that the seeds planted are unhealthy or the climatic condition is unfavourable for the cultivation of the particular plant species. It is in line with the foregoing, that we have reviewed a few research works, to unveil the various ways salinity affects crop productivity in sustainable farming. Hence, a few strategies that have been used by various scientists to address the issue of negative effect of salinity on irrigation water and agricultural land, are hereby uncovered to the agricultural and scientific communities, in order to avoid some draw-backs the phenomenon has on agricultural productivity. In line with this, review on the negative effects of salinity on germination, seedling emergence, crop growth and yield have been considered fundamental. This is to provide a guideline on crop production and soil fertility management to farmers and emerging researchers. This review aligns itself to this agricultural growth course.

4.1.4 The Life of a Green Crop Plant

A crop is a cultivated plant that is harvested for food, clothing, livestock fodder, biofuel, medicine or other uses. A crop comprise of green plants and non-green plants. Example of a green plant include green pepper. Example of a non-green plant include mushroom (fungus). This review only emphasises on green crop plants. A green crop plant is basically a plant that carries out photosynthesis. Photosynthesis is the process whereby all green plants manufacture food using carbon IV oxide (CO_2) and water in the presence of sunlight. The life of a green plant begins with a seed germinating in soil or other appropriate medium containing a means of anchorage, a supply of nutrients and adequate water for growth and development (Jones 1979; Fageria 2009). As pointed out by Jones (1979), plants early stage is strongly influenced by soil moisture, soil structure, soil nutrient status, salt concentration and other factors. From the time the first root begin to penetrate the pore space in the soil and the first protruding leaf exhibits its movement towards the source of light, the rate at which the plant grow and produce yield is set by a host of soil properties. The understanding and modification of these soil properties promote healthy plant growth and high yield at harvest (Jones 1979). In regard to these aforementioned factors, literature dealing with the impact of salinity on concepts evolving from the incorporation of crop seed into the soil environment to the final harvesting stage, that is germination, seedling emergence, crop growth and yield are considered in this chapter.

This section revealed that soil and water salinity stress negatively affects plant growth, development and yield, worldwide. A factor, also having negative impact on the productivity of ruminant animals, which their feeding largely depends on the grazing of grass and forage terra firma. Salinity was indicated as one of the environmental factors causing soil degradation. It was pointed out that the process of soil degradation is related to the dynamics of ground water, which is dependent on soil type, cropping systems and the condition of irrigation networks.

4.2 Salinity, Germination and Seedling Emergence

High salinity decreases the germination rate of crop seeds. It also reduces the rate at which crop seedlings emerges from the soil surface. Salinity has been described in various ways by scientists. Some described it in relation to salt concentration in water while others described it from the view point of salt content in soil. In a broader term, salinity could simply be described as the content of salt available in the body of either water or soil. Soil salinity could be an accumulation of salt introduced from water containing salt, which evaporates from within the soil and reaching the surface, with a consequential production of salt crust. The salt level in water or soil is usually determined in terms of the contents of chemical compounds such as magnesium chloride, sodium chloride, bicarbonates, nitrates and calcium sulphates. The salt level in soil affects the processes of germination and seedling emergence. The stages of germination and seedling emergence are often misinterpreted or confused. To this end, these two terminologies are thereby described as the same mania. According to Bewley and Black (1994) and Kigel and Galili (1995), germination is the developmental change occurring between the time a plant seed is incorporated into the soil or other medium containing a means of anchorage and the time of seedling emergence. Bewley and Black (1994) pointed out that, with the exception of vegetative propagation, the cultivation of most crop species depends on seed germination. Germination begins with water uptake by the seed and ends with the start of elongation by the embryonic axis, usually the radicle.

Hamman et al. (1996) indicated that good seed germination is one of the most important components of yield quality and it is affected by various factors during seed development, maturation and harvesting. Brar and Stewart (1994) reported that rapid germination and seedling emergence are important, and these are considerable factors in crop cultivation. Consequently, these scientists associated soil chemical toxicity to suppression of seedling emergence. Bewley and black (1994) postulated that germination process involves numerous events, which include protein hydration, sub cellular structural changes, respiration, molecular synthesis and cell elongation. None of which is in itself unique to germination. Therefore, it was stated by Bewley and black (1994) that germination exclude seedling growth which commences after germination ends. According to Bewley and Black (1994), it is incorrect to equate germination with seedling emergence from soil as germination ends sometime before the seedling is visible. After the emergence of seedling, the growth of the young plant commences. Precisely, seed germination is the development of a seed to a seedling or the establishment of a seedling from a seed incorporated into the soil or other growth media.

Seedling emergence could be described as the outgrowth on the surface of a plant. Seedling emerges in many plant species through elongation of the hypocotyl. For seedling to be established, the elongation of the hypocotyls must cease and leaf development must begin. Seedling emergence is termed as the establishment of a plant, as seeds open in the soil or growth media. It could also be defined as the appearance of new plant or plant species in the course of evolution.

Seed germination and seedling growth are the most sensitive stages to ecological stresses in general, as the negative impact of these two phenomena significantly reduce the yield of crops and salt stress in particular was pointed out by Kaveh et al. (2011) to be one of the major environmental problems. Lauchli and Grattan's (2007) postulation, which indicated that most crop plants are salt tolerant at germination but salt sensitive during emergence and vegetative development also has similarity with Kevah and others' ideologies. Studies carried out by researchers in various part of the world, revealed that salinity affects germination and seedling growth in different crops (Yo and Shaw 1990; Schimidhlter and Gertli 1990; Singh Mangal 1991; Crucci et al. 1994; Pascale and Barberi 1995; Kowaski and Palada 1995; Sharma et al. 2001; Maiti et al. 2004, 2007a; Bagayoko 2012; Van Averbeke et al. 2012). In an experiment carried out to determine the effect of salinity on the germination of okra, tomato, cabbage, onion and pepper, using white salt soil, black salt soil and normal soil (healthy soil) with specific electrical conductivities of 1.90, 1.63 and 0.08 mmho cm^{-1} 25 °C, respectively. Bagayoko (2012) reported high germination rate of the seeds of all the crops incorporated into the healthy soil. While the germination rates were found to be 87-95 % for okra, 87-97 % for tomato, 77-95 % for cabbage, 90-95 % for onions and peppers seeds planted in the healthy soil, none of the crops germinated in black salt or in white salt soil. Report provided by Bagayoko (2012), showed that in pots containing white salt, some seeds of okra, tomato and cabbage initiated germination, but the seedlings died within 2-3 days after. In contrast, suckers and seeds of onion and red peppers did not develop into seedling in white salt soils. In black salt soil, none of the crops demonstrated any sign of seed germination. According to Bagayoko (2012), similar scenario occurred when seedlings of tomato, onion and cabbage were transplanted. This is shown in Table 4.1, which provides an indication of the level of survival in crops' seedlings that developed from the white salt, black salt and healthy soil types in one of the experiments conducted by Bagayoko (2012).

Germination and seedling emergence are also affected by salinity induced by excessive application of animal manure to plant growth medium. For example, in an incubation experiment carried out to examine the emergence of Chinese cabbage, pumpkin, nightshade and Amaranth seeds to different application rates of goat manure. Also, to determine the effect of application rates of the ruminant manure applied at 30, 60, 90, 150, 180 and 210 t ha^{-1} on the electrical conductivity of soil. The electrical conductivity estimate of 480 mS m^{-1} was suggested to be the threshold point. According to Van Averbeke et al. (2012), found out that at the rate of 90 t ha^{-1} (with EC estimate of 650 mS m^{-1}), soil salinity induced by the amendment of a low fertility soil with goat manure negatively affected the development of Amaranth and pumpkin seeds. Table 4.2 reports on the effect of manure induced

Table 4.1 Crop seedling survival in soils with different salt contents (From Bagayoko 2012)

Soil type	Electrical conductivity (EC) (mmho/cm 25 °C)	Soil pH (KCl)	Percentage crop seedling survival 1 week after transplanting		
			Tomato	Onion	Cabbage
Healthy soil	0.08	4.59	85.0	99.61	86.5
White salt soil	1.90	7.20	3.0	0.0	15.0
Black salt soil	1.63	9.27	0.0	0.0	0.0

The result from this table show differences in the response of crops towards salinity level in the different plant growth media, considering the survival of the crops' seedlings after emergence. In the normal soil, onion seedling exhibited almost 100 % growth, followed by cabbage, which had an increase of 1.5 % than tomato. Despite the fact that the highest level of EC was obtained from the white salt soil, cabbage seedling growth was five times that of tomato while onion, which had almost all its seedling exhibiting adequate growth on the healthy soil, had zero seedling survival. An indication that cabbage is more tolerant to salinity than onion and tomato

Table 4.2 Effect of goat manure induced salinity on vegetable seed germination and or seedling survival (From Van Averbeke et al. 2012)

Application rate of goat manure (t ha^{-1})	Electrical conductivity estimate (mS m^{-1})	Dry above ground biomass (g plant^{-1})			
		Chinese cabbage	Amaranth	Nightshade	Pumpkin
30	270	4.17	6.65	12.97	9.19
60	480	6.25	11.46	16.80	20.96
90	650	7.53	12.76	24.10	12.29
150	750	1.18	0.00	0.00	0.00
180	1,400	0.91	0.00	0.00	0.00
210	1,600	0.35	0.00	0.00	0.00

NB-The estimated EC values were obtained using a mass soil to water ratio of 1:5, which represented an estimated dilution of 18 times the concentration of salts expected to be found in the saturated paste extract of the soil- which is the usual basis for the EC effects on plant growth (Van Averbeke et al. 2012)

Among the four crops, Chinese cabbage seed seemed to be the only vegetable seed that appeared to mostly tolerate high salinity level in soil, from application rates of 150–210 t goat manure ha^{-1}, as far as germination and seedling survival are concerned.

For Amaranth, Nightshade and pumpkin, seed germination failed to occur in soils amended with goat manure at the application rates of 150, 180 and 210 t ha^{-1}, indicating zero tolerance to soil salinity level of 750, 1,400 and 1,600 mS m^{-1}

salinity on vegetable seed germination and or seedling survival, using above ground biomass as the indicator of plant growth and development.

The negative effect of salinity was also reported by Turhan and Ayaz (2004) who planted the seeds of three cultivars of sunflower in soils supplied with seven different concentration of NaCl, which were added in different rates as 0, 0.25, 0.50, 0.75, 1.00, 1.25 and 1.50 % of the salt, and observed the performance of the crop seeds afterwards. In their investigation, it was observed that salinity caused a delay in both seed germination and emergence. According to Turhan and Ayaz (2004), no seed

Table 4.3 Sunflower seedling emergence from saline soils (From Turhan and Ayaz 2004)

Sunflower cultivar	Percentage of Sodium chloride (NaCl%)	8th day after sowing	9th day after sowing
Edirne-87	Control	22	22
Turkuaz	Control	9	9
Dolunay	Control	19	22
Edirne-87	0.25	–	–
Turkuaz	0.25	3	3
Dolunay	0.25	25	25
Edirne-87	0.50–1.50	0	0
Turkuaz	0.50–1.50	0	0
Dolunay	0.50–1.50	0	0

Control: indicates zero level of salt concentration. Turhan and Ayaz (2004) investigation indicates that the germination and seedling emergence of all the cultivars decreased with increasing NaCl concentrations

emergence was observed from soils that had salt concentrations ranging from 0.50 to 1.50 % NaCl, 8–9 days after sowing. Table 4.3 indicates the response of the three cultivars of sunflower to different levels of salt in soil at the 8th and 9th day after sowing, as reported by Turhan and Ayaz (2004).

Studies carried out by researchers in various parts of the world, revealed that salinity affects germination and seedling growth in different crops. This section revealed that good seed germination is one of the most important components of yield quality. Seedling growth decreases with increasing level of salinity. Germination and seedling emergence are important and considerable factors in crop cultivation. Seedling emergence is suppressed by soil chemical toxicity. According to Bewley and Black (1994), it is incorrect to equate germination with seedling emergence from soil as germination ends sometime before the seedling is visible.

4.3 Crop Growth and Plant Development

The growth and development of crops are negatively affected by saline water or soil containing salts above crop tolerant or critical point. Srivastava (2002) defined crop growth as the overall increase in size of various parts of the plant throughout its life (i.e. from seed to seed). According to Tisdale and Nelson (1975), together with Wild (1988), the increase could be in the form of dry weight or in dimensions which arises due to formation of new cells, expansion of the constituent cells, and the production of assimilates. Srivastava (2002) pointed out that plants are able to carry out growth process because of the presence of meristems at certain location in their body. These meristems are composed of stem cells which perpetuate themselves by cell division and give rise to derivative cells which differentiate along new lines. The activity of the meristem results in the formation of fresh quotas of tissues and organs with a consequent continuity of growth in height and in many cases girth throughout its life. Wild (1988) expressed the rate of plant growth as the increase in weight, volume, area or length per unit time and described development as the progress of a plant from germination to maturity through a series of changes. Crop growth is determined by a number of factors including soil condition and cultivation practices such as date of sowing, amount of irrigation and applied nutrients (Overman and Scholtz 2002; Fageria 2009). According to Fageria (2009), adequate supply of nutrients to crop plants is one of the most important factors required to achieve higher yield. Nutrient imbalances in crop may result from several factors. These include the effect of salinity on nutrient ion activity and availability, the uptake or distribution of a nutrient within the plant, and or increasing the internal plant requirement for a nutrient element resulting from physiological inactivation (Läuchli and Grattan 2007). Excessive application of water containing high salt or over application of nutrients disrupts the activities of the plant meristems, which consequently result in negative effect on crop growth and development. For example, in a water culture experiment conducted to study the effect of sodium chloride salinity at the seedling stage of six inbred lines and a salt tolerant variety of rice plant. The seeds of the rice plants were placed for germination in a greenhouse condition at a temperature of 35 °C. In the experiment, the seedlings were transplanted 7 days after emergence; salinised water applied after 14 days, with appropriate amount of sodium chloride to make a concentration of 50 and 75 mM sodium chloride. As reported by Shereen et al. (2005), this corresponds to EC of 6 and 9 dSm^{-1}. In their findings, it was reported that salinity caused a significant reduction in the seedling growth, with various degree of variability among these lines, as the fresh weight of all the lines were observed to decrease with increasing level of salinity from 50 to 75 mM NaCl. According to Shereen et al. (2005), in the rice lines with high sensitivity to salt level, growth decrease became more pronounced with increasing number of days, even at lower level of salinity. After 14 days of saline treatment, the higher salt level (75 mM NaCl) resulted in a significant decrease in seedling growth when compared with the control treatment (treatment without salt).

Similarly, in an experiment conducted by Turhan and Ayaz (2004) who assessed the effect of salinity on sunflower (*Helianthus annus*. L), using seven different levels of sodium chloride (NaCl) concentrations at 0, 0.25, 0.50, 0.75, 1.00, 1.25 and 1.50 %. It was observed that the growth of all the cultivars of the crop that featured in the study decreased with increasing NaCl concentrations. Indications from Turhan and Ayaz's (2004) report showed that there was a drastic reduction in the growth of plants grown in soils treated with salt concentrations that ranged from 0.50 to 1.50 %. Similarly, in a study carried by Saeed and Ahmad (2009), to observe the effects of organic mulch with and without gypsum on the vegetative growth of tomato plant (*Lycopersicon esculentum* Mill. cv. F1 Avinash) under non-saline and saline conditions. A decrease in vegetative growth with increasing salinity levels was reported. Table 4.4 indicates the effect of salinity on the vegetative growth of tomatoes, as reported by Saeed and Ahmad (2009).

In their study, mulch, gypsum and a combination of mulch and gypsum were added concomitantly. This was carried out using a control-0.2 and 0.4 % without

Table 4.4 Growth of tomato as affected by salinity (From Saeed and Ahmad 2009)

Treatment	Electrical conductivity (EC) of irrigation water [dS m^{-1}]	Electrical conductivity of soil-saturated paste extract (ECe) [dS m^{-1}]	Tomato dry biomass (mg $plant^{-1}$)
Control (tap water)	0.5	1.5	38.0
0.2 % sea salt	2.8	5.7	33.5
0.4 % sea salt	5.4	7.1	29.0

Tomato growth determined in terms of plant dry biomass. Dry biomass decreased with increasing salt level in water. Result from the table indicate that the dry biomass of tomato decreased by 4.5 mg $plant^{-1}$ when the electrical conductivity (*EC*) of the irrigation water was increased from 0.5 to 2.8 dS m^{-1} and the vegetable had a decrease of up to 9 mg $plant^{-1}$ when the EC of the irrigation water was increased from 0.5 to 5.4 dS m^{-1}

mulch, neither a combination of mulch and gypsum. It was observed that the addition of these substances reduced the negative effect of salinity on tomato growth, thereby resulting in an increase in the dry biomass of the crop. Tomato grown in soil amended with a combination of mulch and gypsum was observed to contain the highest dry biomass, followed by soil amended with gypsum and then soil amended with mulch. Consequently, Saeed and Ahmad (2009) suggested the use of organic mulches or gypsum or a combination of both to soil irrigated with saline water, as a means of reducing salinity hazards and increasing the growth and yield of fruit vegetables.

Variable responses of crops to salinity at vegetative and reproductive stages were also observed by Shereen et al. (2005). In an experiment carried out to evaluate the effect of different levels of salinity (0, 50, 75 mM NaCl) on the growth of different lines of rice, Shereen et al. (2005) reported that salinity exacerbated a significant reduction in seedling growth with unstable degree of variability among the lines but generally, fresh weight of all the lines decreased as the level of salinity increased from 50 to 75 mM NaCl. Similarly, in a pot experiment carried out to determine the effect of different levels of salt water (3, 6, 9, and 12 dS m^{-1}) on the growth attributes of mutant rice, Islam et al. (2007) reported variable growth responses among the cereal genotypes, using plant height as an indicator of increase in the size of the crop plant. In their investigation, significant variation among the different genotypes in terms of plant height was observed. Islam et al. (2007) findings are presented in Table 4.5.

The investigation carried out by Islam et al. (2007) indicated that there was a significant variation in plant height among the three genotypes of the rice plant, which were namely-Q-31, MR-219 and Y-1281 genotypes. Their report indicated that the genotype Q-31 had the tallest plant height (67.33 cm) from seedling emergence to maturity stage. The genotype MR-219 exhibited an intermediate standing of 54.18 cm throughout the growth period. Y-1281 had the shortest plant height with 41.04 cm. According to Islam et al. (2007), the interaction effects between salinity stresses and genotypes on plant height indicated that at control, all varieties were different for plant height but tolerance capability to salinity varied significantly.

Table 4.5 Rice growth as affected by different water salinity levels (Islam et al. 2007)

Salinity level of irrigated water (dS m^{-1})	Height of rice plant (cm)		
	35 days after sowing	65 days after sowing	At maturity
Control (non saline water-0 salt level)	63.00	81.00	86.00
3	59.00	73.00	81.00
6	52.00	67.00	74.00
9	48.00	57.00	64.00
12	29.00	35.00	–

Plant growth decreased with increasing level of saline water. The result from the table indicates that the growth of the grain crop terminated before after 65 days and before maturity stage, in the case of treatment irrigated with water that had EC level of 12 dS m^{-1}. Indication from the table also shows that the more the salt level in the irrigation water, the higher the level at which the growth of the plant is retarded

Reduction of plant height with increasing salinity was comparatively lower in Y-1281, over the other two genotypes (Islam et al. 2007).

Apart from the reduction in crop growth initiated by the excessiveness or high level of salt in soil, high level of salinity also has negative effect on the nutrient quality of crops. For example, in a field experiment carried out to determine the effect of salinity on the protein content of four genotype of oat leaves and grains, which involved irrigation with different saline water applied at 3, 6, 7.2, 10, 12 and 14 dS m^{-1}, Kumar et al. (2010) reported a decrease in the protein content of the leaves and grains of all the cultivars of oat that featured in the investigation, with increasing level of salt in irrigation water. The negative effect of high level of salt on the protein content of crop leaf and grain, as reported by Kumar et al. (2010) is presented in Table 4.6.

Information from this section revealed that the growth of crops decreased with increasing salt concentration in both soil and water. It could be deduced from the results of investigations carried out on salinity effect on crops that vegetative growth is retarded with increasing salinity levels. The use of organic mulches or gypsum or a combination of both to soil irrigated with saline water were suggested as means of reducing salinity hazards and increasing the growth and yield of fruit vegetables. Among the varieties of a particular crop species, the interaction effects between salinity stresses and genotypes on plant height are different in terms of plant height and the tolerance capability to salinity varies significantly.

4.4 Salinity and Crop Yield

Several research works have shown that most anomalies caused by environmental factors in plant growth pass on to the yield of crops, as reduction in crop growth causes yield losses. Yield was reported to be one of the most important dimensions of a crop plant's economic value (Cooke 1982; Fageria 2009). Fageria (2009) described

Table 4.6 Effect of irrigating saline water on the leaf and grain protein content of oat (*Avena sativa* L.) varieties (Kumar et al. 2010)

Salinity level of water irrigated (dS m^{-1})	Cultivar A		Cultivar B		Cultivar C		Cultivar D	
	Leave protein (mg g^{-1})	Grain protein (mg g^{-1})	Leave protein (mg g^{-1})	Grain protein (mg g^{-1})	Leave protein (mg g^{-1})	Grain protein (mg g^{-1})	Leave protein (mg g^{-1})	Grain protein (mg g^{-1})
Control	79.3	183.6	77.2	181.2	69.5	165.8	75.6	170.3
3	84.0	190.9	81.0	185.3	70.6	172.5	79.0	180.8
6	80.2	182.0	78.0	180.2	69.0	160.2	75.0	165.0
7.2	75.6	170.8	72.5	176.0	64.5	158.8	71.6	162.0
10	72.1	174.0	71.0	168.2	62.3	153.8	70.2	154.4
12	64.5	166.2	64.0	163.4	60.0	150.0	68.4	151.0
14	59.2	154.7	54.0	160.3	53.4	140.2	55.3	145.2

Leaf protein determined at 90 days after sowing; Grain protein determined at final harvest-which was 120 days after planting

Cultivar A-JHO-822; Cultivar B-JHO-851; Cultivar C-KENT and Cultivar D- UPO-94

The protein content in both oat leave and grain increased in the treatment that had its irrigation water with salinity level of 3 dS m^{-1}. The threshold point for protein increase in the crop leaf and grain appeared to be between 3 and 6 dS m^{-1}. Invariably, this implies that the optimum salinity level in irrigation water used for soil cultivated with oat could be within the range of 3–6 dS m^{-1}, as the protein level in the leaves and grains of oat were observed to decrease in irrigation water with salinity level above 6 dS m^{-1}

yield as the amount of specific substance produced (e.g. grain, straw, total dry matter) per unit area. This refers to the weight of cleaned and dried crop plants harvested from a unit area and it is usually expressed in kilogram per hectare (kg ha^{-1}) or in metric tons per hectare (Mg ha^{-1}) at 13 or 14 % moisture (Fageria 2009). Potential yield is the most optimistic estimate of crop yield based on present knowledge, available biological material and ideal management in an optimum physical environment. According to Fageria (2009), potential yield is an estimate of the upper limit of yield increase that could be obtained from a crop plant. The yield of a crop is determined by management practices, which maintain the productive capacity of a crop ecosystem. These practices include the use of efficient water management, use of healthy crop cultivars, control of diseases, insects, weeds and the adequate utilisation of organic and inorganically formed plant nutrients (Chapin et al. 2002).

Similarly, as in the case of seed germination and crop growth, presence of salt in water and soil above plant absorbable threshold points negatively affects the yield of crop plants. Evidence of this occurred in an experiment carried out by Van Averbeke et al (2012). The experiment was carried out to monitor the response of leafy vegetables to different application rates of cattle manure. In their investigation, Van Averbeke et al (2012) found out that the yield of Amaranth-*Amaranthus cruentus* L., Nightshade-*Solanum retroflexum* Dun. and Pumpkin-*Cucurbita maximum* Duchesne, increased when manure from the ruminant was applied from 30 to 180 t ha^{-1}, where the electrical conductivity of the saturated paste extract of soil ranged from 95 to 270 mS m^{-1} but declined when the manure application rate was

Table 4.7 Effect of manure induced salinity on leafy vegetable yield (From Van Averbeke et al. 2012)

Application rate of cattle manure (t ha^{-1})	Electrical Conductivity estimates (mS m^{-1})	Total oven-dry weight of vegetable shoot (g plant^{-1})		
		Amaranth	Nightshade	Pumpkin
30	95	6.05	10.85	10.82
60	135	10.35	17.94	19.83
90	155	17.46	23.51	27.17
150	210	28.27	30.87	31.36
180	270	34.01	37.00	37.83
210	365	26.37	31.69	32.61

Indications from the table show that the electrical conductivity threshold or critical point could be considered to exist between 270 and 365 mS m^{-1}. Van Averbeke et al. (2012) asserts "that the EC values were obtained using a mass soil to water ratio of 1.5, representing an estimate dilution of 18 times the concentration of salts expected to be found in the saturated paste extract of the soil, which is the usual basis for the study of EC effects on crop yield"

increased from 180 to 210 t ha^{-1} (EC-350 mS m^{-1}). The negative effect of high level of salt in the organic nutrient source, as reported by Van Averbeke et al. (2012) is presented in Table 4.7.

While a few studies only pointed out the negative effect of high salinity level in water and soil, a study carried out to determine the effects of salinity on fruit yield and quality of tomato grown in soil-less culture, using greenhouse experimental assay, came up with an additional factor. Considering the research, which was carried out in the Mediterranean by Magan et al. (2008), both the negative and positive effects of increasing level of salinity in soil and water on crop productivity were pointed out. Observation from Magan et al. (2008) indicated that there was a linear decrease in total and marketable yield of tomato, as the salinity level increases from 2.5 to 8.0 dS m^{-1} (2.5; 3.0, 4.0; 5.0; 6.0; 7.0 and 8.0 dS m^{-1}). In their investigation, 3.2 and 3.3 dS m^{-1} were reported to be the average threshold values for the total and marketable fruit yield respectively. As reported by Magan et al. (2008), while increase in salinity level caused a decrease in the fruit size, aspects of fruit quality, which include proportion of extra fruit (high visual quality), soluble solids content and titratable acidity content improved.

The positive effect of saline water could also be seen in the case of irrigation networks where water supplied to farm lands is sourced from irrigation dams close to storage where water treatment activities are carried out. In such dams, the nitrate and phosphate levels of the water are sometimes high enough to provide farm lands and even greenhouses with adequate nutrients which turns out to be a form of fertigation due to the nitrogen from solid wastes and phosphorous from urine, as nutrient may not be added to the water that passes to such irrigation dams Evidence of this is shown in an irrigation dam where the source of water is from Roodeplaat dam in which its source of water comes from Zeekoegat waste water treatment plant. In such irrigation reservoir, invading plants grow rapidly due to the high level of nutrients in the water (Fig. 4.5).

Fig. 4.5 Irrigation reservoir showing its water covered with plant leaves. The dam water contains high level of nitrate. Water surface, totally covered with plants due to its high nitrate and phosphate contents, which encourages the growth of water plants

However, this situation also has draw-back to some aspect of biotic activities. For example, in cases where the level of nitrates and phosphates becomes too high for irrigation purposes or the continuity of marine life, most especially when legumes which may requires low level of nitrogen due to their nitrogen fixing characteristic, surface water plants are introduced to such dams to help absorb and reduce the level of these nutrients available in the water. Carrying out this requires painstaking human efforts. The dam in Fig. 4.6 shows a different scenario with no or zero plant growth activity due to its low nutrient characteristic which habitually encourage marine life that attracts fishers who often visits the irrigation reservoir for fishing and even terrestrials animals drinking the water, from the practical and or visual point of view. This is also consistently used for irrigation activities in various cropping lands.

This section divulged that salinity has positive and negative effects on crop productivity, terrestrials and marine animals.

4.5 Salinity, Crop Breeding and Crop Productivity

Research has shown that salinity induced by high level of salt in irrigation water or excessive application of fertilisers into soil, reduces the yield of crops. The negative effect on crop ranges from a slight crop loss to complete crop failure, depending on

Fig. 4.6 Dam situated 1.5 km apart from the dam in Fig. 4.5. The is clear and has no plants growing on its surface due to its negligible nutrient level

the type of crop and the severity of the salinity problem (Hill and Koenig 1999). Although, several treatments and management practices might have helped in reducing salt levels in the soil, there are some situations where it is either impossible or too costly to attain desirably low soil salinity levels. In some cases, the only practical management option is to plant salt-tolerant crops developed or selected through plant breeding. Breeding has shown extraordinary significance in crop productivity, from the view point of developing crops that adapt to particular environment; crops resistant to diseases or pests; crops with improved or high yield and crops with shorter maturity or harvesting period. Poehlman and Sleeper (1995) defined plant breeding as the purposeful manipulation of plant species in order to produce desired characteristics for specific purposes. This phenomenon could be accomplished through different techniques ranging from simple selection of plants with desirable characteristics for propagation, to more complex molecular techniques. For example, the ideology which emanated from the school of thought of early Ethiopian Agrarian society to replace salt-sensitive crops with more salt-tolerant crops, such as the replacement of wheat (*Triticum aestivum*) with barley (*Hardeum vulgare*), is an evidence that farmers had in the past tried to take in hand the problem of salinity (Marr 1967). In such scenario, the variations among different genotypes of each crop in response to salinity effect were not highlighted. However, recent studies conducted to evaluate the level of genetic diversity among different genotypes of wheat for salinity tolerance at germination and early seedling stage were able to close these gaps. In line with the foregoing, findings from some research endeavours that relates breeding to salinity revealed that some potential genotypes tolerant to salt can be improved and utilised under saline soils (Ahmad et al. 2013; Haq et al. 2003; Sarwar et al. 2003).

As plants vary in their ability to withstand salinity stress induced by saline water or saline soil or excessive application of nutrient sources, plant species have been classified based on their tolerance level to salinity effect. The variability in salt-tolerance within and between plant species, need to be explored and combined with agronomic sound lines in terms of other desirable characteristic, such as high yield, disease and insect pest resistance. The classification varies from sensitive to tolerant plant species (Blaylock 1994). Salt tolerant plants (plants less affected by salinity) possess the ability to adjust internally to the osmotic effects of high salt concentrations than salt-sensitive plants. Also, salt-tolerant plants are more able to absorb water from saline soils. In contrast, salt-sensitive plants have a limited ability to adjust and are injured at relatively low salt concentrations. In response, reductions in yield due to salinity effect appear to be more common in sensitive crop cultivars than in tolerant crop cultivars. Such reduction in yield caused by salinity as reported by Blaylock (1994) is presented in Table 4.8.

In line with the foregoing, a few agronomic crops were classified by (Blaylock 1994; Hill and Koenig 1999) based on their sensitivity or ability to withstand high salt level in soil. The crop classification is presented in Table 4.9.

The differences in the ability of crops to tolerate brackish level in soil continues to encourage the integration of plant breeding into soil fertility management. In sub-Saharan Africa, where agricultural productivity is highly dependent on smallholder farming activities, which is characterised by poor resources and lack of technical know-how about the efficient use of nutrient sources. The technology transfer approach of introducing crops with in-built salt tolerance achieved from breeding research becomes imperative to farmers. This is because the ideology is considered

Table 4.8 Soil salinity level and yield potential of different salt tolerance plant groups (From Blaylock 1994)

Salt tolerance class	Expected relative growth or yield loss (%)			
	0	25	50	100
	Soil salinity level (EC_e-dS m^{-1})			
Sensitive	<1.0	1.0–3.0	2.6–4.2	>8.0
Moderately sensitive	<3.0	3.0–6.0	4.2–9.5	>16.0
Moderately tolerant	<6.0	6.0–11.0	9.5–15.0	>24.0
Tolerant	<10.0	11.0–16.0	15.0–21.0	>32.0

ECe Electrical conductivity of the saturated paste extract
The data presented on table 4.8 indicates that at 10 ECe (dS m^{-1}) salinity level, only the salt tolerant plant will have no reduction in yield while sensitive plants have their yield drastically declined by excessive salt level in soil

Table 4.9 Relative salt tolerances of horticultural and landscape plants (From Blaylock 1994; Hill and Koenig 1999)

Tolerance classes	Crops
Tolerant	Barley, Sugar Beet, Wildrye, Asparagus
Moderately tolerant	Wheat, Wheat Grass, Zucchini, Beet (red), Squash, Carnation
Moderately sensitive	Tomato, Cucumber, Alfalfa, Clover, Corn, Muskmelon, Potato, Cabbage, Lettuce, Celery
Sensitive	Onion, Carrot, Bean, Apple, Carrot, Cherry, Raspberry, Strawberry

As pointed out by Blaylock (1994) and Hill and Koenig (1999), these crops listed also represent other crops species falling in their groups. Hence, these crops could serve as "indicator" plants in determining the salt tolerant level of closely related plants or plants grown in similar climatic conditions

to be a less resource consuming or economical and socially acceptable approach. As a result of the high cost and strategic knowledge required by farmers lacking scientific approach to farming and the resources to optimally provide good growing condition for crops, the need to develop crops suited for salt stress environment through breeding for smallholder use becomes imperative.

This section revealed that salt tolerant plants (plants less affected by salinity) possess the ability to adjust internally to the osmotic effects of high salt concentrations than salt-sensitive plants. Also, salt-tolerant plants are more able to absorb water from saline soils. The breeding of salt tolerant crop is a less resource consuming or economical and socially acceptable approach when consideration is given to how the agricultural system in sub-Saharan Africa functions.

4.5.1 Breeding and Salt Tolerant Crops

Plant salt tolerance level is generally thought of in terms of the inherent ability of the plant to withstand the effects of high salts level in the root zone or on the plant's surfaces without a significant negative effect. Salt tolerant cultivars are capable of maintaining active water uptake by root cells at high salt level in the soil solution (Ashraf 2004). Salt tolerance is a complex trait involving responses to cellular osmotic and ionic stress and their consequent secondary stresses such as oxidative stress and whole plant coordination. The complexity and polygenic nature of salt stress tolerance are important factors contributing to the difficulties in breeding salt-tolerant crop varieties (Zhu 2000). The development of salt tolerant varieties can be advanced in some crop species, provided the genetic control and physiological mechanism of salt tolerance is clearly understood. Conversely, selection for potential breeding parental material with inheritable variations remains a decisive phenomenon, which is determined by the choice of breeding methods.

4.5.2 Genetics and Mechanisms of Salt Tolerance

Genetic diversity for salt tolerance is a prerequisite for developing salt tolerant crop varieties. Empirical studies conducted to evaluate the level of genetic diversity among genotypes of different crop species has revealed enormous response of plants at different growth stages under saline condition, considering reports provided on different crop. These include maize (Khatoon et al. 2010), sorghum (Geressu and Gezahagn 2008), rice (Momayezi et al. 2009) and millet (Yakubu et al. 2010). Additionally, exploiting the genetic variability of the available germplasm to identify tolerant genotype that may sustain a considerable yield under soils that are negatively affected by salinity was reported as one of the major strategies to overcome salinity problem (Ashraf et al. 2006). Genetic variability is a measure of the tendency of individual genotypes in a population to vary from one another. According to a report provided by Ramanatha Rao and Hodgkin (2002), variability is different from genetic diversity, which is the amount of variation seen in a particular population. The variability of trait describes the extent to which a particular trait tends to vary in response to environmental and genetic influences. Genetic variability in a population is important for biodiversity (Sousa et al. 2011), because without variability, it becomes cumbersome for a population to adapt to environmental changes. Consequently, this becomes prone to extinction. In order to enhance the salinity tolerance in crop species; wide range of variability in available germplasm becomes germane.

Existence of genetic variability for salt tolerance within species is of paramount concern in crop improvement programme. Therefore, the choice of germplasm to be used in breeding programme is most crucial, as the success lies on it. Such variation may be between individuals, varieties, or even species that have some degree of sexual compatibility, so that genes may be transferred from one individual to another. A number of studies were conducted to investigate the genotypic variability of salinity tolerance within and between crop species. Maiti et al. (2004) reported significant variability to salinity tolerance among some vegetable crop species. Among these, celery showed higher level of tolerance followed by cabbage, beet root leaves and green tomato. Genotypic variability in salinity tolerance at the stage of seedling stress was reported among tomato genotypes (Maiti et al. 2007b). Similarly, in a study carried out to examine the response of okra, tomato and chill genotypes for salinity tolerance at different concentrations of salt with respect to different seedling parameters. Also, large variability among genotypes was observed by Maiti et al. (2010).

Development of crop plants with salinity stress tolerance requires knowledge of the physiological mechanisms and genetic controls of the contributing traits at different plant developmental stages. Multiple mechanisms in plants was found to be associated with salt tolerance, which target ion selectivity (Shannon 1978), ion exclusion (Noble et al. 1984), ion accumulation (Tal and Shannon 1983), compatible solute production (Grumet and Hanson 1986; Wyn Jones et al. 1977). To this end, none of these physiological features is uniquely related to tolerance, which is assessed as the ability to survive salt as a seedling. However, vigour, sodium uptake and the plant's ability to tolerate the accumulation of sodium, makes important contribution to overall tolerance (Flowers et al. 1997). In additional, breeding and selection for salt tolerance varieties determined by genotypic variability in traits that influence tolerance; in vigour, in sodium uptake, in the distribution of sodium chloride between old and new leaves and in the ability to tolerate accumulated sodium (Yeo et al. 1990). Studies carried out on crop breeding have shown that selection for low sodium chloride uptake is plausible and heritable (Garcia et al. 1997). The ability to tolerate the sodium chloride that is accumulated is also an important trait, but not easily measured (Flowers et al. 1997). Therefore, the presence of molecular genetics plays an instantaneous role-in marker assisted selection (MAS), to ease the assessment of this trait in various crops.

Apart from those general mechanisms identified in most crops' varieties developed in recent times, halophytes have the capacity to sequester NaCl in vacuoles and produce compatible osmotic characteristic in the cytoplasm. Also, halophytes have a diversity of secondary mechanisms to handle excess salt. At the tissue level, some halophytes have salt glands (Liphschitz and Waisel 1982; Balsamo and Thomson 1993), salt bladders (Schirmer and Breckle 1982; Yeo and Flowers 1986) to handle temporary imbalance of NaCl entry into the plant. Halophytes are naturally adapted plants and knowledge in which they survive and maintain productivity on saline water can be used to define a minimal set of adaptations required in tolerant crop germplasm. This knowledge can help to focus the effort of plant breeders and molecular biologist working with conventional crop plants (Bohnert et al. 1995; Glenn et al. 1997; Rausch et al. 1996). Hence, the planting of halophytic plants in saline soil helps in reducing the salinity level of soil.

This section indicates that the exploitation of genetic variability of accessible germplasm to identify tolerant genotype that may sustain a considerable yield under salt affected soils is one of major strategies to overcome salinity problem. As halophytes help to reduce salt level in soil, due to their capability of withholding NaCl in vacuoles and producing compatible osmotic characteristic in the cytoplasm. Their diversity of secondary mechanisms in handling excess salt is also inclusive. The use of halophytic plants was also indicated as one of the ways of reducing salt level in soil.

4.5.3 Screening and Selection for Salt-Tolerance

In order to identify the potential parental material for breeding, having wide genetic variability, several screening and selection schemes were conducted for salt tolerance improvement in cereal and other crops (Dewey 1962; Kingsbury and Epstein 1984; Kelman and Qualset 1991; Karadimova and Djambova 1993; Pecetti and Gorham 1997). Specific stages throughout ontogeny of the plant, such as seed germination and seedling emergence, vegetative growth and reproduction, should be evaluated separately for assessing tolerance and identifying useful genetic material.

However, selection for improved salt tolerance on the basis of seedling stage have been used in various crop species, for example in rice (Shannon et al. 1998), maize (Rao and McNeilly 1999), wheat (Qureshi et al. 1990; Khan et al. 2003), sunflower (Islam et al. 2008), sorghum (Tigabu et al. 2013), and tomato (Solovier et al. 2003). The responses to salinity effect was also evaluated further at germination stage in other crops including pearl millet (Ashraf and Mcneilly 1992; Varma and Poonia 1979), chickpea (Singh and Singh 1980; Dua 1992; Dua and Sharma 1995), groundnut and pigeon peas (Srivastava et al. 2007).

Germination and seedling characteristics were pointed out to be the most viable criteria used for selecting salt tolerant plants due to the fact that the final plant stand of crop primarily depends on seedling characteristics (Tigabu et al. 2013). Therefore, identification of salt tolerant genotypes at both germination and seedling stages is particularly useful (Mano and Takeda 1997). As germination remains a critical stage for plant establishment (Song et al. 2008), poor germination may lead to poor stand establishment, which further results to lower crop yields. The seedling stage is generally the most sensitive phase of plant development and studies on salt tolerance in different crops species have mostly involved plant assessment at this stage (Song et al. 2008; Tlig et al. 2008; Badridze et al. 2009). In addition, selection for salinity tolerance based on seedling response has suggested that variation at this stage is genetically controlled (Maiti et al. 1996).

Generally, breeding crop for any character becomes very efficient mostly if the desirable trait is dominantly controlled by the genotype. Hosseini et al. (2012) emphasised the importance of finding sufficient variation in particular trait and to devise such screening techniques, which are reliable to identify tolerance level in order to enhance crop plants tolerant ability. Development of salt tolerant crop varieties requires an efficient screening technique for assaying genetic variation in adapted, as well as exotic germplasm. Among others, root length and shoot length have proven to be reliable traits used to identify sound parental breeding material tolerant to salt. These traits were reported in many studies carried in different crops, which included sunflower (Islam et al. 2008; Turhan and Ayaz 2004), rice (Hosseini et al. 2012), cotton (Ibrahim et al. 2007), sorghum (Tigabu et al. 2013), okra, tomato and chill (Maiti et al. 2010), to mention a few.

4.5.4 *Breeding Strategies for Salt Tolerance*

Strategies for breeding salt tolerant crops in cross-pollinating species by cycles of recurrent selection were identified long ago (Dewey 1962). The use of different breeding techniques and strategies for improving crop salt tolerant level have been discussed in most recent research works (Flowers 2004; Shannon 1993; Munns et al. 2002). It has been documented that some of the breeding strategies for improving salt tolerance may not be applicable for some crops due to the complexity of the trait responsible for tolerant mechanism of particular crop. Typical examples are presented in Table 4.10.

Table 4.10 Strategies for selection, breeding and development of salt tolerant crops

Approach	Crops	Examples
Conventional breeding	Barley, Lettuce	Ramage (1980)
		Shannon (1980)
Wide crosses	Tomato	Rush and Epstein (1981)
		Tal and Shannon (1983)
		Foolad (1996)
Domestication of wild salt-tolerant species	Tobacco, Chickpea	Nabors et al. (1980)
		Pandey and Ganapathy (1984)
Molecular biology	Wheat	Gulick and Dvorak (1987)

This example is only an indication of breeding strategies of major plant groups. These "indicator plants" could be useful in determining the strategies for selection and breeding of closely related plants or plants having sexual compatibility

Most of the conventional breeding methods have been considered for the development of salt tolerant varieties i.e. introduction, selection, hybridisation, mutation and shuttle breeding approach (Foolad 1996; Richards 1983; Hosseini et al. 2012; Ibrahim et al. 2007). Thereafter, other salt tolerant varieties were developed through recombination breeding. Recombination breeding strategy employs the grouping of genotypes based on the inherent physiological mechanism responsible for salinity tolerance. Subsequently, inter-mating of the genotypes with high degree of expression of the contrasting salinity tolerance mechanism and identifying or screening of the recombinants for pooling of the mechanisms are being followed to enhance further level of salt tolerance (Amaranatha et al. 2014). The genotypes are grouped into different categories based on the physiological mechanism for salt tolerance. Crosses are made between the parents or donors possessing contrasting physiological traits including tissue tolerance, Na^+ exclusion, K^+ uptake and Cl^-exclusion to pyramid the genes leading or contributing to salinity tolerance, which consequently become integrated into one agronomic superior background.

Molecular breeding through Marker assisted selection (MAS) has been considered as one of the strategies used to supplement conventional breeding programme in recent times (Linh et al. 2013). According to Linh et al. (2013), it has also been applied in breeding for salt tolerance to a considerable extent. Moreover, in some crops such as rice and soyabean, markers have been found as link to some specific traits of interest and used as the tools of biotechnology. Therefore, it is plausible to transfer valuable gene of salt tolerance stress in such crops (Cregan et al. 1999; Mackill 2007).

4.6 Conclusion

Worldwide, soil and water salinity stress causes reduction in plant growth, development and yield. The impact of salinity on crop production indirectly reflects on animal production level, particularly grazing animals. Salinity was indicated as one of the environmental factors causing soil degradation due to its relationship with the dynamics of ground water, which is dependent on soil type, cropping systems and the state of irrigation networks. Increases in salinity level in both soil and water used for irrigation decreases seed germination rate, seedling emergence, crop growth and development.

The use of organic mulches or gypsum or a combination of both to soil irrigated with saline water were all indicated as means of reducing salinity hazards and increasing the growth and yield of crops cultivated in saline soils. Other means of alleviating salinity problems pointed out include the exploitation of genetic variability of accessible germplasm to identify tolerant genotype that may sustain a considerable yield under salt affected soils. Also, the planting of halophytes to reduce salt level in soil due to their capability of withholding NaCl in vacuoles and producing compatible osmotic characteristic in the cytoplasm. Also considered among the attributes of the halophytic plants, is their diversity of secondary mechanisms, which makes them to be capable of handling excess salt. Henceforth, it could be recommended that the determination of the electrical conductivity of any water envisaged to be used for irrigation purpose is important to avoid the negative effect of water salinity on crop productivity.

As a result of the high costs of inputs, high cost of equipments and the labour intensiveness involved in today's farming and agricultural research activities. Factors, all known to be inciting the interest in earmarking strategies to manage the various problems home gardeners, smallholders, emerging and commercial farmers encounter in their day to day operational activities. The need to pull together information from various scientists involved in research aimed at the provision of food, for today and future generations remain imperative for use as guide in crop cultivation and research purposes. This is important for crop growers and agronomic researchers to succeed and achieve their goals. This review was aimed at such development. We hope this review work will fulfill the role of making sure the information revealed on salinity reach those in need of it.

Acknowledgement The authors of this chapter are grateful to all the various agriculturists, researchers, scientists and academics to which the information retrieved from their various research findings, reassessed works and books have contributed to this review work. The relentless supports provided to the authors by the management of the Vegetable and Ornamental Plant Institute (VOPI)-ARC, South Africa and the Department of Water Affairs (DWA), South Africa, are greatly acknowledged. Opinions expressed and conclusions arrived at, are those of the authors and are not necessarily to be attributed to VOPI or DWA.

References

Ahmad M, Shahzad A, Iqbal M, Asif M, Hirani AH (2013) Morphological and molecular genetic variation in wheat for salinity tolerance at germination and early seedling stage. Aust J Crop Sci 7(1):66–74

Ali Y, Aslam Z, Ashraf MY, Tahir GR (2004) Effect of salinity on chlorophyll concentration, leaf area, yield and yield components of rice genotypes grown under saline environment. Int J Environ Sci Technol 1(3):221–225

Amaranatha RM, Francies RM, Rasoo SKN, Prakash VR (2014) Breeding for tolerance to stress triggered by salinity in rice. Int J Appl Pharm Biotechnol 5(1):167–176

Ashraf M (2004) Some important physiological selection criteria for salt tolerance in plants. Flora 199:361–376

Ashraf M, Mcneilly TM (1992) The potential for exploiting variation in salinity tolerance in pearl-millet [*Pennisetum-americanum* (L) Leeke]. Plant Breed 108(3):234–240

Ashraf MY, Akhtar K, Hussain F, Iqbal J (2006) Screening of different accession of three potential grass species from Cholistan desert for salt tolerance. Pak J Bot 38:1589–1597

Badridze G, Weidner A, Asch F, Borner A (2009) Variation in salt tolerance within a Georgian wheat germplasm collection. Genet Resour Crop Evol 56:1125–1130

Bagayoko M (2012) Soil salinity alkalinity effects on germination and seedling growth of vegetable crops in the Office du Niger zone. J Res Environ Sci Toxicol 1(12):328–337

Balsamo R, Thomson WW (1993) Ultrastructural features associated with secretion in the salt glands of *Frankenia grandifolia* (Frankeniaceae) and *Avicennia germinans* (Avicenniaceae). Am J Bot 80:1276–1283

Bewley DJ, Black M (1994) Seeds: physiology of development and germination, 2nd edn. Plenum Press, New York

Blaylock AD (1994) Soil salinity, salt tolerance and growth potential of horticultural and landscape plants. University of Wyoming, Department of Plant, Soil and Insect Sciences. B-988, Laramie, February

Bohnert H, Nelson DE, Jensen RG (1995) Adaptations to environmental stresses. Plant Cell 7:1099–1111

Brar GS, Stewart BA (1994) Crop quality and utilization: germination under controlled temperature and field emergence of 13 sorghum cultivars. J Crop Sci 34:1336–1340

Chang C, Sommerfeldt TG, Entz T (1991) Soil chemistry after eleven annual applications of cattle feedlot manure. J Environ Qual 20:475–480

Chapin FS, Matson PA, Mooney HA (2002) Principles of terrestrial ecosystem ecology. Springer, New York

Cooke GW (1982) Fertilizing for maximum yield, 3rd edn. Suffolk Granada Publishing Ltd., London

Cregan PB, Mudge J, Fickus EW, Danesh D, Denny R, Young ND (1999) Two simple sequence repeat markers to select for soybean cyst nematode resistance conditioned by the *rhgl* locus. Theor Appl Genet 99:811–818

Crucci G, Caro A, de Ciciretti L, Leoni B (1994) Salinity and seed germination of some vegetable crops. II. Note. Acta Hortic 362:305–309

Dewey DR (1962) Breeding crested wheatgrass for salt tolerance. Crop Sci 2:403–407

Dua RP (1992) Differential response of chickpea (*Cicer arietinum*) genotypes to salinity. J Agric Sci 119:367–371

Dua RP, Sharma PC (1995) Salinity tolerance of Kabuli and Desi chickpea genotypes. Int Chickpea Pigeonpea Newsl 2:19–22

El-kharbotly A, Maghloub O, Al-subhi A, al-halhali A (2003) Indigenous grass species with potential for maintaining rangeland and livestock feeding in Oman. Agric Ecosyst Environ 95:623–627

Evangelou VP (1994) Influence of sodium and soils of humid regions. In: Pessarakli M (ed) Handbook of plant and crop stress. Marcel Dekker Inc, New York, pp 31–62

Fageria NK (2009) The use of nutrients in crop plants. CRC Press/Taylor and Francis, Boca Raton

Flowers TJ (2004) Improving crop salt tolerance. J Exp Bot 55(396):307–319

Flowers TJ, Garcia A, Koyama M, Yeo AR (1997) Breeding for salt tolerance in crop plants-the role of molecular biology. Physiol Plant 19(4):427–433

Foolad MR (1996) Response to selection for salt tolerance during germination in tomato seed derived from PI 174263. J Am Soc Hortic Sci 121(6):1006–1011

Garcia A, Rizzo CA, Ud-Din J, Bartos SL, Senadhira D, Flowers YAR (1997) Sodium and potassium transport to the xylem are inherited independently in rice, and the mechanism of sodium: potassium selectivity differs between rice and wheat. Plant Cell Environ 20:1167–1174

Geressu K, Gezahagn M (2008) Response of some lowland growing sorghum (*Sorghum bicolor* L. Moench) accessions to salt stress during germination and seedling growth. Afr J Agric Res 3(1):044–048

Glenn E, Miyamoto M, Moore D, Brown JJ, Thompson TL, Brown P (1997) Water requirements for cultivating *Salicornia bigelovii* Torr. with seawater on sand in a coastal desert environment. J Arid Environ 36:711–730

Gosh PK, Ramesh P, Bandyopadhyay KK, Tripathi AK, Hati KM, Misra AK, Acharya CL (2004) Comparative effectiveness of cattle manure, poultry manure, phosphocompost and fertilizer-NPK on three cropping systems in vertisols of semi-arid tropics. Bioresour Technol 95:77–84

Grumet R, Hanson AD (1986) Genetic evidence for an osmoregulator function of glycinebetaine accumulation in barley. Aust J Plant Physiol 13:353–364

Gulick P, Dvorak J (1987) Gene induction and repression by salt treatment in roots of the salinity-sensitive Chinese Spring wheat and the salinity-tolerant Chinese Spring Elytrigia elongatum amphiploid. Proc Natl Acad Sci U S A 84:99–103

Hamman B, Kong G, de Venter V (1996) Germination of South African soya bean seed lots. S Afr J Plant Soil 13(1):5–8

Haq T, Mahmood K, Shahzad A, Akhtar L (2003) Tolerance potential of wheat cv. LU-26S to high salinity and water logging interaction. Int J Agric Biol 5:162–165

Hill R, Koenig RT (1999) Water salinity and crop yield. Utah Water Quality AG-425.3. Utah State University, Logan, UT, 84322, May

Horton ML, Wiersma JL, Schnabel RR, Beyer RE, Carlson CG (1981) Animal waste effects upon crop production, soil, and runoff waters. Rep. EPA-600/S2-81230. USEPA, Ada, OK

Hosseini SJ, Tahmasebi Z, Pirdashti H (2012) Screening of rice (*Oryza sativa* L.) genotypes for NaCl tolerance at early seedling stage. Int J Agron Plant Prod 3(8):274–283

Ibrahim M, Akhtar J, Younis M, Riaz MA, Anwar-ul-Haq M, Tahir M (2007) Selection of cotton (*Gossypium hirsutum* L.) genotypes against NaCl stress. Soil Environ 26(1):59–63

Islam MZ, Mia MAB, Aktar A (2007) Effect of different saline levels on growth and yield attributes of mutant rice. J Soil Nat 1(2):18–22

Islam M, Afzal S, Ahmad I, Ahsan -ul-Haq, Hussain A (2008) Salt tolerance among different sunflower genotypes. Sarhad J Agric 24(2):241–250

Jamin JY, Doucet MJ (1994) La question foncière dans les perimeters irrigués de l'Office du Niger (Mali). Les Cahiers de la Recherche Développement 38:65–82

Jones US (1979) Fertilizers and soil fertility. Prentice-Hall Company, Reston

Karadimova M, Djambova G (1993) Increased NaCl-tolerance in wheat (*Triticum aestivum* L. and *T. durum* Desf) through in vitro selection. In Vitro Cell Dev Biol 29:180–182

Kaveh H, Nemati H, Farsi M, Jartoodeh SV (2011) How salinity affect germination and emergence of tomato lines. J Biol Environ Sci 5(15):159–163

Kelman M, Qualset CO (1991) Breeding for salinity-stressed environment: recombinant inbred wheat lines under saline irrigation. Crop Sci 31:1436–1442

Khan AS, Asad MA, Ali Z (2003) Assessment of genetic variability for NaCl tolerance in wheat. Pak J Agric Sci 40:33–36

Khatoon T, Hussain K, Majeed A, Nawaz K, Nisar MF (2010) Morphological variations in maize (*Zea mays* L.) under different levels of NaCl at germinating stage. Warsaw Agric Sci J 8(10):1294–1297

Kigel J, Galili G (eds) (1995) Seed development and germination. Marcel Dekker Inc, New York

Kihanda FM, Warren GP, Atwal SS (2004) The influence of goat manure application on crop yield and soil nitrate variations in semi–arid eastern Kenya. In: Bationo A (comp) Managing nutrient cycles to sustain soil fertility in sub-Saharan Africa. Academy Science Publishers, Nairobi, pp 174–186

Kingsbury RW, Epstein E (1984) Selection for salt resistant in spring wheat. Crop Sci 24:310–315

Kowaski JA, Palada MC (1995) Response of selected vegetable crops in saline water in the U.S. Virgin Islands. In: Proceeding of the thirteenth annual meeting of the Caribbean, Food Crops Society, St. Thomas, 5 August

Kumar A, Agarwal S, Kumar P, Singh A (2010) Effects of salinity on leaf and grain protein in some genotypes of oat (*Avena sativa* L.). Recent Res Sci Technol 2(6):85–87

Läuchli A, Grattan SR (2007) Plant growth and development under salinity stress. In: Jenks MA et al (eds) Advances in molecular breeding toward drought and salt tolerant crops. Springer, New York, pp 1–32

Linh LH, Linh TH, Xuan TD, Ham LH, Ismail AM, Khanh TD (2013) Improving submergence tolerance of Vietnamese rice cultivar by molecular breeding. J Plant Breed Genet 01(03):157–168

Liphschitz N, Waisel Y (1982) Adaptation of plants to saline environments: salt excretion and glandular structure. In: Sen D, Rajpurohit K, Junk W (eds) Tasks for vegetation science. Dr. W. Junk, The Hague, pp 197–214

Mackill DJ (2007) Molecular markers and marker-assisted selection in rice. In: Varshney RK, Tuberosa R (eds) Genomic assisted crop improvement, genomics applications in crops, vol 2. Springer, New York, pp 147–169

Maerere AP, Kimbi GG, Nonga DLM (2001) Comparative effectiveness of animal manures on soil chemical properties, yield and growth of *Amaranthus* (*Amaranthus cruentus* L). Afr J Sci Technol 1(4):14–21

Magan JJ, Gallardo M, Thompson RB, Lorenzo P (2008) Effects of salinity on fruit yield and quality of tomato grown in soil-less culture in greenhouses in Mediterranean climatic conditions. Agric Water Manage 95:1041–1055

Maiti RK, Amaya LED, Cardona SI, Dimas AMO, De La Rosa IM, Castillo HD (1996) Genotypic variability in maize cultivars (*Zea mays* L.) for resistance to drought and salinity at the seedling stage. J Plant Physiol 148:741–744

Maiti RK, Rio Zavala GJF, Singh VP, Pena RP, Arreola ES, Hernandez P (2004) Evaluation of germination and seedling establishment of some vegetable crops species for tolerance to salinity in Tlaxcala, Mexico. Crop Res 27:258–265

Maiti RK, Aruna Kumari CH, Kalpana K, Patil BS (2007a) Variability in salinity tolerance and osmotic stress among vegetable crop species. Crop Res 33(1):100–106

Maiti RK, Aruna Kumari CH, Prasad H, Kalpana K, Agsthi A (2007b) Genotypic variability in tomato hybrids to salinity at emergence and osmotic stress at germination stage. Crop Res 33(1):107–112

Maiti RK, Vidyasagar P, Umashankar P, Gupta A, Rajkumar D, González-Rodríguez H (2010) Genotypic variability in salinity tolerance of some vegetable crop species at germination and seedling stage. Int J Biochem Mol Biol 1(3):204–209

Mano Y, Takeda K (1997) Mapping quantitative trait loci for salt tolerance at germination and the seedling stage in barley (Hordeum vulgare L.). Euphytica 94:263–272

Marr PD (1967) The social context of irrigation. In: Hagan RM, Haise HR, Edminster TW (eds) Irrigation of agricultural lands. American Society of Agronomy, Madison, pp 12–23

Mathers AC, Stewart BA (1974) Corn silage yield and soil chemical properties as affected by cattle feedlot manure. J Environ Qual 3(2):143–147

McWilliam JR (1986) The national and international importance of drought and salinity effects on agricultural production. Aust J Plant Physiol 13:1–13

Momayezi MR, Zaharah AR, Hanafi MM, Razi IM (2009) Seed germination and proline accumulation in rice (*Oryza sativa* L.) as affected by salt concentrations. Pertanika J Trop Agric Sci 32(2):247–259

Munns R (2002) Comparative physiology of salt and water stress. Plant Cell Environ 25(2):239–250

Munns R, Husain S, Rivelli AR, James RA, Condon AG, Lindsay MP, Lagudah ES, Schachtman DP, Hare RA (2002) Avenues for increasing salt tolerance of crops, and the role of physiologically based selection traits. Plant Soil 247:93–105

Murphy LS, Wallingford GW, Powers WL, Manges HL (1972) Effects of solid beef feedlot wastes on soil conditions and plant growth, pp 449–464. In: Waste management research Proceedings of the Cornell agricultural waste management conference of the mangrove, *Avicennia germinans* L. Trees (2005), 19:721–727

Nabors MW, Gibbs SE, Bernstein CS, Meis ME (1980) NaCl-tolerant tobacco plants from cultured calls. Z Pflanzenphysiol 97:13–17

Noble CL, Holloran GM, West DW (1984) Identification and selection for salt tolerance in lucerne (*Medicago sativa* L.). Aust J Agric Res 35:239–252

Overman RA, Scholtz RV III (2002) Mathematical models of growth and yield. CRC Press/Taylor and Francis, New York

Pandey R, Ganapathy PS (1984) Isolation of sodium-tolerant callus line of Cicer arietinum L. cv BG-203. Plant Cell Rep 3:45–47

Pascale SD, Barberi G (1995) Effects of soil salinity from long-term irrigation with saline sodic water on yield and quality of vegetable crops. Sci Hortic 64:145–257

Pecetti L, Gorham J (1997) Screening of durum wheat germplasm for Na-22 uptake under low and moderate salinity. Cereal Res Commun 25(4):923–930

Poehlman JM, Sleeper DA (1995) Breeding field crops. Panima Publishing Corporation, New Delhi, p 278

Pratt PF, Davis S, Sharpless RG (1976) A four year field trial with animal manures: I Nitrogen balances and yields of forage. Hilgardia 44:99–112

Pratt PF, Lund LJ, Rible JM (1978) An approach to measuring leaching of nitrate from freely drained irrigated land. In: Nelson DR et al (eds) Nitrogen in the environment. Academic, New York, pp 223–273

PSI (Pôle régional de recherche sur les Systèmes Irrigués) (1999) Actes du séminaire de clôture (CD-ROM), Dakar, Sénégal

Qureshi RH, Aslam M, Nawaz S, Mahmood T (1990) Saline agriculture research in Pakistan. In: Proceedings of the India and Pakistan workshop on soil salinity and water management. PARC, Islamabad, pp 409–423

Ramage RT (1980) Genetics methods to breed salt tolerance in plants. In: Rains DW, Valentine RC, Hollaender A (eds) Genetic engineering of osmoregulation: impact on plant productivity for food, chemicals and energy. Plenum Press, New York. Basic Life Sci 14:311–318

Ramanatha Rao V, Hodgkin T (2002) Genetic diversity, conservation and utilization of plant genetic resources. Plant Cell Tiss Org Cult 63:1–19

Rao SA, McNeilly T (1999) Genetic basis of variation for salt tolerance in maize (*Zea mays* L). Euphytica 108:145–150

Rausch T, Kirsch M, Low R, Lehr A, Viereck R, Zhigang A (1996) Salt stress responses of higher plants: the role of proton pumps and Na+/H+-antiporters. J Plant Physiol 148:425–433

Rhoades JD, Chanduvi F, Leach S (1999) Soil salinity assessment: methods and interpretation of electrical conductivity measurements. FAO irrigation and drainage paper 57. Food and Agriculture Organization of the United Nations, Rome, Italy

Richards RA (1983) Should selection for yield in saline regions be made on saline or non-saline soils? Euphytica 32:431–438

Rush DW, Epstein E (1981) Breeding and selection for salt tolerance by the incorporation of wild germplasm into a domesticated tomato. J Am Soc Hortic Sci 106:669–670

Saeed R, Ahmad R (2009) Vegetable growth and yield of tomato as affected by the application of organic mulch and gypsum under saline rhizosphere. Pak J Bot 41(6):3093–3105

Sarwar G, Ashraf MY, Naeem M (2003) Genetic variability of some primitive bread wheat varieties to salt tolerance. Pak J Bot 35:771–777

Schimidhlter U, Gertli JJ (1990) Comparative investigations of the effects of salinity and moisture stress on germination and seedling growth of carrot. Acta Hortic 278:213–220

Schirmer U, Breckle S (1982) The role of bladders for salt removal in some Chenopodiaceae (mainly *Atriplex*) species. In: Sen D, Rajpurohit K, Junk W (eds) Tasks for vegetation science. Dr. W. Junk, The Hague, pp 215–232

Shannon MC (1978) Testing salt tolerance variability among tall wheatgrass populations. Agron J 70:719–722

Shannon MC (1980) Differences in salt tolerance within 'Empire' lettuce. J Am Soc Hortic Sci 105:944–947

Shannon MC (1997) Adaptation of plants to salinity. Adv Agron 60:75–120

Shannon MC (1993) Development of salt stress tolerance - screening and selection systems for genetic improvement. In: Maranville JW, Baligar BV, Duncan RR, Yohe JM (eds) INTSORMIL, Pub No. 94-2, University of Nebraska, Lincoln, NE, pp 117–132

Shannon MC, Grieve CM, Francois LE (1994) Whole-plant response to salinity. In: Wilkinson RE (ed) Plant environment interactions. Springer, Berlin, pp 199–244

Shannon MC, Rhoades JD, Draper JH, Scardaci SC, Spyres MD (1998) Assessment of salt tolerance in rice cultivars in response to salinity problems in California. Crop Sci 38:394–398

Sharma PC, Mishra B, Singh RK, Singh VP (2001) Variability in the response of spinach and coriander to alkalinity and salinity stresses. Indian J Plant Physiol 6:329–333

Shereen A, Mumtaz S, Raza S, Khan MA, Solangi S (2005) Salinity effects on seedling growth and yield components of different inbred rice lines. Pak J Bot 37(1):131–139

Shortall JG, Liebhardt WC (1975) Yield and growth of corn as affected by poultry manure. J Environ Qual 4(2):186–191

Singh Mangal JL (1991) A note on the response of some vegetable crops to underground water having different concentrations of residual sodium carbonate (RSC) in field condition. Haryana J Hortic 20:149–151

Singh G, Singh H (1980) Effect of growth-regulators on the growth-parameters of chickpea grown under different salinity levels. Indian J Agric Sci 50:23–30

Solovier AA, Kuklev MY, Karnauakhova T (2003) Functional status of some genotypes of tomato under salinity conditions in green house. Acta Hortic (ISHS) 609:47–50

Song J, Fan H, Zhao Y, Jia Y, Du X, Wang B (2008) Effect of salinity on germination, seedling emergence, seedling growth and ion accumulation of a euhalophyte *Suaeda salsa* in an intertidal zone and on saline inland. Aquat Bot 88:331–337

Sousa P, Froufe E, Harris DJ, Alves PC, Meijden A (2011) Genetic diversity of Maghrebian *Hottentotta* (Scorpiones: Buthidae) scorpions based on CO1: new insights on the genus phylogeny and distribution. Afr Invert 52(1):135–143

Srivastava LM (2002) Plant growth and development: hormones and environment. Elsevier, Orlando

Srivastava N, Vadez V, Nigam SN, Aruna R (2007) Genetic variability for salinity tolerance in groundnut (*Arachis hypogaea*). In: International conference on sustainable agriculture for food and bio-energy and livelihoods security, Jabalpur, 14–16 Feb 2007

Sutton AL, Nelson DW, Mayrose VB, Nye JC, Kelly DT (1984) Effects of varying salt levels in liquid swine manure on composition and corn yield. J Environ Qual 13:49–59

Sutton AL, Nelson DW, Kelly DT, Hill DL (1986) Comparison of solid vs. liquid dairy manure applications on corn yield and soil composition. J Environ Qual 15:370–375

Tal M, Shannon MC (1983) Salt tolerance in the wild relatives of the cultivated tomato: responses of *Lycopersicon esculentum L. cheesmanii, L. peruvianum, Solanum pennellii* and F1 hybrids to high salinity. Aust J Plant Physiol 10:109–117

Tas B, Basar H (2009) Effects of various salt compounds and their combinations on growth and stress indicators in maize (*Zea mays* L.). Afr J Agric Res 4(3):156–161

Tigabu E, Andargie M, Tesfaye K (2013) Genotypic variation for salinity tolerance in sorghum (*Sorghum bicolor* (L.) Moench) genotypes at early growth stages. J Stress Physiol Biochem 9(2):253–262

Tisdale SL, Nelson WL (1975) Soil fertility and fertilizers, 3rd edn. Macmillan, New York

Tlig T, Gorai M, Neffati M (2008) Germination responses of *diplotaxis harra* to temperature and salinity. Flora 203:421–428

Turhan H, Ayaz C (2004) Effect of salinity on seedling emergence and growth of Sunflower (*Helianthus annuus* L.) Cultivars. Int J Agric Biol 1:149–152

Turner JC, Hattey JA, Warren JG, Penn CJ (2010) Electrical conductivity and sodium adsorption ratio changes following annual applications of animal manure amendments. Commun Soil Sci Plant Anal 41:1043–1060

Van Averbeke W, Chabalala MP, Okorogbona AOM, Ramusandiwa TD, Azeez JO, Slabbert MM (2012) Plant nutrient requirements of African leafy vegetables. In: Oelofse A, Van Averbeke W (eds) Nutritional value and water use of African leafy vegetables for improved livelihoods. WRC, Pretoria

Varma SK, Poonia SR (1979) Germination and early seedling growth of pearl- millet (Pennisetum-typhoides L) as affected by salinity. Indian J Plant Physiol 22(2):144–146

Wild A (1988) Plant nutrients in soil. In: Wild A (ed) Russell's soil condition and plant growth, 11th edn. Longman Scientific and Technical, Harlow, pp 652–779

Wyn Jones RG, Leigh SR, Ahmad RA, Pollard A (1977) A hypothesis on cytoplasmic osmoregulation. In: Marre E, Ciferri E (eds) Regulation of cell membrane activities in plants. Elsevier Press, Amsterdam, pp 121–136

Yakubu H, Ngala AL, Dugje IY (2010) Screening of millet (*Pennisetum glaucum* L.) varieties for salt tolerance in semi-arid soil of Northern Nigeria. Warsaw J Agric Sci 6(4):374–380

Yeo A, Flowers T (1986) Ion transport in *Suaeda maritima*: its relation to growth and implications for the pathway of radial transport of ions across the root. J Exp Bot 37:143–159

Yeo AR, Yeo ME, Flowers SA, Flowers TJ (1990) Screening of rice (*Oryza sativa* L) genotypes for physiological characters contributing to salinity resistance and their overall performance. Theor Appl Genet 79:377–384

Yo SA, Shaw RJ (1990) Salinity tolerance of various crops. Department of Primary Industries, Brisbane

Zhu J (2000) Genetic analysis of plant salt tolerance using Arabidopsis. Plant Physiol 124:941–948

Chapter 5
Vegetable Breeding Industry and Property Rights

João Silva Dias and Rodomiro Ortiz

Abstract Plant genetics and breeding are long-term endeavor that require dedicated expertise and infrastructure plus substantial and stable funding. The development of new vegetable cultivars or breeding techniques requires time, effort and funding. Likewise, access to technology and to crop diversity remains essential for the development of vegetable cultivars. Vegetable breeding is characterised by continuous innovations and the development of new cultivars that meet the requirements of growers and consumers. The driving force behind this innovation is acquiring or increasing seed market share. However, breeding new vegetable cultivars requires high investments that can only be recouped if the breeding companies can commercialise the cultivar for a certain period. Intellectual property rights on cultivars are regarded by some in the private sector as the ultimate guardian of plant breeding entrepreneurs. They are viewed as the opportunity to control as many aspects of the invention as possible, thereby strangling the innovative capacity of the competition. As a result, a few multinationals dominate the global seed trade, while public sector plant breeding and local, small- and medium-size seed enterprises have a marginal role. Plant variety rights through patents may affect both vegetable diversity and the progress of plant breeding research, except within the company holding the patent. While obviously benefiting that company, it is a big step backwards for the plant breeding community and by extension, for horticulture itself. Some vegetable breeding programs were merged to reduce costs, which could lead to growers being dependent on a narrow genetic background that could contribute to biodiversity reduction and food insecurity. Access to "advanced" genetic resources is an important condition for a healthy and innovative vegetable breeding sub-sector and food security. We argued that the private sector relies on fundamental research and proof-of-concept demonstrations of feasibility from the public sector, and the public sector expects their discoveries to be expanded and implemented commercially by

J.S. Dias (✉)
Instituto Superior de Agronomia, University of Lisbon,
Tapada da Ajuda, 1349-017 Lisboa, Portugal
e-mail: mirjsd@gmail.com

R. Ortiz
Department of Plant Breeding, Swedish University of Agricultural Sciences,
Sundsvagen 14, Box 101, Alnarp SE 23053, Sweden
e-mail: rodomiro.ortiz@slu.se

E. Lichtfouse (ed.), *Sustainable Agriculture Reviews*,
Sustainable Agriculture Reviews 17, DOI 10.1007/978-3-319-16742-8_5

the private sector. Hence, we advocate effective and synergistic private-public partnerships for enhancing the use of vegetable diversity, accelerating its breeding and increasing genetic gains.

Keywords Biodiversity • Breeding • Food insecurity • Horticulture • Hybrids • Patenting

5.1 Introduction

Every vegetable product we see on the market has benefited from plant breeding in one way or another. The total genetic information available in a gene pool was used to breed new cultivars (Fig. 5.1). The first suggestion to exploit hybrid vigor or heterosis in vegetables was made by Hayes and Jones (1916) for cucumber. Commercial hybridization of vegetable species began in the United States in the middle 1920s with sweet corn, followed by onions in the 1940s. Since that time, private breeding companies have been placing more and more emphasis on the development of vegetable hybrids, and many species of vegetables have been bred as hybrid cultivars for the marketplace. Besides heterosis, hybrids also allow plant breeders to combine the best horticultural traits and multiple host plant resistance to pathogens and pests plus adaptation to stressful environments. Furthermore, if the parents are

Fig. 5.1 Each vegetable cultivar on the market has benefited from plant breeding

homozygous, the hybrids will be uniform, an increasingly important trait in commercial vegetable market production. The development of vegetable hybrid cultivars requires homozygous inbred parental lines, which provide a natural protection of plant breeders' rights without legal recourse and ensure a market for seed enterprises.

Since the 1970s plant breeders' rights protection has been provided by the International Union for the Protection of New Varieties of Plants (UPOV), which coordinates an international common legal regime for plant variety protection (UPOV 1994). Protection was granted for those who develop or discover cultivars that are new, distinct, uniform, and stable. Cultivars may be either sexually or asexually propagated. The UPOV coverage lasts 20 years for herbaceous species. Protective ownership was extended by UPOV in 1991 to include essentially derived cultivars. At the same time, the farmer's exemption (that permitted farmers to save seed for their own use) was restricted, but giving member states the option to allow farmers to save seed. In addition, after 1998 in Europe, and 2001 in the United States of America plant breeding companies can take advantages of patent laws to protect not only the cultivar itself but all of the plant's parts (pollen, seeds), the progeny of the cultivar, the genes or genetic sequences involved, and the method by which the cultivar was bred. The seed can be used only for research that does not include development of a commercial product (i.e., another cultivar) unless licensed by the patent holder. Patents are the ultimate protective device allowing neither a farmers' exemption nor a plant breeders' exemption (that permitted protected cultivar(s) to be used by others in further breeding to develop new cultivars).

Research and development (R&D) for improved seed development is expensive. Such product protection has presented a business incentive to corporations to invest in the seed industry, which supported an enormous increase in private R&D leading to strong competition in the marketplace between the major seed companies. The majority of current vegetable cultivars being sold nowadays are proprietary products developed by private R&D. A significant consequence of this increase in R&D has been a reduction of public plant breeding programs. As a result, the cost for R&D to develop new cultivars is shifting from the publicly supported research programs to the customers of the major seed companies (Dias and Ryder 2011).

One of the main factors that determine success in vegetable production is biodiversity and genetic capacity. No practical breeding program can succeed without large numbers of lines (genotypes) to evaluate, select, recombine and inbreed (fix genetically). This effort must be organized, so valid conclusions can be reached and decisions made. Scientists, plant breeders, support staff, facilities, budgets, and good management are requirements to assure success in the vegetable seed business. Science must be state-of-the-art to maximize success in a competitive business environment. Since the continued need for fundamental plant breeding research is critical to support development of new technology and expansion of the knowledge base that supports cultivar development, competition among proprietary cultivars results in owner-companies striving to do the best possible research to develop their own products and to compete on genetic and physiological quality of vegetable seed in the marketplace. Reasonable profit margins are necessary to pay back the R&D

costs to the owner and to fund future research on developing even better vegetable cultivars to stay competitive. There is considerable genetic variation within the various vegetable species, which can be exploited in the development of superior proprietary cultivars. The consequences of this dynamic situation will mean relatively short-lived cultivars replaced by either the owner of the cultivar or a competitor seed company. This intense competition means constantly improved and more sophisticated cultivars for the vegetable industry. Seed companies are in the business of manipulating genes to improve cultivar's performance for a profit. The success of the research is judged by the success of the product in making a reasonable profit. The research must improve economic performance starting with the seed production costs and include the grower-shipper/processor and the end user. If any link in this sequence of events is weak or broken, the new cultivar will likely fail (Dias 2010a).

Biotechnology is a new, and potentially powerful, tool that has been added by all the major seed corporations to their vegetable breeding research programs, and is part of ongoing public research for developing transgenic vegetable projects. It can augment or accelerate cultivar development by saving time, providing better products, delivering genetic uniformity, or getting results that are not possible through conventional crossbreeding (Dias and Ortiz 2012a, b). There are new challenges being face by vegetable breeding. This article provides an overview of these challenges and highlights the importance of biodiversity, plant breeding and improved cultivars to modernize vegetable production and to alleviate some protective measures that can create obstacles for innovation, and risks for biodiversity and food security.

5.2 Vegetable Breeding Industry, Biodiversity and Food Security

Plant breeders play a key role in determining what we eat, since the cultivars they develop begin the dietary food chain. Vegetable breeding is the development of vegetable cultivars with new proprieties (or traits). Innovation in vegetable breeding depends on biodiversity and access to genetic resources, on specific knowledge, on the development and application of new technologies, and capital to utilise them. Access to genetic biodiversity as well as to technology is essential for the development of new vegetable cultivars. Selection is impossible without genetic variation and new cultivars cannot be bred without it, thereby making access to this variation essential for vegetable breeders.

The impact of plant breeding on vegetable production rests on the complex relationships involving growers, available cultivars, and the developers of these cultivars. Vegetable growers consist of commercial producers with varying size land holdings ranging from moderately small farms to very large ones, and poor growers many of them subsistence farmers with small farms often on marginal lands. The subsistence farmers are usually also poor. Several types of cultivars are available.

The least sophisticated in terms of the method of development are landraces, also known as local or farmers' cultivars. Modern vegetable cultivars are bred by crossing and selection alone and F_1 hybrids between desirable inbred lines. The developers of landraces are usually the farmers themselves, and are obtained by repeated simple selection procedures generation after generation, while public sector plant breeders or seed companies develop improved cultivars and hybrids. Farmers in some cases can plant and save their own vegetable seeds, but there are real problems in this system in commercial production, where typically many different species may be grown. In farmer-grown seeds, viability may be low, due to poor seed storage environment, pollination is often uncontrolled, genetic improvement is lacking and seed born pathogens including virus are constraints. Hence, in modern vegetable production the seed business is most efficiently conducted by a distinct industry dominated today by multinational seed corporations (Dias 2010b; Dias and Ryder 2012).

Vegetable breeding has to address and satisfy the needs of both consumers and growers. The general objectives for farmers are high yield, host plant resistance to pathogens and pests, uniformity and abiotic stress adaptation. The main attributes sought by consumers are produce quality, appearance, shelf life, taste, and nutritional value. Thus, color, appearance, taste, shape, are usually more important than productivity. The priority goal of vegetable breeding programs is then to release new cultivars combining many desirable horticultural characteristics. Consumers want more vegetable diversification in a continuous supply. Vegetables are purchased based partially on eye appeal, which means that the development of desire to consume increases market demand. Diversification also tends to increase consumption (Dias 2014).

In our view, product differentiation, including new or renewed product introductions, is a key strategy for expanding sales in vegetable markets. To exploit such an opportunity, it is important therefore to continue research in biodiversity and to disseminate information regarding the benefits of vegetables, develop new improved vegetable cultivars and processed products, evaluate the economic opportunities and the market scope of these new products, and identify marketing trends and alternatives. Furthermore, the increasingly more wealthy and healthy people will demand greater vegetable dietary diversity in a global bio-based economy, which means that biodiversity will be crucial for the future of vegetable farming (Dias 2012a). Likewise, biodiversity remains the main raw material for vegetable agricultural systems to cope with climate change because it can provide traits for plant breeders and farmers to select resilient climate-ready crop germplasm and release new cultivars. Hence, collecting samples of endangered vegetables to be preserved in genebanks is the first step, but also protecting the agricultural systems where those vegetables are produced is also important to ensure the *in situ* evolutionary processes remain in place. The consequences of all these relationships may be quite profound for the farmers at each level, the seed producers, the consumers and the availability of food worldwide. It is therefore worthwhile to examine the commercial breeding industry and the future of crop biodiversity and sustainability to assess our expectations for food security.

5.3 The Commercial Breeding Industry

In the case of most vegetable crops, biodiversity and genetics are delivered in a marvelous package known as the seed. Individual growers cannot carry on the special techniques of seed production, such as seed treatment for the control of planting pathogens or the development of hybrids, as well as the incorporation of biotechnology. Vegetable seed production is often a business undertaken by a distinct industry. High tech seed industry is a key part of modern horticulture that combines plant breeding, seed production, storage, and distribution (Dias and Ryder 2011).

The private breeding sector emphasizes the development of hybrids to exploit heterosis as well as to combine multiple host plant resistance and abiotic stress adaptation. The vegetable seed business aims that growers purchase seed for each of their planting. Control of the parents prevents others to reproduce the hybrid seed. Farmers pay all the breeding work and seed marketing costs when purchasing improved or hybrid vegetable seed. International seed companies are mainly interested in the breeding and production of vegetable seeds with a high commercial value. Seed companies, policymakers, and researchers have neglected vegetable landraces, whose production often takes place under low-inputs, Nonetheless, these vegetable landraces still contribute significantly to household food and livelihood security, particularly for small resource-poor farmers (Weinberger and Msuya 2004). For example, in Africa landraces constitute an important source of micronutrients, contributing between 30 % and 50 % of iron and vitamin A consumed, respectively, in poor households (Gockowski et al. 2003; Weinberger and Msuya 2004).

Although hybrid seed technology has a significant impact on most vegetable crops in the industrialized agriculture, the unavailability of high quality seeds remains a limiting factor to vegetable farming in the developing world. Hybrid seed production is a high-level technology and cost-intensive venture. Only a well organized seed company with scientific manpower and a well-equipped research facility can afford hybrid seed production. The public sector in the developing world often does not have sufficient capacity to supply adequate quantities of good quality vegetable seed to poor growers and at present, there are few private sector seed companies adapting cultivars to local environments, especially in the poorest countries (Rohrbach et al. 2003). Farmers themselves often produce seeds of locally preferred landraces, as the individual markets are too small and the private seed sector has little interest in producing open pollinated cultivars (Weinberger and Msuya 2004). Without proper seed production, processing technology, quality assurance, and management supervision, locally produced seeds are often contaminated by seed-transmitted viruses and other seed-borne pathogens, and are genetically diverse. Lack of proper storage facilities and an effective monitoring mechanism often leads to low or uncertain seed viability and vigor. Moreover, low capital resources and poor market information discourage the development of seed-related agribusinesses. Seed quality and treatment are keys to product quality, and there is a need for upgrading quality control laboratories to meet international standards.

The global seed trade dominated by a few international corporations has effectively marginalized public plant breeding and local, small- scale seed companies. About 30 years ago there were thousands of seed companies in the world, most of which were small and family owned. Today, the top six global seed companies control almost 50 % of the commercial seed trade. Some of these companies belong to multinational corporations that also own other agri-business, for producing or selling pesticides and biotechnology-derived products. A large number of acquisitions of small and big seed companies happened between 1996 and 2008 and these companies have increased their turnover both in conventional and in organic vegetable production (Dias and Ryder 2011).

There are five company's business models in the vegetable breeding: (i) those traditionally integrating cultivar development, production and marketing of seed; (ii) others undertaking plant breeding and producing seed in their home country but licensing their cultivars to companies in other countries; (iii) those developing their own capacity in applied biotechnology; (iv) some specialized in plant biotechnology only, without being active in practical plant breeding, cultivar development, and seed production; and (v) a few operating globally and having a strategic research capacity. Some of these companies belong to worldwide corporations that are also involved with pesticides and biotechnology. In the traditional vegetable breeding companies (i and ii) their income is primarily the selling of seeds. Although even these traditional companies are now also increasingly using biotechnology in their breeding programmes. In the companies that have still developed their own capacities in applied biotechnology (iii) their income remains by selling seed and not by generating income via licences on patents. This group of companies comprises some companies originated from the agrochemical sector and that later became breeding companies via acquisitions and mergers. These last companies are combining two businesses: selling seeds and acquiring market positions via licences on their patents. Biotechnology companies (iv) are focusing on income from contract research for seed companies and on licence income from their biotechnological findings based on patent rights. This in particular concerns patents on molecular breeding techniques, marker platforms and on properties or "traits" of the plants, and marketing of traits. The value of such patents will in the end have to be paid at the level of the market for the seeds and planting materials by the end users (farmers and growers). The companies under (v) combine a large biotechnological capacity with the production and marketing of seed while at the same time licensing technologies to other plant breeding companies. This category comprises most multinationals in the seed sector that are also active in agrochemicals or pharmacy, but also larger traditional plant breeding companies with a significant biotechnology capacity. For these companies the income from seed sales is the most important but some also generate income from licences.

Commercial vegetable breeding has brought a paradigm shift in the agricultural cropping system by developing superior and productive vegetable crop cultivars in a short period. The vegetables attracting the most breeding attention vary considerably between small enterprises and huge seed corporations. Small seed companies have a tendency to specialize in a few vegetable crops. In large international companies the breeding activity is more diverse, but is concentrated on the more economically important crops. In these companies, the application of modern biotechnologies such as the use of molecular marker has become an integral component of many commercial vegetable breeding programs (Dias 1989). The access to modern tools of plant breeding such as genomic information to develop markers for important traits and genetic resources are the key drivers of successful modern vegetable plant breeding. In an era of continuous change, vegetable plant breeding is contributing towards fulfilling requirements of producers and consumers as well as in assessing climate or growing conditions, through continuous innovations to develop new and better cultivars. The vegetable breeding strategy and targets are dependent on market trends. Successful breeders anticipate changes in the market by developing new cultivars that are ready to be released to the growers when their demand increases. It will be therefore interesting to see how breeding companies react to changes in vegetable consumption and to evaluate the potential influence that the vegetable market and growing systems may have on breeding targets and priorities.

The commercial vegetable breeding sector produces a continuous flow of innovative new cultivars for a number of vegetables. Breeding focuses on the following most important properties: host plant resistances against pathogens and pests, increasing yield, and improving quality such as shelf life or taste, and enhancing production efficiency. Companies that are introducing a new cultivar with a new trait usually have a lead of about 4 years, after which the competitors can introduce their own new cultivars with the same trait. In such cases they make use of the "breeder's exemption". This is how this "open innovation" system leads to a wide availability of such an innovation. Investments in R&D by the top companies in this sector are between 15 and 25 % of their turnover, which keeps track with the annual increase in very high turnover. Most of the top companies show an annual growth of 5-7 % with net profits exceeding 10 %. Such growth can be realised in two different ways: by mergers and acquisitions or by autonomous growth. Enterprises with autonomous growth have to spend more on innovative R&D since they have to breed new cultivars and new technology themselves.

Plant breeding is a long-term and therefore costly activity. It was, at its beginnings, merely an empirical activity where plant breeders, on the basis of much knowledge and experience about traits of the reproductive material made crosses and select the most suitable plants. This process was strongly affected by growing season, length of the generation cycle, growing conditions, and available space. This meant that the development of a new cultivar or a new hybrid took 10-24 years, depending on the species. This development period decreased to 4-11 years in the last three decades through the use of a wide range of biotechnological methods, such as *in vitro* tissue culture, *in vitro* haploidization, mutation breeding, recombinant DNA technology and DNA marker-aided breeding, among others. The application of modern technology has made plant breeding less time and space-depend and breeding processes have become much more efficient. These advances led to reducing the time by a factor 2.5 for developing a new cultivar. Even though the research and development (R&D) costs increase strongly by about 10 % annually, the return for such investments was ensured by the faster production of new cultivars.

We conclude that a breeding company tries to maintain, or preferably expand, its market share by developing good cultivars. A company can therefore only continue breeding new cultivars if a good "return on investment" is ensured. The long time needed for the development of a new cultivar entails high risks and costs. This situation requires an adequate protection against the misuse of cultivars developed by the breeder with a lot of creativity and professionalism. In Europe, Plant Breeder's Rights provide, depending on the vegetable crop, a protection of 25 or 30 years. This time is long enough because the success period of a cultivar is usually 3-7 years. Seed companies can recover their investments by increasing the price of innovative seeds. This is possible in view of the usually fairly low price elasticity of vegetable seeds caused by the seed price being only marginal in comparison to the total production costs of a plant, by seeds being essential as basic material for production, and by innovations giving the seed a worthwhile added value. Currently, its also possible to protect a new trait in a cultivar via patent rights, provided that the new trait does at least meet the criteria of novelty, inventiveness and industrial applicability, and if this "invent" is not restricted to one cultivar. The exclusivity for the patent holder means that these innovative traits cannot be used in plant breeding without permission such as a licence of the patent holder.

5.4 Intellectual Property in Vegetable Breeding

To encourage innovations, compensate and reward innovators, and to protect the rights of the plant breeder the legislator has developed systems to be used to protect the "discoverer" against the risk that others without permission simply copy, imitate and commercialise own results, the new cultivar or the new finding. In the pre-protection era, most of the innovators were compensated in terms of their professional growth. In private breeding, the 'first mover advantage' and 'trade secrets' built in hybrid seeds gave sufficient compensation to innovators, but after the enactment of intellectual property right laws related to agriculture, in most countries, private research increased and research companies rushed to gain as much intellectual property rights or patents to obtain commercial benefits. The rapid development in biotechnologies has led to "breeding by design". The knowledge ensuing from molecular biology and genomic research keeps increasing and soon access to genetic information of the complete genome of all major crops will be available. These approaches in plant breeding are anticipated to produce lot of alternative processes like "breeding by chromosomes" resulting in patentable products. Presently, big corporations are earning income by selling products and from royalty on their patents. The OECD (2009) predicts widespread use of the technologies based on high-throughput sequencing, proteomics, metabolomics and phenotyping, new types of genetic markers and new genetic engineering system by 2030. Transgenic vegetable plants will include genes for producing pharmaceuticals and other valuable products.

Whether these technologies will become commercial successes depends upon costs related to research, market introduction and regulations, public acceptance and balanced intellectual property policies that stimulate innovations and competition.

Plant breeders' rights have a few weaknesses but they were written specifically for plants, and thereby implicitly recognize the differences between plants and inanimate objects. This is a saving grace. Much more egregious is the application to plants of the patent laws, which do not recognize these differences and therefore creates serious problems. The patent laws were written and amended over the years to protect inter alia a process, a machine or a manufacture, but not for living organisms. It became therefore necessary to apply the criteria of the patent laws to living entities, for which they were not intended, which has had some interesting consequences. Consider the bases for granting a patent. Under the patent law, an invention must be novel, non-obvious, and useful. The use of the term novel in intellectual property right laws may be confusing. For plant breeder's rights protection, as explained above, it means new in a commercially available sense. Under patent law, it means: "of a remarkably new and different kind". As stated by Ryder (2005), this criterion is badly abused in plant patents. For example in 12 lettuce patents involving lettuce found in an Internet search, eight are for new cultivars. All are unequivocally obvious. Hundreds of lettuce cultivars have been bred and released over the years. These eight lettuce cultivars are not remarkable in any way. The concept, breeding methods, and characteristics claimed are all ordinary. Most plant cultivars are bred after shuffling known genes in various combinations. These genes code for obvious known traits. The other four patents were for characteristics or procedures. One was for aphid resistance transferred by traditional breeding crosses from a related wild species. The resistance was closely linked with a deleterious character. They were separated through crossing-over and the recombinants were identified by molecular methods. The overall process was clever but obvious: breeders often find it necessary to break undesirable linkages. The second patent was for a trait called "multi-leaf characteristic" and refers to lettuce plants subject to fasciation, a flattening of the stem due to a wide meristematic apex. The trait was selected to occur very early in the life of the plant and resulted in the production of many leaves within a relatively narrow size range. This trait would be advantageous in producing cut leaves for packaging. This innovation may be considered non-obvious. The third trait is an elongated iceberg type lettuce produced by crossing iceberg lettuce and romaine lettuce. Iceberg lettuce is normally spherical. The head leaves are closely oppressed and cup-shaped and are therefore hard to separate. Romaine lettuce has elongated leaves that remain separated. The claimed trait specifies iceberg type leaves (characteristic texture and taste) in an elongated head where the leaves also separate easily. This combination of traits is non-obvious. The fourth patent is for a chemical treatment that inhibits head formation of iceberg or butter head lettuce, so that the leaves remain upright and open. Interior leaves are exposed to light and therefore are green instead of white. This presumably increases the content of certain nutrients, for example, beta-carotene, of these leaves. This may qualify as a non-obvious invention, although the idea of producing all green head leaves has been proposed before. The last criterion for protection is utility. The meaning of this is straightforward: the invention is marketable and therefore has potential economical use. This criterion is particularly important to the inventor, because the driving purpose of the invention is to sell it and make money. The difficulties noted above stem from a failure to properly apply two of the three basic requirements that qualify an invention as patentable under the law.

Much of the above discussion leads to the inevitable conclusion that the patent laws are inadequate for plants and should be replaced. Intellectual property rights for plants must be framed in different terms than for inanimate objects. Many patent applications are granted broad claims on traits and processes that are essential in nature. So, in patents, the essential processes like crosses, segregations and recombinant selections that are used for developing new cultivars should be excluded. However the term "essentially biological" processes are not well defined. In the European Biotechnology Directive, these are defined as entirely natural phenomenon of crossing and selection. A technology step in plant breeding seems sufficient to make whole process not entirely a natural phenomenon, thus patentable.

Patents allow elevation of the profit motive far above the good-of-society and biodiversity requirements. There are two major products of plant biotechnology: traits and methods. Traits such as a host plant resistance or product quality (e.g. increase antioxidant content) create value in the process of vegetable breeding (Dias 2012b, c). For vegetable breeding enterprises, specialized in plant breeding biotechnology that have based their business model on the development and marketing of traits or marker platforms the protection through patents is essential. For them patent system is the only way to create freedom to operate for further innovation. Patents are also necessary to enter into public-private partnerships, to maintain freedom to operate for scientists, assist in the downstream utilisation of public inventions, and to obtain cash benefits for a public institute facing increasing difficulties to secure funding.

Intellectual property rights have provided an essential contribution to the innovation and the success of plant breeding until now but breeder's exemption that allows them to benefit from the availability of the competitor's genetic resources and to use protected cultivars for further breeding seems crucial for the future of biodiversity and food security. Breeder's exemption plays therefore an essential role in innovation in practical plant breeding whose motivation is to find creative solutions for problems in vegetable farming and in the value chain that can capture a market segment. It should also be noted that nowadays no breeder's rights are requested for many vegetable crops because the economic life of a new cultivar is no more than few years and that most income can be generated during the time required to register such cultivars; i.e., 1 or 2 years. Another reason is that most vegetable cultivars are hybrids than cannot be reproduced as "true breeding lines".

We believe that a patent is however a means to slow the flow of progress of plant breeding research, except within the company holding the patent. While obviously benefiting that company, it is a big step backwards for the plant breeding community and by extension, for agriculture itself. Theoretically, if each seed company could obtain a patent on a new cultivar with certain favorable traits, each would do further breeding only with its own protected cultivar. So there would be parallel lines of research without the enrichment to each program that comes from crossing those lines with cultivars in other programs. The owner of a patented cultivar can share it by licensing its use in breeding to other companies. The cost of the license, in outright payment or in royalty fees, may be quite steep. This would certainly limit the interest in using that cultivar, since the cost may negate any profit from a new cultivar.

5.5 Trends in Genetic Diversity and Vegetable Breeding

About 52 % of vegetables grown in the world receive commercial breeding attention by seed companies and, of those, only 17 % are by large scale breeding programs, fostering a need for serious attention to maintenance of vegetable crop biodiversity (Dias and Ryder 2011). There has been a severe decline in the vegetable cultivar genetic base, as evidenced by the significant reduction, especially within the last 50 years, in the number and range of vegetable cultivars grown. During this period vegetable genetic diversity has been eroding all over the world and vegetable genetic resources are disappearing, on a global scale, at an unprecedented rate of 1.5-2 % per annum (Dias 2010b).

Widespread adoption of simplified vegetable systems with low genetic diversity carries a variety of risks including food insecurity. In the short term, such systems risk potential crop failure. In the long term, they encourage the reduction of the broad genetic base that contributes to high yields, quality traits, or host plant resistance to pathogens and pests. This compromises the future genetic wealth of vegetables. Especially prominent among the "enemies" of genetic diversity are the commercial markets and economic social pressures promoting breeding methods leading to uniformity, encouraging extensive cultivation of preferred improved and hybrid vegetable cultivars with insufficient diversity (Fig. 5.1; Dias 2010b). In addition, globalization has stimulated the consolidation of vegetable seed companies into huge corporations and the decline of small seed enterprises that serve local and regional markets. In consequence some vegetable breeding programs have been merged or eliminated to reduce costs. Thus fewer and fewer companies and corporations are making critical decisions about the vegetable research agenda, and the future of vegetables worldwide. Inevitably, two things will happen. There will be fewer vegetable breeders in the future and growers will be dependent on a narrower genetic background that could lead in the near future to food insecurity for poor growers and consumers (Dias 2010b). Likewise, with the advent of genetic engineering, these huge seed corporations are also assuming ownership of a vast array of living organisms and biological processes. Of equal concern are expanded uses of legal mechanisms, such as patents and plant breeder's rights that are removing vegetable plant germplasm from general public use (Ryder 2005). IPR for plants were intended as a defensive mechanism to prevent the loss of invented cultivars to competitors. However, with the more stringent enforcement of plant breeding rights, and particularly with the application of the utility patent law in the USA to protect all forms of an innovation, this has become an offensive weapon to stifle competition and inhibit the flow of germplasm and information. This situation can have serious implications for the future conservation of vegetable genetic resources and for world food security (Dias 2010a; Dias and Ryder 2011).

Some landraces and old open-pollinated cultivars of vegetables have existed for long periods outside the commercial and professional plant breeding circles because they have been kept alive within communities by succeeding generations of seed savers. Unfortunately, there are fewer and fewer active seed savers among the millions of vegetable growers, due to the demand of commercial markets and the professionalization of the sector. This is an additional threat to genetic diversity. Hence, the continued survival of landraces and open-pollinated cultivars of vegetables depends largely on popular interest and initiative as well as preservation in genebanks. We should be alerted and concerned about the loss of genetic diversity in vegetables and about its impact on food security (Dias 2010b).

Vegetable growers have an important role in conserving and using vegetable genetic diversity. The future of world food security depends not just on stored vegetable genes, but also on the people who use and maintain crop genetic diversity on a daily basis. In the long run, the conservation of plant genetic diversity depends not only on a small number of professional plant breeders and genebanks, but also on the vast number of growers who select, improve, and use vegetable genetic diversity, especially in marginal farming environments. That is why we should be also alerted and particularly alarmed by the current trend to use improved and hybrid vegetable cultivars exclusively. Growers do not just save seed, they also act as plant breeders who are constantly adapting their vegetable crops to specific farming conditions and needs. For many generations, vegetable growers have been selecting seeds and adapting their plants for local use. This genetic diversity is the key to maintaining and improving the world's food security and nutrition. No plant breeder or genetic engineer starts from scratch when developing a new cultivar of tomato, pepper, cabbage or lettuce. They build on the accumulated success of generations of growers, who have selected and improved vegetable seeds for thousands of years. If poor small-scale growers in marginal areas stop saving seeds, we will lose genetic diversity (Dias 2010b). Growers will lose the means to select and adapt vegetable crops to their unique farming conditions, which are characterized by low external inputs. Hybrid seed technology is designed to prevent growers from saving seed from their harvest, thus forcing them to return to the commercial seed market every year. Hybrid vegetable seeds alone, and used globally, can be a dead-end to biodiversity. If growers abandon completely their traditional vegetable landraces in the process of adopting only hybrids, crop genetic diversity achieved over centuries will be lost forever (Dias 2010b). Many horticultural benefits will be lost to worldwide growers and thus to consumers.

The exclusive adoption of hybrid cultivars in marginal areas may restrict the vegetable producing capacity of growers. It will also destroy biodiversity, and it may contribute in the long-term to food insecurity (Fig. 5.2). For example, a study by Daunay et al. (1997) points out that the release of F_1 hybrids (in Europe and some Asian countries such as China and Japan), which had high productivity but poor

Fig. 5.2 Commercial markets promote breeding methods leading to uniformity

phenotypic variability, contributed to the losses of eggplant landraces, thus inevitably leading to genetic erosion of *S. melongena*. Moreover, some African cultivated eggplants have been lost following social, economic, and political changes (Lester et al. 1990). Hence, the eggplant cultigen pool has been considered a priority for the preservation of vegetable genetic resources since 1977. As a result, research has been carried out in Asia and Africa (Lester et al. 1990; Gousset et al. 2005), and collections built up (Bettencourt and Konopka 1990), particularly in China (Mao et al. 2008).

Fortunately, in the industrialized world new independent seed companies, offering unique collections of regionally adapted landrace vegetable cultivars, have recently emerged. Furthermore vegetable hobbyist groups, mainly from organic horticulture, are thriving and maintaining old vegetable landraces, in organizations known as "seed savers." In this way traditional landraces are being restored to native growers and urban and peri-urban growers. Some of these traditional landraces display combinations of traits that make them especially responsive to local or regional conditions, or are well-suited to particular growing methods, such as those used in organic horticulture or low-external-input systems, or are tolerant to local pests and diseases or other stresses and constraints. Organic growers, who seek to grow "full-cycle" or seed-to-seed, are also working to ensure the continued availability of organically grown seeds. There are also considerable ongoing efforts by national governments and international organizations to preserve plant vegetable germplasm in genebanks. This is a valuable but static approach, as further evolutionary changes and improvements will not occur until the seeds are planted, and selection takes place. It is also an activity that relies heavily on continued political stability and support, including sustained governmental funding. Active and positive connections between the private breeding sector and large-scale genebanks are required to avoid possible conflict involving breeders' rights and gene preservation. The genetic diversity of vegetable species will be promoted by the maintenance of crop genebanks by governmental and non-governmental organizations, the continued use of diverse sources by plant breeders, especially in the public sector, and by the use of local cultivars and landraces by farmers (Fig. 5.3).

We argued that issues indicated above related to biodiversity, plant breeding and intellectual property rights are not confined to either the public or the private sector because addressing them will require partnerships and collaboration for success. The private sector relies on fundamental research and proof-of-concept demonstrations of feasibility from the public sector, and the public sector expects their discoveries to be expanded and implemented commercially by the private sector. Issues such as intellectual property, competition and privacy can complicate public–private interactions, but effective partnerships between the two sectors have proven to be highly synergistic (Spielman et al. 2007). With a set of unified priorities (above) and new models for cooperation and collaboration, the strengths of both sectors can be brought to bear on the significant challenges we face. Pre-competitive research in

Fig. 5.3 Local market in Africa promoting biodiversity

the public sector, jointly funded in some cases by private resources, can "lift all boats" and provide tools and resources for accelerating plant breeding improvements. Mechanisms are needed to allow the significant private investments in fundamental research, such as genome sequences and genetic maps, to be available to public researchers. Cost sharing programs, with public funds matching private investments in public research, are excellent models for encouraging direct public/private research collaborations (Yarkin and Murray 2003). A number of public sources of research funding now require matches from private industry or commodity groups, making private partnerships even more critical for public research. Creating a shared vision that supports systemic change increases the opportunities for success. These new approaches need to focus on leveraging the potential for synergy between the collaborators and set the foundation early in the arrangement to manage the risks and dangers of food security that are of greatest concern (Rausser et al. 2000).

5.6 Food Security and Prospects for Developing Countries and Poor Vegetable Farmers

Food security exists when all people, at all times have access to sufficient, safe and nutritious food to meet their dietary needs and preferences for an active and healthy life. Vegetable breeding in the developing world is reduced and focused on a very limited number of crops. The general lack of private investment in developing countries can be explained by the dominance of the public sector on the one hand and the low purchasing power of the majority of the farmers. Besides in some of these developing countries the market is too small to generate the interest of the international breeding companies for specific programmes.

Nearly half of the world's vegetable farmers are poor and cannot afford to buy hybrid seed every growing season. What are the prospects for these growers since they produce 15-20 % of the world's vegetables and they directly feed almost one billion people in Asia, Latin America, and Africa? Capital and risk factors are the key constraints that limit the adoption of improved vegetable cultivars by small and poor farmers, because these vegetables generally are much more costly to produce per hectare than traditional landrace cultivars (Key and Runsten 1999; Ali and Hau 2001; Ali 2002), and most farmers require credit to finance their production. While landraces are usually cultivated using a level of input intensity appropriate to the financial resources available within a household, improved vegetable cultivars often require an intensive input regime, including large labor inputs for planting and harvest that cannot be met with family labor alone (Weinberger and Genova 2005). For small and poor farmers improved vegetable cultivars also tend to be riskier than landraces, since the higher costs associated with seeds and production impose a greater income risk. Small farmers may have lower production costs with landraces, because they achieve adequate yields with fewer inputs. In addition, the profits from improved cultivars or hybrids tend to vary because yields are often higher but prices fluctuate. From another perspective variable prices and yields increase the variability in market supply (Key and Runsten 1999).

The lack of capital available to small and poor farmers denies them the opportunity to invest in vegetable production inputs. Without collateral help these farmers are usually unable to secure a loan from a bank or moneylender. For those who can get a loan, rates are often unmanageably high, with strict penalties for late repayments. Similarly, a lack of awareness, education, resources, skill training, and support prevent these farmers from using improved cultivars and then to generate a stable income from their production. In addition, governments usually do not regulate the price of vegetable crops or even provide market information, unlike for field crops. Improving market information systems for vegetable crops and facilitating farmers' access to credit are then essential components of a strategy to enable poor farmers to grow improved vegetable cultivars and to overcome the insecurity of their food supplies. The problem of food insecurity in this situation, like that of poverty, is thus frequently traceable to macroeconomic conditions and market failures due to actions of exploitative intermediaries, including landowners, moneylenders, and traders.

We strongly argue that a major obstacle to success in vegetable production is the shortage of affordable credit. In some cases vegetable farmers must pay high interest rates of 15-25 % per 100 days. Desperate for cash, subsistence farmers are forced to sell their crops immediately after the harvest to middlemen or their creditors at unfavorable prices. As pointed by HKI (2010) low cost quality seeds are essential for these farmers. Hence, credit facilities and other inputs must be also part of these vegetable production systems, so that the use of improved vegetable cultivars can help subsistence vegetable growers to overcome their poverty and food insecurity.

5.7 Conclusion

Vegetable breeding is the development of new vegetable cultivars with new proprieties. In this era of changes, vegetables will play a major role in well-balanced diets and in the current global battle against malnutrition. There will be continuous need of biodiversity and new and performing cultivars for sustainability of vegetable production. Biodiversity is the basis for vegetable breeding and for the introduction of new cultivars to improve quality and productivity. As advances in breeding are dependent upon genetic diversity, preserving and characterizing existing germplasm resources and expanding collections are essential to future crop improvement. Changing agricultural practices, including adoption of improved cultivars, can result in loss of genetic diversity that exists in native landraces.

Breeding vegetable hybrids is a key means towards the development of cultivars for modern vegetable production. Hybrid seed production is high technology and a cost intensive venture. Only well organized seed companies with good scientific manpower and well-equipped research facilities can afford seed production. Due to globalization, most vegetable breeding research and cultivar development in the world is presently conducted and funded in the private sector, mainly by huge multinational seed companies. Few companies are controlling a large part of the world market. Public vegetable breeders and public sector cultivars development are disappearing worldwide. It is therefore imperative that national governments and policymakers, as part of a social duty, invest in plant breeding research and development of traditional open-pollinated cultivars and in the minor and so-called "forgotten" vegetables. Smaller seed companies, which are usually specialized in few vegetable crops, must be supported, possibly through autonomous affiliation with the larger companies. More investments in this area will mean less expensive seed for growers to choose from, and increased preservation of vegetable diversity. The accomplishment of this goal may require new approaches to vegetable breeding research and development by both the public and private sector. We must ensure that society will continue to benefit from biodiversity and from the vital contribution that plant breeding offers, using both conventional and biotechnological tools, because improved and hybrid vegetable cultivars are, and will continue to be, the most effective, environmentally safe, and sustainable way to ensure global food security and healthy human nutrition.

References

Ali M (2002) The vegetable sector in Indochina: a synthesis. The vegetable sector in Indochina countries: farm and household perspectives on poverty alleviation. The World Vegetable Center, Bangkok

Ali M, Hau VTB (2001) Vegetables in Bangladesh. Technical bulletin 25. The World Vegetable Center, Shanhua

Bettencourt E, Konopka J (1990) Vegetables. In: Directory of germplasm collection. IBPGR, Rome, pp 204–220

Daunay MC, Lester RN, Ano G (1997) Les aubergines. In: Charrier A, Jacquot M, Hamon S, Nicolas D (eds) L'amélioration des Plantes Tropicales. Repéres. CIRAD-ORSTOM, Paris, pp 83–107

Dias JS (1989) The use of molecular markers in selection of vegetables. Acta Hortic 3:175–181

Dias JS (2010a) Impact of improved vegetable cultivars in overcoming food insecurity. In: Nath P, Gaddagimath PB (eds) Horticulture and livelihood security. Scientific Publishers, Jodhpur, pp 303–339

Dias JS (2010b) Impact of improved vegetable cultivars in overcoming food insecurity. Euphytica 176:125–136. doi:10.1007/s10681-010-0237-5

Dias JS (2012a) Vegetable breeding for nutritional quality and health benefits. In: Carbone K (ed) Cultivar: chemical properties, antioxidant activities and health benefits. Nova Science Publishers, Inc., Hauppauge, pp 1–81

Dias JS (2012b) Major classes of phytonutriceuticals in vegetables and health benefits: a review. J Nutr Ther 1:31–62. doi:10.6000/1929-5634.2012.01.01.5

Dias JS (2012c) Nutritional quality and health benefits of vegetables: a review. Food Nutr Sci 3:1354–1374. doi:10.4236/fns.2012.310179

Dias JS (2014) Guiding strategies for breeding vegetable cultivars. Agric Sci 5:9–32. doi:10.4236/as.2014.51002

Dias JS, Ortiz R (2012a) Transgenic vegetable crops: progress, potentials and prospects. Plant Breed Rev 35:151–246

Dias JS, Ortiz R (2012b) Transgenic vegetable breeding for nutritional quality and health benefits. Food Nutr Sci 3:1209–1219. doi:10.4236/fns.2012.39159

Dias JS, Ryder EJ (2011) World vegetable industry: production, breeding, trends. Hortic Rev 38:299–356

Dias JS, Ryder E (2012) Impact of plant breeding on the world vegetable industry. Acta Hortic 935:13–22

Gockowski J, Mbazo'o J, Mbah G, Moulende T (2003) African traditional leafy vegetables and the urban and peri-urban poor. Food Policy 28:221–235

Gousset C, Collonnier C, Mulya K, Mariska I, Rotino GL, Besse P, Servaes A, Sihachakr D (2005) *Solanum torvum*, as a useful source of resistance against bacterial and fungal diseases for improvement of eggplant (*S. melongena* L.). Plant Sci 168:319–327

Hayes HK, Jones DF (1916) First generation crosses in cucumber. Republic Conf Agric Exp Stat 5:319–322

HKI (2010) Homestead food production model contributes to improved household food security, nutrition and female empowerment-experience from scaling-up programs in Asia (Bangladesh, Cambodia, Nepal and Philippines). Nutr Bull 8:1–8

Key N, Runsten D (1999) Contract farming, smallholders, and rural development in Latin America: the organization of agroprocessing firms and the scale of outgrower production. World Dev 27:381–401

Lester RN, Jaeger PML, Spierings BHB, Bleijendaal HPO, Hooloway HLO (1990) African eggplant: a review of collecting in West Africa. Plant Gen Resour Newsl 81–82:17–26

Mao W, Yi J, Sihachakr D (2008) Development of core subset for the collection of Chinese cultivated eggplant using morphological-basal passport data. Plant Genet Resour 6:33–40

OECD (2009) The bioeconomy to 2030. Designing a policy agenda. Organization for Economic Cooperation and Development, Paris

Rausser G, Simon L, Ameden H (2000) Public-private alliances in biotechnology: can they narrow the knowledge gaps between rich and poor? Food Policy 25:499–513

Rohrbach DD, Minde IJ, Howard J (2003) Looking beyond national boundaries: regional harmonization of seed policies, laws and regulations. Food Policy 28:317–333

Ryder EJ (2005) Intellectual property rights for plants. The case for a new law. Chronica Hortic 45(2):5–11

Spielman D, Hartwich F, von Grebmer K (2007) Public-private partnerships in international research. International Food Policy Research Institute, Washington, DC

UPOV (1994) International Convention for the Protection of New Varieties of Plants of 2 December 1961, as revised at Geneva on 10 November 1972, 23 October 1978, and March 1991. UPOV publication 221. International Union for the Protection of New Varieties of Plants, Genéve, Switzerland

Weinberger K, Genova C (2005) Vegetable production in Bangladesh: commercialization and rural livelihoods. Technical bulletin 33. The World Vegetable Center, Shanhua

Weinberger K, Msuya J (2004) Indigenous vegetables in Tanzania: significance and prospects. Technical bulletin 31. The World Vegetable Center, Shanhua

Yarkin C, Murray A (2003) Assessing the role of the University of California in the State's biotechnology economy: heightened impact over time. University of California, Berkley

Chapter 6
Local Plants for Rural Food Security

Ruparao T. Gahukar

Abstract Local plants appear to be of minor importance as they are either neglected or underutilized by the farmers and consumers. Some of these crops are disappearing, even lost or became extinct, because of preference for high value cash crops or plants of industrial importance. Consequently farmers have lost their traditional knowledge and skill for growing local crops raised mostly on marginal lands with little farm inputs. With availability of easy to cook and tastier foods, consumers have changed their food habits. Poor people cannot however, afford these modern but costly foods and have to rely solely on local crops particularly during time of distress and natural calamities. There is environmental degradation, loss in genetic resources and changes in crop sustainability. Unsteady climate, paucity of resources and developmental facilities, inappropriate government policies hamper the cause of food security. Content of poor nutrients and/or toxic substances in local crops are the additional challenges.

I review here the issues of current food security in poor countries in the context of local crops. Great potential exists for exploitation and utilization of this natural asset as these crops have ability to grow and thrive in adverse conditions for growth and development. Likewise, existing crops are being rejuvenated by modern breeding techniques. Conservation of natural resources becomes a priority especially in ecologically degraded regions. When food is inaccessible due to high prices and low paying capacity, local crops and wild plants enhances household security not only by availability by substituting normal crops but also by cheap processing. Food is inexpensive yet provides nutritional elements, supports family livelihood and economy of poor people. Therefore, due attention has to be given by research scientists, planners, policy makers, administrators and consumers. In near future, creating community institutions and enhancement of local agricultural practices may help marginal and small farmers to cope with disastrous poverty and to eradicate hunger. Pro-poor policies launched by government can also improve food and nutrition security.

Keywords Neglected and underutilized crops • Wild plants • Local participation • Food security • Climate change • Initiatives and government policies

---

R.T. Gahukar (✉)
Arag Biotech Pvt. Ltd., Plot 220, Reshimbag, Nagpur, 440009, India
e-mail: rtgahukar@gmail.com

E. Lichtfouse (ed.), *Sustainable Agriculture Reviews*,
Sustainable Agriculture Reviews 17, DOI 10.1007/978-3-319-16742-8_6

6.1 Introduction

Since centuries, local plants had been used for food purposes by native ancestors for their survival whenever they had to face difficult situations such as, drought, crop failure, erratic rainfall, floods, human disease epidemics, ethnic war or social conflict. Even now people would use them for food or beverages rather than to starve. Out of the 300,000 plant species currently available on earth, some 10,000 are used as food. Of this, 150–200 species are commercially exploited with 30 crops providing food, particularly wherever the food insecurity prevails, by way of malnourishment and underfeeding, at least in the low-income families (Anthony et al. 1993). In India alone, out of 45,000 species of wild plants, 7,500 species are practically used in medicinal and indigenous health practices (Kamble and Jadhav 2013). Also, about 3,900 plant species including 145 species of root and tubers and 521 species of leafy vegetables are used by tribal communities as food (Kamble and Jadhav 2013). In reality, there is no distinction between plants used for nutrition and therapeutic qualities (Das 2010; Akhtar et al. 2012; Arivalagan et al. 2012).

Apart from food, few plants provide fodder for domestic animals in places where meat is frequently eaten in hard times especially by the desert inhabitants and tribal communities. Despite these natural foods, there are around two billion people lacking essential micronutrients (McDermott et al. 2013) and again, human population is expected to peak at 9–10 billion by 2050. How to feed the ever-increasing human population is a big challenge particularly in a situation when productivity of major food crops is nearly stagnant (Gregory and George 2011; Smith 2013). The global food security may therefore remain a worldwide concern for the next 50 years and beyond. Ultimately, there is need to stimulate production of local food crops in order to produce sustainable diets locally and unlock their promising contributions to household livelihood. The local produce is inflation neutral as people can collect it from the wild or procure from natural resources.

Local plants include neglected and underutilized often regarded as minor plants. The neglected crops are grown primarily by traditional methods but yet important for the subsistence of local communities. The underutilized plants are widely grown but are less used by the farmers and consumers though they play important roles in supporting livelihood and economy of poor people.

Currently, statistical data are not available on plant species (including minor crops) that serve as food in the period of distress and food shortage, and are protective of biodiversity and ecosystems in developing and less-developed countries (Koohafkan 2012). Similarly, census of the wild plants is rather difficult due to their inaccessible locations, seasonal availability, local taboos, unpleasant taste, intoxication and health complications arising from consumption (Ahmad and Javed 2007; Vincent et al. 2013).

From the review of literature it appears that sufficient information on local food crops is available (Wickens et al. 1989; Anthony et al. 1993; Hu 2005; van Wyk 2005). But the scientists, development workers and policy makers unintentionally overlooked the potential of these crops and instead, focused research on new sources

Photo 6.1 Biodiversity in marginal farmer's field (adoption of sustainable agriculture and cultivation of underutilized legumes and cereals)

of staple food crops (Ahmad and Javed 2007). Further, documentation on local food plants is lacking probably because the collection of scattered and rare literature is not easy. In this paper, I have compiled pertinent information and discussed the role of local food plants in the context of climate change, participatory development and the initiatives launched by national and international organizations to rejuvenate and build up local food systems that can ensure family livelihood to poor people (Photo 6.1).

6.2 Issues and Challenges

6.2.1 Crop Production

A great genetic variability among local food plants exists in developing and less-developed countries where sustainable improvement in food and nutrition security is important (Frison et al. 2011). Genetic improvement can therefore help to overcome some of the constraints that hamper enhanced uptake of underutilized crops (Mayes et al. 2012).

Local crops are grown on poor soils without much water and farm inputs (fertilizers and pesticides), are comparatively resistant to pests and diseases and can adapt to local climate (Abukutsa-Onyango 2003). Consequently, existing local food crops have been improved for productivity and quality and some new crop varieties have been developed and marketed worldwide. In the same period, many of the local land races have been overlooked and even vanished from the market place or were neglected to the extent that they are currently regarded as 'lost or extinct' crops. Otherwise, they were present in a variety of agricultural and pastoral landscapes.

Major reason for this unwarranted situation is that the area planted under minor crops is continuously decreasing year after year or replaced by high value commodities, hybrids or cash crops. For example, area under cotton in India increased from 7.44 m ha in 1990–1991 to 12.88 m ha in 2011–2012 (Ramnath 2014). Soybean, sorghum, millets are used as source of biofuel feedstock or raw material for agro-industries (Gahukar 2009; Tirado et al. 2010). In fact, with industrial development promoted by the accelerated trade liberalization and privatization and dependence on global commodity markets, there had been increased hunger in poor countries (Watkins 2008). Also, strong competition in marketing and trade of minor crops with major cereals resulted in the reduced amount of food produced for the family, decreased control over family income and the food supply and often, provided less food for poor families (Gahukar 2011).

Cropping systems including mix cropping, strip cropping and intercropping in minor crops with short duration legumes and condiments as per family needs are generally practised by small farmers even on slopes in the mountainous areas (Das 2010). By this way, the research and development activities could have solved some of the problems of food penury in down trodden communities. However, limited infrastructure, lack of technical facilities, loss of traditional knowledge, low income generation from minor crops, unavailability of genetic material for breeding, lack of information on perspectives and inadequate policies to promote these crops appear to be the major constraints (McDermott et al. 2013).

There are other problems for cultivating minor crops. For example, most of the farmers do not plan the crop cultivation based on the climate, family needs and market demand. Only marginal or degraded land is used for cultivation. Mechanization is not available to compensate for labour scarcity and improved technology is rarely applied to crop cultivation. Moreover, the exotic invasive plants are responsible for the loss in biodiversity and diminishing genetic pool. Once they enter into the local ecosystem, thrive by competing with local food plants for nutrients, sunlight and water. Subsequently, the spread of these plants becomes a threat to the survival of local flora and thereby to food availability and biodiversity (Mooney and Hobbs 2000). The increasing number of these plants in all agro-ecosystems is now problematic not only in cultivated areas but also in forest areas even at high altitude (Bhat et al. 2012). The recent technique of field spectro-radiometer and simulated hyperspectral remote sensing can be useful for detecting presence of the invasive plants through survey of geographical distribution (Ghosh et al. 2013).

With recent breeding technology, domestication of wild edible food plants can be achieved because firstly, they are derived from a genetically diverse assemblage of different traits such as, greater tolerance to drought, heat and frost; increased input-efficiency and enhanced pest and disease resistance (Maxted and Kell 2009).

Photo 6.2 Food production through improved breeding technology (productivity of pearl millet has been increased)

Secondly, overall time required to domesticate a species has decreased since the earliest domestication events. The frequencies of some domestication syndrome traits have also decreased over time as Meyer et al. (2012) discussed the species complexity in 203 crops from 75 families. This is a viable contribution to plant-based local diets. Therefore, a system of the Systematic Management of Agricultural Resources and Technology (SMART) can play a pride role in improving current agricultural systems. Nutritional improvement through diverse germplasm resources will provide balanced food and nutritional security (Ramya et al. 2013).

It is ascertained that production of local food plants is hampered by man-made and natural factors. Some of plant species are now considered as 'lost or extinct'. With modern breeding techniques, existing crops can be rejuvenated (Photo 6.2).

6.2.2 Food Availability and Access

Among underutilized plants, 13 species of cereals and pseudocereals, 33 species of fruits and nuts, 33 species of vegetables and pulse crops and 17 species of root and tuber crops are currently used in various food recipes (Williams and Haq 2002). But in the modern era, consumers changed their food habits with preference to hybrid crops. For example, people in China have stopped eating wild amaranth (*Abelia engleriana* Rehder) as soon as they were able to cultivate new high yielding genotypes (Kang et al. 2012). In India, wild plants used as vegetables in rainy season such as, *Cassia tora* (L.) Roxb., *Commelina benghalensis* L., *Achyranthes aspera* L., *Boerhavia diffusa* L., *Celosia argentea* L. are no more collected by tribal women.

Collecting and drying of leafy vegetables in the period of plenty and then cooking in the period of non-availability of food had been a regular practice of poor communities. This custom progressively disappeared because fresh improved vegetables were available round the year.

In Mali, 60 % of local varieties of sorghum have disappeared due to expansion of cotton area, introduction of maize cultivation and the saturation of the available cropping area by a specific improved sorghum variety (FAO 2011a). In Nepal, there is gradual disappearance of landraces and native cultivated species of azuki bean, *Vigna angularis* (Willd.) Ohwi & Ohashi and grass pea, *Lathyrus sativus* L. (FAO 2011a). These indigenous and traditional food (often considered as 'poor people's food') enhanced household security not only by availability but also by cheap and easy processing and preservation for home consumption. The known examples include pearl millet in the sub-Saharan Africa, underutilized millets in North-East India, and tuber crops in South-East Asia and South America (FAO 2010a). By this way, family capacity to survive is enhanced significantly and farmers become near self-sufficient in their food requirements even though taking food in a dose recommended by the World Health Organization remains problematic.

Access to food is generally a function of income-generating livelihood through farm activities based on locally available resources and market demand. Local food plants are readily available in abundance (often in the wild state) and affordable to poor people. Locals have skill to identify edible food plants and to know their palatability and method of food preparations. Knowledge and expertise for the crop cultivation also exist in villages. For example, whenever cultivated plants are not available, farmers grow domesticated wild plant species, particularly vegetables, in the kitchen gardens or on degraded lands near villages. Vegetables are easy to cook even without oil and food additives, and are consumed with no spices and condiments. They are inexpensive yet high quality sources of nutrition especially for low income people who do not have sufficient food stock. Also, they have limited or no option during times of crisis. Forest food sources are not always available for two reasons: (1) the federal/state forest laws for reservation of forest areas are in force and people are denied access to food and wood, (2) forest cover is diminishing due to large scale mining and other purposes.

Currently, due to recurring drought, erratic rainfall, small land holding, volatility in market price and declining productivity of long duration crops including cash crops and fruit crops, farmers are turning to short duration local cereals, pulses and vegetables. But the traditional knowledge is not passed down from one generation to the next. Therefore, local foods are not popular. On the contrary, high yielding hybrids or improved varieties of cultivated crops were provided during the era of green revolution to traditional farmers whose culture, food habits and family expenses have been largely affected. On the social front, there is tremendous discouragement for traditional story tellers who otherwise could rejuvenate the importance of minor food crops and wild flora, especially for their cultural and nutritional value. Several countries have initiated food supply projects through public distribution systems. Even when there is enough food for existing population, difficulties are encountered through faulty distribution system that hinders food access (Gahukar 2011).

Photo 6.3 Changes in cropping systems (intercropping/mix cropping in cotton)

Generally, people are ready to purchase foods, but their purchasing power is influenced by the expenditure on other family needs, food price, availability at fair price stores/shops and food allocation by government agencies (Gahukar 2011). The meagre space for food storage in the family house is also an impediment and makes food inaccessible whenever need arises.

Finally, when food is inaccessible due to high prices and low paying capacity, local plants enhances household security as food prepared from them are easily available and can be processed cheaply (Photo 6.3).

6.2.3 Nutritional Value

Governments in developing and less-developed countries have focused too much on crop productivity leaving behind the concern of nutrition. Nutrition is much more related to diversified diets consisting of lots of vegetables and other plant-based sources. Currently, the loss in food biodiversity and ecosystem degradation and their impact on poverty and health are major issues in the poor world. These are compelling reasons for re-examining food-agriculture systems vis-à-vis quality of human diet. In any case, emphasis is needed to increase household income as well as educating masses on diet and nutrition as being practised in China (Li and Shangguan 2012).

In urban areas, consumers do not like food prepared from coarse cereals and wild plants as they are unaware of nutritional quality of the local foods and prefer processed, packaged or 'ready to serve' foods. With rapid changes in the life style, modern cooking is evolved to suit everybody's preference. Generally, the main dish is mixed with starchy staple foods for the taste and enhancing iron absorption in human body (Smith and Ezyaguirre 2007). Otherwise, lack of iron and fibre in meal results in anaemia and diseases. Local crops are therefore essential in areas where protein and micronutrient deficiency is quite common.

Nowadays, bio-fortified foods are provided to areas with nutrient deficiency, For example, rice and wheat diets are supplemented with zinc and iron where corn, cassava, rice, banana and sweet potato are main foods. Additional vitamin A is required in areas where pearl millet and beans are main diets. Iron needs to be supplied in other areas (Padulosi et al. 2008; Jain and Dutta 2013; McDermott et al. 2013). In reality, root and tuber vegetables particularly, livingstone potato (*Plectranthus esculentus* N.E. Br.), Chinese potato (*Solenostemon rotundifolius* (Poir.)), Rogon daji (*Ampelocissus grantii* (Baker) Planch) having high calorific value, and high content of protein and carbohydrates are considered as potential new crops for domestic consumption (Kona et al. 2012). Underutilized vegetables such as, broad bean, *Vicia faba* L., winged bean, *Psophocarpus tetragonolobus* (L.) DC. and rice bean, *Vigna umbellata* (Thunb.) contain 25 % higher protein than many cultivated legumes (Katoch 2013). Finger millet is superior to wheat and rice for the content of calcium, iron, manganese, copper, vitamins and essential amino acids (Das 2010). Buckwheat, *Fagopyrum esculentum* Moench, *F. tartaricum* (L.), contains more protein than cultivated cereals (Bonafaccia et al. 2003). Millet grains have high content of bioavailable minerals, vitamins, dietary fibres and phytochemicals. These unique physico-chemical characteristics render them slowly digestible and with low glycemic index. Long shelf life is also an appreciated attribute preferred by consumers (Bala Ravi et al. 2010). Unfortunately, foods prepared from certain plants once popular in rural areas have now become rare or unavailable. Some of these plants are: *Glyceria maxima* (Hartm.) (Holmb.), *Ziziphus macronata* (Willd.), *Eleusine coracana* (L.), *Echinochloa crus-galli* L., *Echinochloa frumentacea* Link, *Panicum miliaceum* L., *Paspalum scrobiculatum* L, *Setaria italica* L., *Panicum sumatrense* (Roth & Schult.), *Amaranthus* spp., *Cardamine* spp., *Caryopteris divericata* Maxim., *Celastrus orbiculatus* Thunb., *Chenopodium giganteum* Don, *Cirsium vulgare* (Savi) Ten., *Hewingia japonica* (Thunb.), *Pimpinella* sp., *Pteridium aquilinum* (L.) Kuhn, *Staphylea bumalda* DC, *Staphylea holocarpa* Hemsl. (Bala Ravi et al. 2010; Kang et al. 2012).

In order to satisfy growing demand of nutritious food, besides other measures, food crops are genetically modified for improving food and nutritional security. However, consumption of food from these crops without risk to human health is questionable. As such, success of these crops has been mired in controversy with Non- Government Organizations, social institutions and citizen groups. Questions of safety to human health, profitability to farmers and environmental benefits are still not clearly answered (Gahukar 2004; Cressey 2013). Instead, the present agriculture should be improved with all possible good agricultural practices that are based on locally adapted sovereignty principles. By this way, farmers will not be deprived of their traditional vocation and would contribute to nutrition-sensitive food systems (Smith 2013).

The kitchen gardening is an example that is quite practical and profitable with perspective in easy availability and ready access to food. Vegetables such as, *Amaranthus* spp., *Portulaca* spp. can be grown with little money, in a small patch in the field, house roof or around dwelling huts as kitchen garden. Over the past few years, this system has been introduced in tribal areas by Non-Government Organizations and governments have supported it to fight anaemia in local populations (Kona et al. 2012). Farmers can get enough money for family livelihood by selling fresh vegetables at local markets. It is therefore imperative to create a model for community-based participatory research on food and nutrition that can be replicated wherever needed. This system may be useful subsequently to build food security network by examining food assets and access. These traditional home gardens can then be considered as 'Health and nutrition gardens' as suggested by McDermott et al. (2013).

When purchased from the market, the raw material and processed foods are often found adulterated or contaminated making them unfit for human consumption (Gahukar 2014). Additionally, certain types of food are not suitable for home consumption since they contain anti-nutritional or toxic substances that reduce uptake of micronutrients and may be hazardous to human health. These foods may therefore pose health hazards (Akubugwo et al. 2007; Kang et al. 2012; Katoch 2013). For example, Beta-N-oxalyl-amino-L-alanine in seeds of *Lathyrus sativus* L., Beta-oxalyl-L-alpha-beta diamino propionic acid in seeds of *Lathyrus japonicus* (Willd.), alkaloids in tubers of *Fagopyrum esculentum* Moench or *Dioscorea* spp., leaves of purple amaranth (*Amaranthus cruentus* L.), phytolaccotoxin and phytolaccine in *Phytolacca acinosa* Roxb., oligosaccharides in rice bean, phylate, phytic acid and polyphenols in wild beans, glycoalkaloids (solanine, solasonine, solamargine) and oxalate in unripe green berries of *Solanum nigrum* L. However, it has been possible to lower or eliminate the toxic content by simple processing. Soaking rice bean seeds in water and then sprouting or autoclaving and cooking (Kaur and Kapoor 1990), parboiling of grass pea split grains followed by washing with fresh water (Gahukar 2014) and boiling leaves of purple amaranth and disposing of water (Kang et al. 2012) have been successful to make food consumption safe. Now, it is a challenge for scientists to search detoxification process for other natural toxic contents.

Conclusively, those who fail to buy diverse foods adequately suffer from malnutrition. Food prepared from local crops and wild plants is inexpensive yet provides nutritional elements to poor people. However, the anti-nutrients and toxic substances pose a challenge (Photo 6.4).

6.2.4 Conservation of Biodiversity

Among developing countries, China could retain maximum number of minor crops with 222 taxa occupying 16 % of the world's acreage (FAO 2011a). The biodiversity is important for improving genotypes for various traits especially tolerance to drought, soil salinity/alkalinity, climate change etc. In this context, the recent

Photo 6.4 Local foods produced for family consumption (assuring livelihood in difficult times)

development is rather disappointing and may impact poverty due to alarming pace of the food plant diversity loss and ecosystem degradation.

In reality, shrinking food diversity makes people vulnerable to malnutrition which is an important local and national issue. Ecotypes are disappearing or extinct due to either commercial exploitation or unfavourable plant growth conditions in ecosystems. Also, rapid rate of intensification of agricultural practices has led farmers to abandon traditional crops resulting in substantial reduction in the genetic diversity of domesticated plants in agricultural systems (FAO 2010a). In an effort to maintain biodiversity, indigenous knowledge and participatory plant breeding has been successfully used to regenerate crops (Brocke et al. 2012). Minor food plants are excellent source of useful genes that can be exploited by using biotechnological techniques such as, micro-propagation, molecular biology and genetic transformation (Jain and Dutta 2013) and therefore, it is now possible to rejuvenate local cultivars/land races along with exotic high yielding genotypes by employing new breeding techniques. For example, inter-specific and inter-varietal hybridization of local gourds, including wild species, has facilitated to transfer traits like pest and stress resistance and nutritional elements (Arivalagan et al. 2012).

Food plants presenting non-timber forest products help family livelihood throughout the year. Therefore, conservation of this natural resource of diversified foods cannot be neglected. In order to prevent further genetic erosion, *in-situ* and *ex-situ* conservation strategies for buffer forest plant species (neglected and underutilized) (Ghorbani et al. 2012) and minor food plant species (Stapit and Padulosi 2012) are ideal solutions. In all suitable agrosystems, cultivation of minor crops facilitated by collection of local seeds may be introduced to conserve the genetic diversity of indigenous plants. In central places, the gene banks may be established in the crop growing areas (Vidyadharan and Tiwari 2009). Also, threatened plants can be equated to plant species that evolve low dispersal habits and render them

Photo 6.5 Food diversity exploited for nutritional security (availability of local plants)

more vulnerable to extinction when facing anthropogenic disturbance, habitat loss and climate change (Kotiaho et al. 2005). To overcome these constraints, there is urgent need for inventory of all neglected and underutilized plants including their wild relatives. Intensive studies on clarity over the identity and systematic conservation of local plants would also be helpful (Photo 6.5).

In all, reduction or loss in genetic diversity of local crops and degradation in agro-ecosystems are adversely affecting food production. Therefore, conservation of natural resources becomes a priority.

6.2.5 Climate Change

Changes in the climate and global environment factors such as, drought, floods, unseasonal rainfall, CO_2 concentration, solar radiation, rise in temperature, water depletion etc. and interaction of these factors, affect crop growth and productivity, and diminish the options of food crops to most small land holding farmers (Gahukar 2009; Jaggard et al. 2010; Tirado et al. 2010). In this regard, the recent report of the United Nations Intergovernmental Panel on Climate Change mentioned about likely rise in temperature by 0.3–4.8 °C and sea level by 26–82 cm at the end of the century (IPCC 2013).

Local crop species and varieties, including wild crop relatives, constitute a vulnerable genetic pool for supporting food security under climate change (Padulosi et al. 2008). Minor millets and rare crops would therefore be a boon to farming communities because by virtue of fast growth, they show extreme high resilience to the harsh agro-climatic conditions or are well adapted to semi-arid and hilly regions particularly in marginal soils and under aberrant rainfall (Bala Ravi et al. 2010). For example, tuber and root crops are tolerant to poor soils and seasonal drought; the ulluco plant, *Ullucus tuberosus* Caldas is frost resistant; bitter cucumber, *Citrullus colocynthis* (L. Schrad.) is well adapted to sandy soils in desert areas; the miner's lettuce, *Montia perfoliata* Willd. and hausa groundnut, *Macrotyloma geocarpum* (Harms) are well suited to arid climate and drought conditions (Enete et al. 2012; Kona et al. 2012). Moreover, since farmers do not generally use synthetic fertilizers and pesticides for cultivation, emission of the Green houses gases such as, nitrous oxide, nitric oxide, is quite low (Gahukar 2009).

Considering current global situation, it would be advisable to examine the models based on crop yield simulations that can lead to further research on genotypic adaptation to climate change (Challinor et al. 2009). Other mitigation strategies include alternate land use systems and management, agroforestry, crop diversification, intercropping and crop rotations, resource conservation, resistant varieties, biotechnological tools and risk management through early warning and crop insurance (Padulosi et al. 2008).

Finally, changes in the climate and global environment can probably be studied for mitigation strategies and risk management in the productivity of local food crops (Photo 6.6).

Photo 6.6 Challenge of climate change (drought affecting plant development)

6.2.6 *Local Initiatives and Government Policies*

Although local governments in some countries have launched pro-poor and pro-farmer measures such as, food provision at subsidized prices, rural employment guarantee schemes, bank loan at low interest rate, loan waiving etc., there had been little improvement in the food security and living standard of poor families (Gahukar 2011) probably due to administrative reasons. Lack of promotional efforts for dissemination of appropriate technologies and ignorance to clear cut government policies can hinder the utilization of neglected and underutilized food plants.

Women are a backbone of household food security management in poor countries. Empowerment of women and disadvantaged communities can therefore help to enhance food security (McDermott et al. 2013). In villages, the socio-cultural constraints, lack of technical information delivery, poor social participation and inadequate extension support hinder their active roles in family livelihood (Mishra and Mishra 2012). In reality, women have little say on how income should be used for livelihood and have less control over food supply. To earn extra money, the tribal women in India collect fresh flowers of mahua (*Madhuca indica* J.F. Gmel) during the period of peak flowering and harvest vegetables in the wild areas during rainy season. This system generates some income to support family livelihood (Patel and Naik 2010). Value-addition of local food plants by creating community institutions in rural areas also led to more income to women and helped them to safeguard their families from hunger (Patel and Naik 2010).

The government encouraged and appreciated the development of social capital in the form of Self help groups of men and women. These groups could innovate, adapt and rebuilt local systems by encouraging and supporting small farmers who made the traditional small-scale farming reasonably sustainable and profitable. These groups are closely and strongly associated with local authority and have traditional knowledge and skill for cultivating local crops (Pimbert 2011; Ikerd 2011). For example, in the Sub-Saharan Africa, community cultivation assured food access and availability to the landless and helped to eradicate hunger through enhancement of local agricultural practices. With this method, marginal farmers rediscovered their lost dignity and enhanced significantly the household livelihood (Pimbert 2011). Earlier, these systems have been abandoned or undermined by farmers but nowadays, they are moving from conventional to sustainable farming to assure food security to all villagers irrespective of social and economic strata, and can achieve hunger alleviation and environmental protection (Pimbert 2011).

Creating community institutions and enhancement of local agricultural practices can help marginal and small farmers to cope with disastrous poverty and to eradicate hunger. Pro-poor policies launched by government can also improve food and nutrition security (Photo 6.7).

Photo 6.7 Field demonstration to popularize recommended practices (farmers' participatory approach)

6.3 Perspectives

Improving crop productivity is needed to secure food security with proper knowledge on plant breeding, understanding and application of management practices, increased communication and execution of laboratory to field techniques (Fan et al. 2012). Besides, efficient and diversified cultivation of minor food crops would provide nutritious food, empower local communities through participatory approach particularly for selecting crops/varieties as per their preference and family needs (Bordoni 2012). This system encourages preservation and innovation- based traditional knowledge and new information, influences public policy and enforces further understanding of the links between farming, food, health and local economy.

Informal seed systems and social networks are important to strengthen farmers' capacity through research and development to cope with climate change (Stapit and Padulosi 2012). Further improvement in food and nutrition security is possible by providing highly nutritious and biofortified foods, improving nutritional quality and food safety, increased investments and better policy reforms (Rosegrant and Cline 2003; Gahukar 2011). Of course, sustainability in food production is more important in long term than external financial aid to reduce poverty in developing countries (Pretty et al. 2003; Shreeran 2011; FAO 2011b). These steps would possibly ensure people to get easy access to nutritious food at least for few years in future so as to be free from hunger.

At regional level, international symposiums on neglected and underutilized plant species have been organized by various national organizations in Ghana (November

2006, September 2013), Tanzania (March 2008) and the USA (August 2013). The recent international quinoa research symposium was scheduled to coincide with the United Nations International year of the quinoa. Thus, national policies executed by the local governments and global efforts initiated by international organizations including voluntary, non-government and federal agencies, play a vital role in food and nutrition security (Tirado et al. 2010). Earlier the Global Facilitation Unit of Underutilized Species of the Global Forum on Agricultural Research established in 2002 emphasized the role of neglected and underutilized plants in raising income of the rural poor and pointed out the need of the international research and development in cultural and local cuisine diversity (Padulosi et al. 2008). The nutrition-sensitive foods can reduce the deficiency of micronutrients as demonstrated by the South Africa Food and Nutrition Security Initiatives. Also, a group named 'Crops for the Future' has been formed by merging International Centre for Underutilized Crops and the Global Facilitation Unit of Underutilized Species is based in Malaysia. Recently, the International Plant Genetic Resources Institute prepared a strategic action plan with objectives of increasing purchasing power of rural poor and maintaining ecological stability and cultural diversity.

The International Plant Genetic Resources Institute and the recently established International Centre for Underutilized Crops (Sri Lanka) are jointly promoting the conservation and use of neglected and underutilized plants worldwide. Similarly, the International Treaty on Plant Genetic Resources for Food and Agriculture will facilitate diversification of minor crops and sustainable use of biodiversity by enhancing country compliance with mandate of 1992 Convention on Biological Diversity proposals. For preservation of the genetic resources to be used for crop production by the small farmers, international efforts are being made by the Food and Agriculture Organization by promoting 'Save and grow campaign' (FAO 2011b).

The FAO planned to invest higher than $10 million in adaptation of food crops to ensure sustainable food supply and assist farmers to stay ahead of the climate change. The Second call of International Treaty will ensure fair distribution of benefits for the conservation and sustainable use of plant genetic resources that are base of food security (FAO 2010b). Finally, efforts of International Union for Conservation of Nature, the International Treaty on Plant Genetic Resources for Food and Agriculture and Convention on Biological Diversity can certainly prompt the awareness of minor crops and strengthen the new global mandate of food and nutrition security.

Social network, increased investments, better policy reforms and international cooperation would certainly boost the production of local crops and would strengthen the global food and nutrition security.

6.4 Conclusions

Poor families can cultivate local crops of their preference or needs. Crops cultivated in diversified soils and ecologically sound cropping systems can thus support family livelihood through nutritious food in difficult periods. For achieving this goal, better supportive policies, increased investments by the way of government initiatives and international cooperation are needed to build local capacity to adapt and innovate proper solutions for upcoming climate change, future food demand and changing social systems in developing and less-developed countries.

References

Abukutsa-Onyango MO (2003) Unexploited potential of indigenous African vegetables in Western Kenya. Maseno J Educ Arts Sci 4:103–122

Ahmad SS, Javed S (2007) Exploring the economic value of underutilized plant species in Ayubia National park. Pak J Bot 39:1435–1442

Akhtar S, Karak C, Biswas P, Chattopadhyay A, Hazra P (2012) Indigenous leafy vegetables: a potential source of B-carotene and ascorbic acid. Int J Veg Sci 18:370–375. doi:10.1080/19315260.2011.649163

Akubugwo IE, Obasi AN, Ginikas SC (2007) Nutritional potential of leaves and seeds of black nightshade, *Solanum nigrum* L. var. *virginicum* from Afikpo, Nigeria. Pak J Nutr 6:323–326

Anthony K, de Groot P, Haq N (eds) (1993) Underutilized fruits and nuts of Asia. Commonwealth Science Council, London and International Centre for Underutilized Crops, Southampton

Arivalagan M, Prasad TV, Bag MK (2012) Role of underutilized crops for combating iron deficiency in Indian population. Curr Sci 103(2):137

Bala Ravi S, Hrideek TK, Kishore Kumar AT, Prabhakaran TR, Mal B, Padulosi S (2010) Mobilizing neglected and underutilized crops to strengthen food security and alleviate poverty in India. Indian J Plant Genet Resour 23(1):110–116

Bhat JA, Kumar M, Negi AK, Todaria NP (2012) *Acacia dealbata* Link (Silver wattle): an invasive species growing in high altitudes of the Himalaya. Curr Sci 103:130

Bonafaccia G, Marocchini M, Kreft I (2003) Composition and technological properties of the flour and bran from common and tartary buckwheat. Food Chem 80:9–15

Bordoni P (2012) Coping with climate change: using agrobiodiversity to enhance resilience and adaptability. In: Padulosi S, Bergamini N, Lawrence T (eds) On-farm conservation of neglected and underutilized species: status, trends and novel approaches to cope with climate change. Proceedings of an international conference, Frankfurt, Germany, 14-16 June 2011. Biodiversity International, Rome, pp 101

Brocke KV, Zougmore R, Barro C, Trouche G, Weltzieri E (2012) Using indigenous knowledge and participatory plant breeding to improve crop production and regenerate crops and tree species in West Africa. In: Padulosi S, Bergamini N, Lawrence T (eds) On-farm conservation of neglected and underutilized species: status, trends and novel approaches to cope with climate change. Proceedings of an international conference, Frankfurt, Germany, 14-16 June 2011. Biodiversity International, Rome, pp 103-122

Challinor AJ, Wheeler T, Hemming D, Upadhyaya HD (2009) Ensemble yield simulations: crop and climate uncertainties, sensitivity to temperature and genotypic adaptation to climate change. Climate Res 38:117–127. doi:10.3354/cr00779

Cressey D (2013) Transgenics: a new breed. Nature 497(7447):27–29. doi:10.1038/497027a

Das A (2010) Millets for food and nutritional security. ICAR Research Complex for NEH Region, Umiam

Enete AA, Madu IA, Onwubuya EA (2012) Climate change and the profitability of indigenous adaptation practices in smallholder agriculture in South- East Nigeria. Outlook Agric 41(3): 179–185

Fan M, Shen J, Yuan L, Jiang R, Chen X, Davies WJ, Zhang F (2012) Improving crop productivity and resource use efficiency to ensure food security and environmental quality in China. J Exp Bot 63:13–24

FAO (2010a) State of diversity of major and minor crops. Biodiversity International & FAO, Rome

FAO (2010b) Call for proposals 2010: benefit-sharing fund. FAO, Rome

FAO (2011a) The second report of the state of the world's plant genetic resources for food and agriculture. FAO, Rome

FAO (2011b) Save and grow: a policy maker's guide to the sustainable intensification of smallholder crop production. FAO, Rome

Frison EA, Cherfad J, Hodgkin T (2011) Agricultural biodiversity is essential for a sustainable improvement in food and nutrition security. Sustainability 3:238–253

Gahukar RT (2004) Are Indian foods from genetically modified crops safe? J Food Agric Environ 2(2):11–13

Gahukar RT (2009) Food security: the challenges of climate change and bioenergy. Curr Sci 96:26–28

Gahukar RT (2011) Food security in India: the challenge of food production and distribution. J Agric Food Inform 12:270–286. doi:10.1080/10496505.2011.588932

Gahukar RT (2014) Food adulteration and contamination in India: occurrence, implication and safety measures. Int J Basic Appl Sci 3(1):47–54. doi:10.14419/ijbas.v3il.1727

Ghorbani A, Langenberger G, Liu JX, Wehner S, Sauerborn J (2012) Diversity of medical and food plants as non-timber forest products in Naban River Watershed National Nature Reserve (China): implications for livelihood improvement and biodiversity conservation. Econ Bot 66:178–191

Ghosh A, Joshi PK, Garg RD, Mukherjee S (2013) Identification of invasive plant species using field spectro-radiometer and simulated hyperion spectra: a rapid mapping of invasiveness. Curr Sci 104:1148–1151

Gregory PO, George TS (2011) Feeding nine billion: the challenge for sustainable crop production. J Exp Bot 62(15):5233–5239. doi:10.1093/jxb/err232

Hu SY (2005) Food plants of China. Chinese University Press, Hong Kong

Ikerd JE (2011) Local food revolution and reality. J Agric Food Inform 12:49–57. doi:10.1080/10496505.2011.540557

IPCC (2013) Climate change 2013: the physical science basis. Intergovernmental Panel on Climate Change, IPCC Secretariat, Geneva

Jaggard AW, Qi A, Ober ES (2010) Possible changes to arable crop yields by 2020. Philos Trans R Soc B 365:2835–2851

Jain SM, Dutta GS (eds) (2013) Biotechnology of neglected and underutilized crops. Springer, Berlin, 459 pp

Kamble VS, Jadhav VD (2013) Traditional leafy vegetables: a future herbal medicine. Int J Agric Food Sci 3(2):56–58

Kang Y, Tuczaj T, Ye S, Zhang S, Kang J (2012) Wild food plants and wild edible fungi of Heihe valley (Qinling mountains, Shannxi, Central China): herbophilia and indifference to fruits and mushroom. Acta Soc Bot Pol 81:405–413 (summary in English)

Katoch R (2013) Nutritional potential of rice bean (*Vigna umbellata*): an underutilized legume. J Food Sci 78(1):C8–C16. doi:10.1111/j.1750-3841.2012.02989

Kaur D, Kapoor AC (1990) Some antinutritional factors in rice bean (*Vigna umbellata*): effects of domestic processing and cooking methods. Food Chem 37(1):171–179

Kona HA, Aliyu IA, Chammang HB (2012) Review of neglected and underutilized root and tube crops as food security in achieving the millennium development goals in Nigeria. J Agric Vet Sci 4:27–33

Koohafkan P (2012) Sustainable diets and biodiversity: directions and solutions for policy, research and action. FAO, Rome

Kotiaho JS, Kaitala V, Komonen A, Paivinen J (2005) Predicting the risk of extinction from shared ecological characteristics. Proc Natl Acad Sci U S A 102:1963–1967

Li J, Shangguan Z (2012) Food availability and household food security. Outlook Agric 41:57–63

Maxted N, Kell SP (2009) Establishment of a global network for the *in-situ* conservation of crop wild relatives: status and needs. FAO Commission on Genetic Resources for Food and Agriculture, Rome

Mayes S, Massawe FJ, Alderson PG, Roberts JA, Azam-Ali SN, Hermann M (2012) The potential for underutilized crops to improve security of food production. J Exp Bot 63:1075–1079. doi:10.1093/jxb/err396

McDermott J, Ait-Aissa M, Morel J, Rapando N (2013) Agriculture and household nutrition security: development practices and research needs. Food Secur 5:667–678. doi:10.1007/s12571.013-0292-6

Meyer RS, du Val AE, Jensen HR (2012) Patterns and processes in crop domestication: an historical review and quantitative analysis of 203 global food crops. New Phytol 196(1):29–48. doi:10.1111/j.1469-8137.2012.04253

Mishra M, Mishra SK (2012) Participation of rural women in household food security management. Ann Agric Res (New Ser) 33(1–2):58–61

Mooney HA, Hobbs RJ (2000) Invasive species in a changing world. Island Press, Washington, DC

Padulosi S, Hoeschie-Zeledon I, Bordoni P (2008) Minor crops and underutilized species: lessons and prospects. In: Maxted N, Ford-Llyod BV, Kell SP, Iriondo JM, Dulloo ME, Turok J (eds) Crop wild relative conservation and use. CAB International, Wallingford, pp 605–624

Patel M, Naik SN (2010) Flowers of *Madhuca indica* J.F. Gmel: present status and future strategies. Indian J Nat Prod Resour 1(4):438–443

Pimbert M (2011) Supporting locally determined food systems: the role of local organizations in farming, environment and people's access to food. Leisa India 13(3):12–15

Pretty JN, Morison JIL, Hine RE (2003) Reducing food poverty by increasing agricultural sustainability in developing countries. Agric Ecosyst Environ 95(1):217–234

Ramnath M (2014) Call of the wind. Down Earth 22(19):48–50

Ramya KT, Abdul FR, Yasin JK (2013) SMART agriculture for nutritional security. Curr Sci 105(11):1458

Rosegrant MW, Cline SA (2003) Global food security: challenges and policies. Science 302:1917–1919

Shreeran J (2011) Preventing hunger: sustainability not aid. Nature 479(7374):469–470. doi:10.1038/479469a

Smith P (2013) Delivering food security without increasing pressure on land. Global Food Secur 2(1):18–23

Smith FI, Ezyaguirre P (2007) African leafy vegetables: challenges and opportunities in the Kenyan context. Afr J Food Agric Nutr Dev 17:4–12

Stapit B, Padulosi S (2012) On farm conservation of neglected and underutilized crops in the face of climate change. In: Padulosi S, Bergamini N, Lawrence T (eds) On-farm conservation of neglected and underutilized species: status, trends and novel approaches to cope with climate change. Proceedings of an international conference, Frankfurt, Germany, 14-16 June 2011. Biodiversity International, Rome, pp 31-48

Tirado MC, Cohen MJ, Aberman N, Meerman J, Thompson B (2010) Addressing the challenge of climate change and biofuel production for food and nutrition security. Food Res Int 43(7):1729–1744

Van Wyk BE (2005) Food plants of the world. Timber Press Inc., Portland

Vidyadharan V, Tiwari MK (2009) Tribal women attain food sovereignty through seed banks. Leisa India 11(3):13–14

Vincent H, Wiersema J, Kell S, Fielder H, Dobbie S, Castaneda-Alvarez NP, Guarino L, Eastwood R, Leon B, Maxted N (2013) A prioritized crop wild relative inventory to help underpin global food security. Biol Conserv 167:265–275

Watkins K (2008) Agricultural trade, globalization and the rural poor. In: Van Braun J, Diaz-Bonilla E (eds) Globalization of food and agriculture and the poor. University of Oxford Press, Oxford, pp 155–180

Wickens GE, Haq N, Day P (eds) (1989) New crops for food and industry. Chapman and Hall, London, 436 pp

Williams JT, Haq N (2002) Global research on underutilized crops: an assessment of current activities and proposals for enhanced cooperation. International Centre for Underutilized Crops, Southampton, 46 pp

Chapter 7
Phytoremediation Crops and Biofuels

M.N.V. Prasad

Abstract Environmental decontamination is an integral part of sustainable development. In recent years there has been growing interest in using plants for decontamination. On the other hand, water, soil and air are increasingly contaminated. Large amounts of toxic waste have been dispersed in thousands of contaminated sites spread all over the globe. These pollutants belong to two main classes: inorganic and organic. The challenge is to develop innovative and cost-effective solutions to decontaminate polluted environment. Phytoremediation is emerging as an invaluable tool for environmental cleanup. Various strategies are being applied to reduce the accumulation of toxic metals in plants. Cultivation of edible crops in contaminated soils is a subject of human health concern if the contaminant concentration in the edible parts of crops plant exceed the permissible level. In such cases non-food crop production viz. value chain and value additions appears profitable. In this review: (1) the contamination due to industrial effluents in peri urban region of greater Hyderabad, and (2) the strategies to use contaminated soil and water for raising phytoremediation crops, and generation of value products. Crops and products include medicinal and aromatic plants, ornamental plants, biofuels, tree crops, fiber crops, dyes, and plants for carbon sequestration.

Keywords Biodiesel • Bioeconomy • Biomass conversion • Biorefinery • Carbon sequestration • Circular economy • Constructed wetlands • Energy crops • Environmental cleanup • Industrial crops • Ornamental crops • Value additions • Value chain products

7.1 Introduction

Environmental (Soil, water and pollution) (conflict between agro-ecosystems and urbanization) is not only a subject of pan-Indian concern, but also for the entire world. Consequently, developing eco-innovative remediation technologies to re-use

---

M.N.V. Prasad (✉)
Department of Plant Sciences, University of Hyderabad, Hyderabad 500 046, India
e-mail: prasad_mnv@yahoo.com; prasad.heavymetal@gmail.com

E. Lichtfouse (ed.), *Sustainable Agriculture Reviews*,
Sustainable Agriculture Reviews 17, DOI 10.1007/978-3-319-16742-8_7

contaminated substrates (land and water) in line with biomass production for the bioeconomy are gaining considerable global attention (Prasad 2015). Rapid industrialization and extraction of large quantity of natural resources is the main cause of environmental contamination and pollution. Thus, every one of us are being exposed to contamination from past and present industrial practices, emissions in natural resources (air, water and soil) even in the most remote regions. The risk to human and environmental health is rising and there is evidence that this cocktail of pollutants is a contributor to the global epidemic of cancers, and other degenerative diseases.

Plant based technologies are emerging as viable alternatives to conventional remediation techniques when time constraint is low, being more energy-efficient and less disruptive to contaminated sites. Based on small-scale studies, annual and perennial accumulators are suitable candidates for these plant based technologies (= Phytotechnologies). Plant-associated microbes are also implicated in enhancing plant performance, reducing contaminant phytotoxicity or modifying (degrading) (organic) contaminants. These innovative phytotechnologies can be *in-situ* or *ex-situ* (Conesa et al. 2012), remediate soil layers explored by roots and at the same time provide plant biomass, contributing towards achieving envisaged targets on the use of renewable plant-based feedstock for various purposes (renewable energy sources, ecomaterials, biomass for bioreffineries, green fine chemistry, bioplastics, etc. (Prasad 2014b)) in substitution to fossil fuels and other non-renewable raw materials. They can also reduce the diversion of croplands to bioenergy and other non-food crops (Liu et al. 2011, 2012; Sheng et al. 2012; Zhang et al. 2014a). Development of economically sound valorisation pathways for complete chain of phytoproducts of value addition and value chain from phytoremediated plants would go a long way (Grison 2015; Prasad 2015; van der Ent et al. 2015) (Fig. 7.1).

Energy and environment are interlinked complex issues. Population explosion (~7.2 billion) on plant earth and excessive use of natural resources resulted in environmental pollution and contamination with inorganic and organics (Prasad 2011; Xu et al. 2006). Pollution in environment is an ever increasing phenomenon and often regulatory systems and cleaning operations are not commensurate with waste

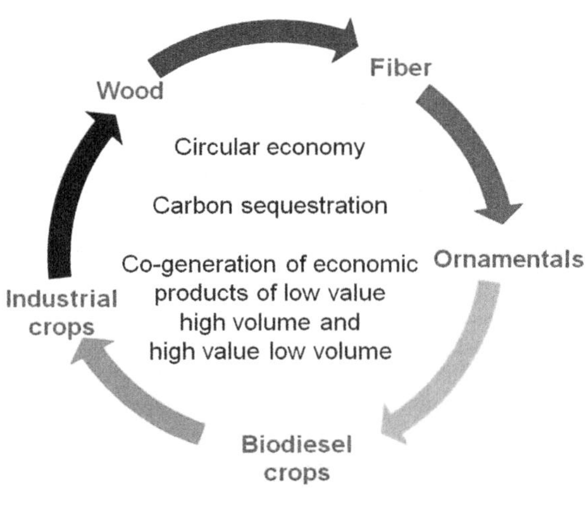

Fig. 7.1 Environmental crops boosting circular economy, carbon sequestration and co-generation of phytoproducts

generation. It is therefore, necessary to search for effective technologies for decontaminating the natural resources. Biodiversity has immense potential not only for monitoring and abatement of environmental pollution but also in generating a wide variety of by-products. Breakthrough research innovations that are amenable for field applications of pollution abatement are discussed with specific examples in several papers (Conesa et al. 2012; Prasad 2011; Prasad and Freitas 2003; Prasad and Prasad 2012; Prasad et al. 2010; Prasad 2013; Puhui et al. 2011).

The growing need for biomass for conversion to biofuels, require lignocellulosic-rich raw materials. Phytoremediated phytomass is one such option used to produce fuel like methanol, biodiesel, synthetic gas, and hydrogen (using thermal and thermo-chemical processes by direct or indirect liquefaction or gasification) and ethanol (through hydrolysis and subsequent fermentation). Biorefinery processes (the sustainable processing of biomass to a spectrum of marketable products and energy) is an absolute necessity and it is the key to meet this vision towards bio-based economy.

Brassica juncea (Indian mustard) *Helianthus annus* (Sunflower), *Prosopis juliflora*, bamboo, and *Pistia stratiotes* (water lettuce), could be grown for different purposes including energy generation Annual and perennial crops, including algae in wastewater ponds are potential candidates. Management and production of phytoremediation crops in contaminated substrates serves as a sink for contaminants with possibility for co-generation of economic products. Economics and byproduct generation with the overall success level of integration in the bioremediation fostering circular economy (Prasad 2015) (Fig. 7.2).

The challenge for environmental decontamination of inorganic and organic pollutants and contaminants is to develop innovative and cost-effective solutions to decontaminate polluted environments. Biodiversity is the raw material for environmental cleanup and is an invaluable tool box with wider application in the field of pollution abatement (Witters et al. 2012). Phytoremediation includes variety of technologies using plants and microbes to remediate or contain contaminants in soil, groundwater, surface water, and/or sediments including air. These technologies have become attractive alternatives to conventional cleanup technologies due to relatively low capital costs and the inherently aesthetic nature (Fig. 7.3).

Phytoremediation is a sustainable environmental cleanup technology. This approach is based on a wide range of plants that include terrestrial, aquatic, weeds, annuals, perennials (agricultural, tree and field crops) and ornamentals (Poonam

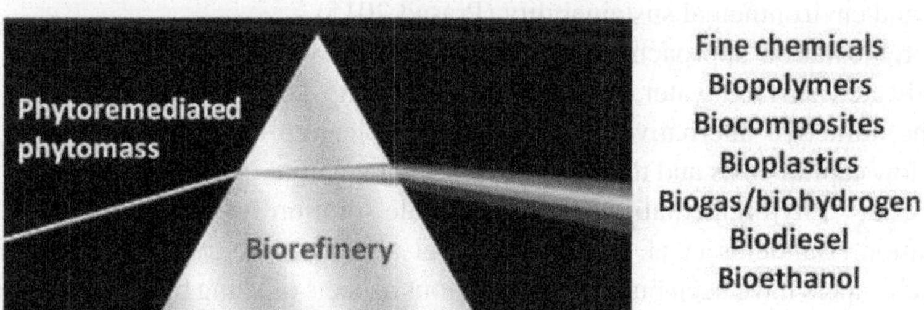

Fig. 7.2 "Biorefinery" is the sustainable processing of conversion of phytoremediated phytomass into a spectrum of marketable products for e.g. chemicals, composite materials, and energy

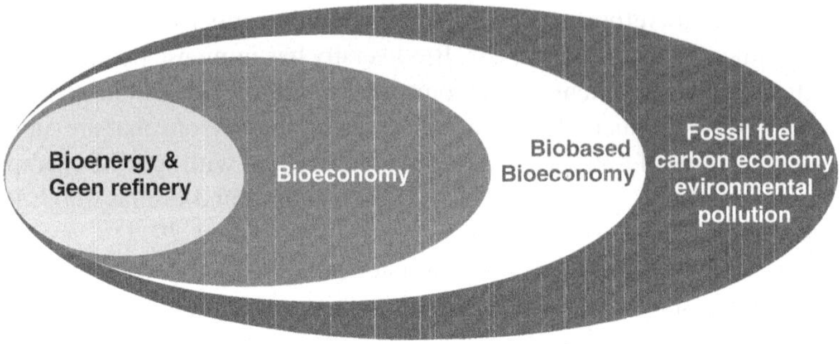

Fig. 7.3 Bioremediation and biobased economy. A biobased economy is a sustainable economy, optimising economic value through valorization of bimass. The new wave in this direction is to use phytoremediated phytomass as substrate in biorefinery for replacing fossil resources and to produce a variery of economic products

et al. 2014; Vamerali et al. 2011, 2014). Plants applied in phytoremediation have several beneficial uses. Nature cure mechanisms for natural resources and environmental protection viz. phytosequestration, rhizodegradation, phytohydraulics, phytoextraction, phytodegradation and phytovolatilization are well established with suitable examples (Shekhar 2012).

Arsenic, Mercury, Chromium, Fluoride, Cyanide, abandoned mines, fly ash disposed sites, engineered phytotreatment technologies, biological permeable barriers and degradation of Organics viz., petroleum hydrocarbons, pesticides and explosives are some of the contaminants subjected to bioremediation and promising results have been obtained.

Quite a variety of plants, natural, transgenic, and/or associated with rhizosphere micro-organisms are extraordinarily active in these biological interventions and cleaning up pollutants by removing or immobilizing. While diverse microbes are the most active agents, fungi and their strong oxidative enzymes are key players as well in recycling recalcitrant polymers and xenobiotic chemicals following systems biology and biotechnology interventions (de Lorenzo 2008; Dowling and Doty 2009; Wood 2008). Constructed wetlands are the result of human skill and technology integrating the geology, hydrology and biology (Prasad 2004a, b, 2007a, b). Constructed wetlands have been designed centuries ago by mankind to treat wastewater (Bi et al. 2007; Türker et al. 2013). Bioeconomy is a solution for clean air, water and environmental sustainability (Prasad 2015).

Bioremediation approach is currently applied to contain contaminants in soil, groundwater, surface water, or sediments including air. These technologies have become attractive alternatives to conventional cleanup technologies due to relatively low capital costs and the inherently aesthetic nature. Nature's cure using plant resources (= Phytoremediation) is a sustainable solution for environmental decontamination (Bañuelos et al. 1997; Bitterli et al. 2010). As of now about 20,000 research papers have been published on various aspects of using biological resources for environmental cleanup starting with only 11 in 1989. Environmental sources of heavy metals are shown in Table 7.1.

Table 7.1 Environmental sources of heavy metals (Prasad 2011)

Cadmium (Cd)	Production of stainless steel, alloys, storage batteries, spark plugs, magnets and machinery, Cadmium (Cd) Cd-Ni battery production, pigments for plastics and enamels, fumicides, and electroplating and metal coatings.
Molybdenum (Mo)	Spent catalyst.
Zinc (Zn)	Brass and bronze alloys production, galvanized metal production, pesticides and ink. Zinc smelting, waste batteries, e-waste, paint sludge, incinerations, fuel combustion and electroplating.
Tin (Sn)	Soft drink, beer and beverage can production.
Cobalt (Co)	Steel and alloy production, paint and varnish drying agent and pigment and glass manufacturing.
Chromium (Cr)	Corrosion inhibitor, dyeing and tanning industries, plating operations, alloys, antiseptics, defoliants and photographic emulsions. Mining, industrial coolants, chromium salts manufacturing, and leather tanning.
Lead (Pb)	Lead battery industry, fuel additives, paints, herbicides manufacturing of ammunition, caulking compounds, solders, pigments, and insecticides, Smelting operations, coal- based thermal power plants, lead acid batteries, paints, and E-waste.
Mercury (Hg)	Electrical apparatus manufacture, electrolytic production of Cl and caustic soda, paints, pharmaceuticals, plastics, paper products, batteries, pesticides and burning of coal and oil chlor-alkali plants, thermal power plants, fluorescent lamps, hospital waste (damaged thermometers, barometers, sphygmomanometers), electrical appliances etc.
Arsenic (As)	Production of pesticides, veterinary pharmaceuticals and wood preservatives, geogenic/natural processes, smelting operations, thermal power plants, and fuel burning.
Copper (Cu)	Textile mills, cosmetic manufacturing and hardboard production sludge, mining, electroplating, and smelting operations.
Vanadium (Va)	Spent catalyst, sulphuric acid plant.
Nickel (Ni)	Smelting operations, thermal power plants, and battery industry.

Biodiversity and ecosystem services offer means and ways to render the usage of natural resources more sustainably. This can be achieved by using the soil and ecosystem to generate solutions for modern-day challenges such as industrial pollution and the growing demand for natural resources (Aparna et al. 2010). Thus, there is a need to develop appropriate tools with like minded parties and allied disciplines. This kind of action plan would assist planners in calculating the social benefits of green missions in urban environments. The nexus of 'Biodiversity *vs* Industry' consortium and knowledge based systems for better understanding of production chains would contribute to sustainable development.

Industrial and urban activities impact our environment, especially in terms of polluting soil, air and water. Large numbers of sites are nowadays contaminated/polluted by inorganic and/or organics. Treating these pollutants represents an economic need, which often remains unanswered by conventional civil/chemical engineering methods, due to their inappropriateness, their environmental impact and costs (notably for large sites) (Figs. 7.4 and 7.5).

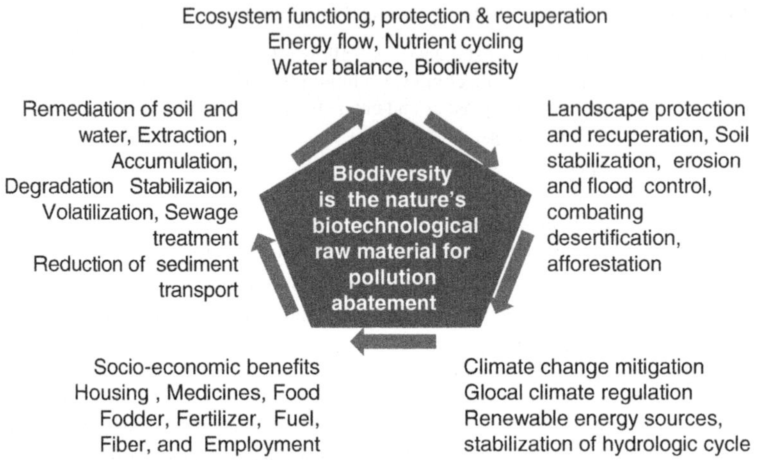

Fig. 7.4 Nature's biotechnology involves biodiversity as raw mterial for pollution abatement. During the last two decades, we have witnessed the emergence of gentle soil remediation techniques using various plant species and the combination of microbial biotechnologies

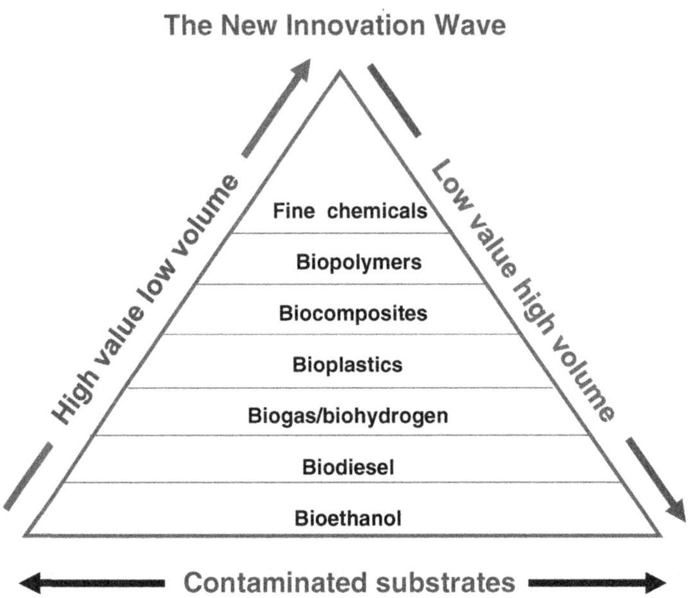

Fig. 7.5 Utilization of contaminated substrates for boosting bioeconomy through cascading approach is a new innovation wave. The added value is the highest at the *top* of the pyramid and the lowest at the *bottom*. On the contrary, the volume of biomass needed for the application is the lowest at the *top* of the pyramid and the highest at the *bottom* of the pyramid

7.2 Study Area

Major contaminated sites in Greater Hyderabad = Tri cities (Hyderabad, Secunderabad and Cyberabad) are selected for this study (Figs. 7.6–7.12).

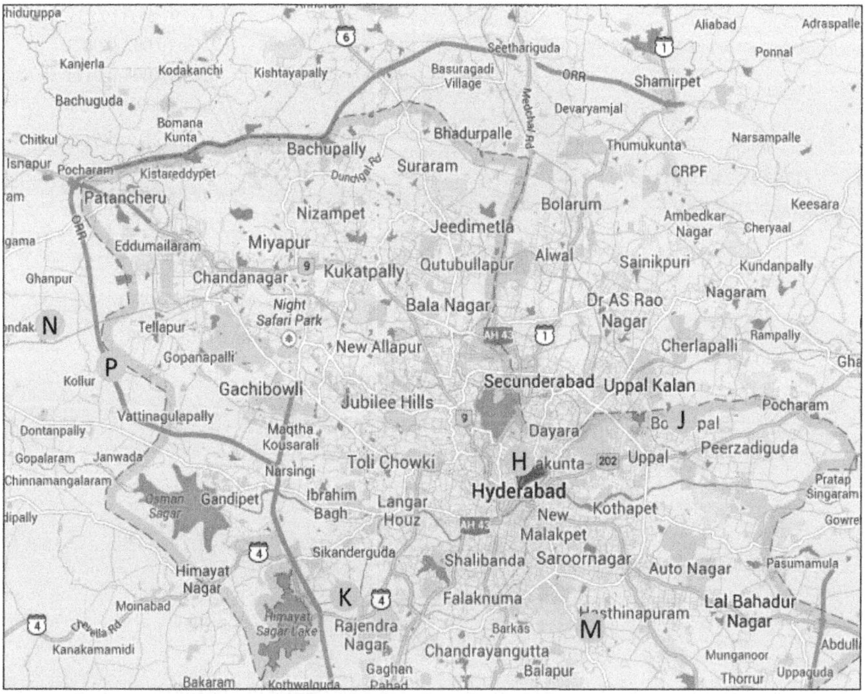

Fig. 7.6 Major contaminated sites in Greater Hyderabad = Tri cities (Hyderabad, Secunderabad and Cyberabad)
H Hussain Sagar, *J* Jawaharnagar, *K* Katteda
M Musi river, *N* Nakkavagu, *P* Patancheru
Bholakpur area – Abdul et al. (2012)
Municipal solid waste dumping sites – Jawaharnagar – Ahmed et al. 2011; Parth et al. 2009, 2011
Industrial waste contaminated areas – Ahmed et al. (2011), Balanagar, Machender et al. (2011, 2012)
Kattedan – Sekhar et al. (2006), Govil et al. (2008, 2012)
Hussainsagar lake sediments – PTE – Gurunadha Rao et al. (2004, 2008), Jain et al. (2010), Suneela et al. (2008), Vikram Reddy et al. (2012)
Industrially contaminated sites of Hyderabad – Sekhar et al. (2003, 2005)
Patancheru Industrial Area – Dasaram et al. (2011), Govil et al. (2001)

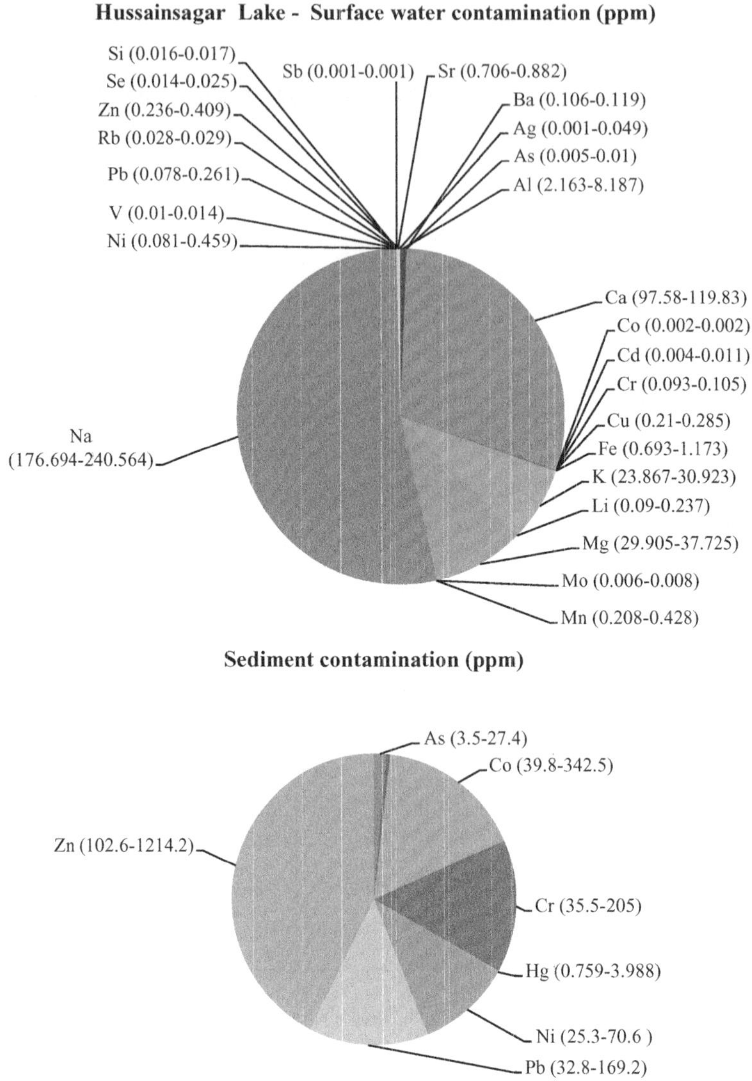

Fig. 7.7 Hussainsagar metal contamination budget (**a**) Surface water and (**b**) sediment

7.3 Products from Plants Applied in Phytoremediation

The objective of this paper aims at highlighting the sustainable valorization of phytoremediated phytomass for production of value additives and value chain products for boosting bio-economy (Bañuelos 2002, 2006) (Fig. 7.13).

India is forging ahead to attain high GDP growth rate to achieve the status of industrialized and developed nation by 2020. In order to achieve this, heavy

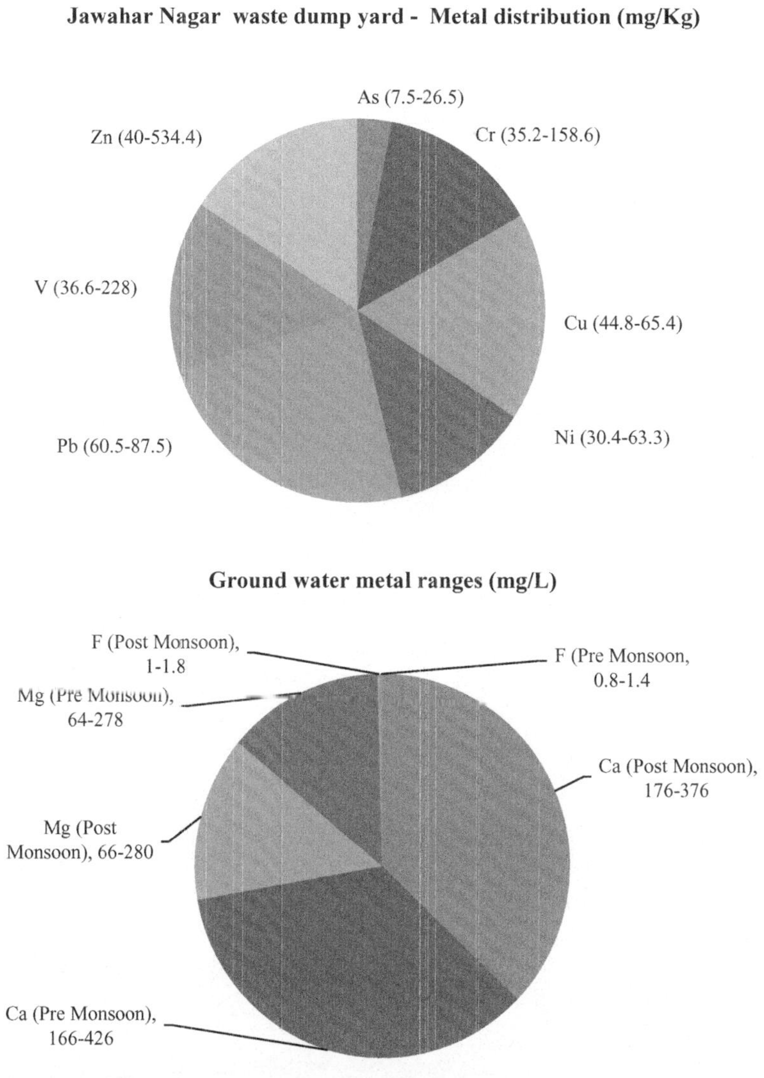

Fig. 7.8 Jawaharnagar waste dump site (**a**) Soil and (**b**) ground water

consumption of natural resources is inevitable. The other side of the coin of this task is enormous waste generation. The waste management summit, held in 2012 has estimated that the poised industrial growth would generate 100 million tons of non-hazardous solid waste, 6–7 million tons of hazardous waste annually. Thus, soil pollution, land-use and land cover change (conflict between agro-ecosystems and urbanization) are of pan-Indian concerns. Consequently, developing eco-innovative soil remediation technologies to re-use contaminated lands in line with biomass production for the bioeconomy are priority objectives.

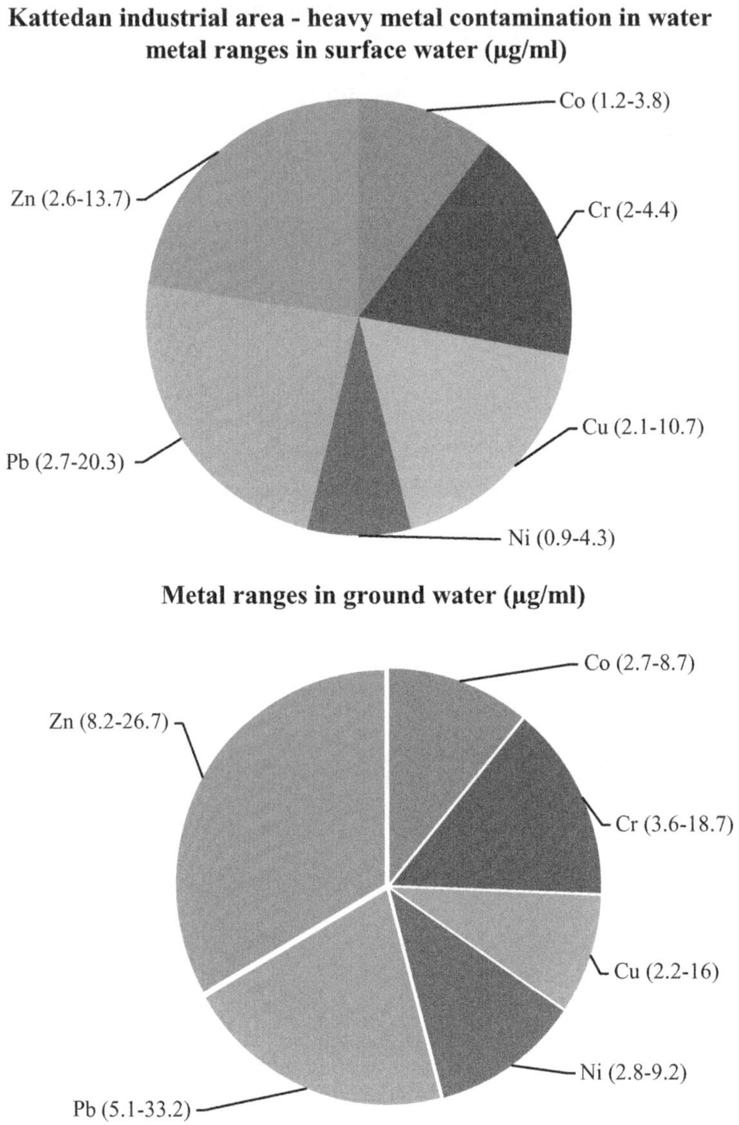

Fig. 7.9 Kattedan industrial area metal contamination (**a**) Surface water and (**b**) ground water

Land resources on a global perspective are under immense pressure. Soil remediation is a solution to reduce the pressure on land resources. Land degradation is due to: (a) Natural processes, (b) erosion, (c) nutrient depletion, (d) loss of organic matter, (e) structural losses, (f) Induced land degradation, (g) irrigation with waste waters, (h) atmospheric deposition of pollutants, (i) poor agricultural practices and (j) climate change increases the intensity of land degradation.

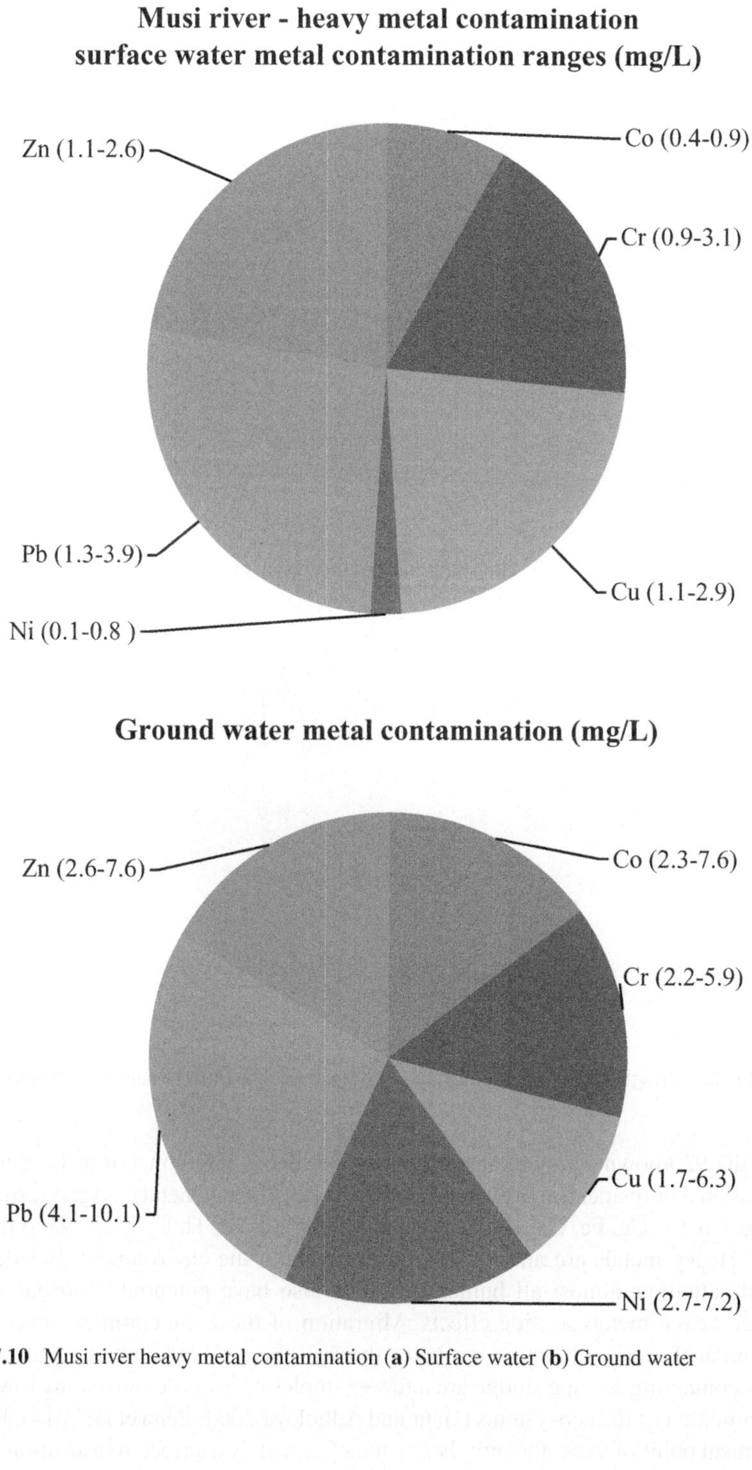

Fig. 7.10 Musi river heavy metal contamination (**a**) Surface water (**b**) Ground water

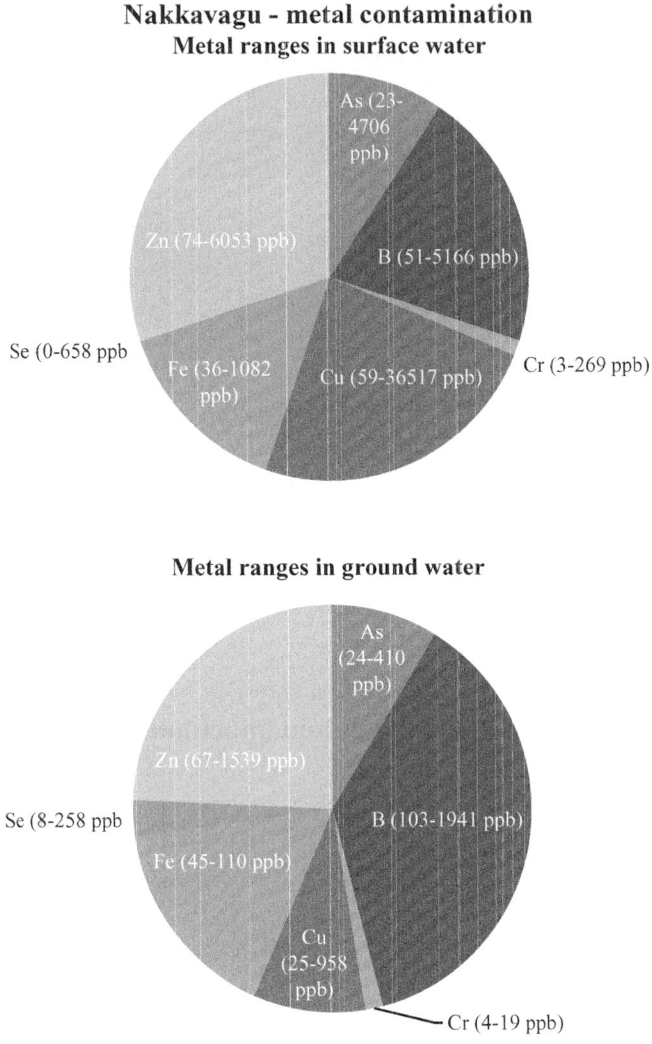

Fig. 7.11 Nakkavagu – Metal contamination (**a**) Surface water and (**b**) Ground water (ppb)

Of the 92 known elements in earth crust, 35 metals that concern us because of occupational or residential exposure, 23 of these are "heavy metals": Ag, As, Au, Bi, Cd, Ce, Co, Cr, Cu, Fe, Ga, Hg, Mn, Ni, P, Pb, Sb, Sn, Te, Tl, U, V, and Zn (Glanze 1996). Heavy metals are among the contaminants in the environment. Beside the natural activities, almost all human activities also have potential contribution to produce heavy metals as side effects. Migration of these contaminants into non-contaminated areas as dust or leachates through the soil and spreading of heavy metals containing sewage sludge are a few examples of events contributing towards contamination of the ecosystems (Gaur and Adholeya 2004; Peña et al. 2014). From a chemical point of view, the term heavy metal is strictly ascribed to transition met-

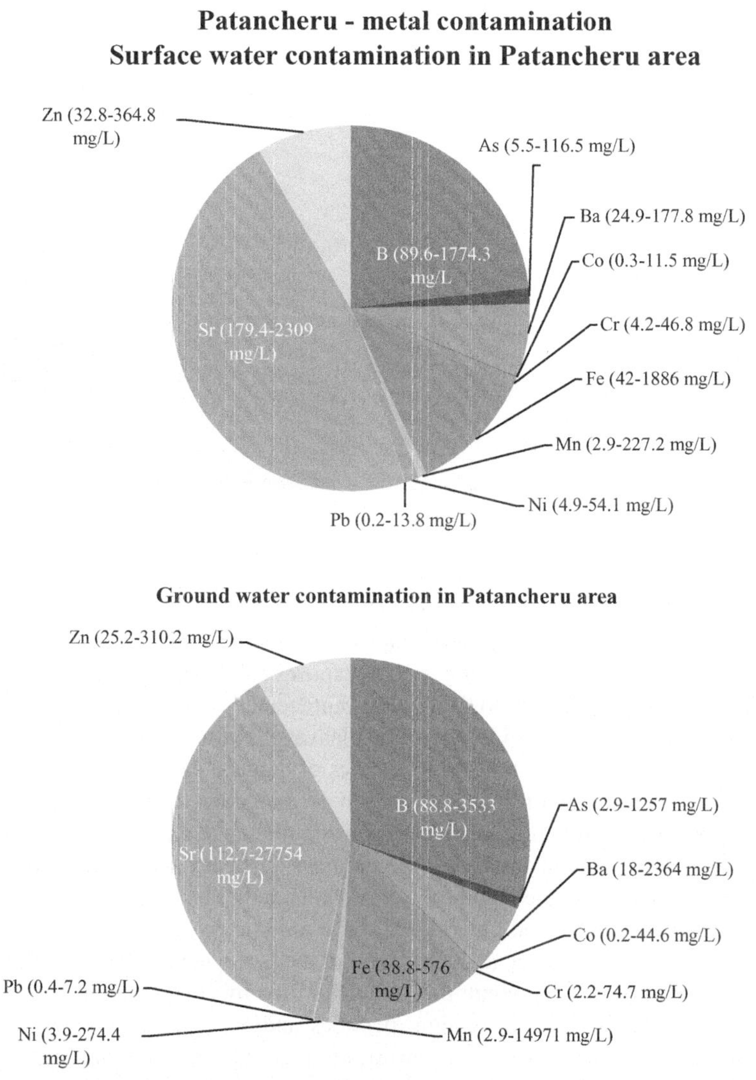

Fig. 7.12 Patancheru – Metal contamination (**a**) Surface water (**b**) Ground water

als, metalloids, with atomic mass over 20 and specific gravity >5. Here the term "heavy metals" will be for these potentially phytotoxic elements. Some of these heavy metals do not perform any known physiological function in plants. However, others such as micronutrients and beneficial elements are required for plant growth and metabolism, but these elements can easily lead to phytotoxicity when their concentration rises above optimal ranges. Heavy metal phytotoxicity may result from alterations of numerous physiological processes caused at cellular/molecular level by inactivating enzymes, blocking functional groups of metabolically important molecules, displacing or substituting for essential elements and disrupting membrane

Fig. 7.13 Products from phytoremediated plants

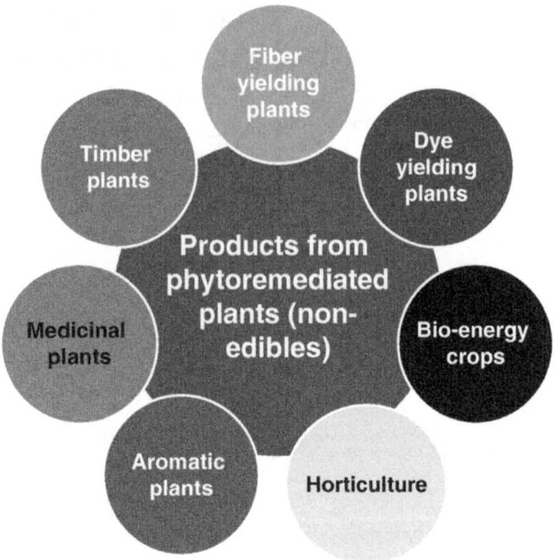

integrity. A rather common consequence of heavy metal poisoning is the enhanced production of reactive oxygen species (ROS) due to interference with electron transport activities, especially that of chloroplast membranes (Pagliano et al. 2006; La Rocca et al. 2009). This increase in ROS exposes cells to oxidative stress leading to lipid peroxidation, biological macromolecule deterioration, membrane dismantling, ion leakage, and DNA-strand cleavage (NavariIzzo et al. 1998, 1999; Quartacci et al. 2001). Plants resort to a series of defence mechanisms that control uptake, accumulation and translocation of these dangerous elements and detoxify them by excluding the free ionic forms from the cytoplasm. One commonly employed strategy lies in hindering the entrance of heavy metals into root cells through entrapment in the apoplastic environment by binding them to exuded organic acids (Watanabe and Osaki 2002) or to anionic groups of cell walls (Dalla Vecchia et al. 2005; Rascio et al. 2008). Most of the heavy metals that enter the plant are then kept in root cells, where they are detoxified by complexation with amino acids, organic acids or metal-binding peptides and/or sequestered into vacuoles (Hall 2002). This greatly restricts translocation to the above-ground organs thus protecting the leaf tissues, and particularly the metabolically active photosynthetic cells from heavy metal damage. A further defence mechanism generally adopted by heavy metal-exposed plants is enhancement of cell antioxidant systems which counteracts oxidative stress (Navari-Izzo et al. 1998; Sgherri et al. 2003).

Several methods are already being used to clean up the environment from these contaminants, but most of them are costly and far away from their optimum performance. The chemical technologies generate large volumetric sludge and increase the costs (Rakhshaee et al. 2009). Both chemical and thermal methods are technically difficult and expensive, and can also degrade the valuable component of soils (Hinchman et al. 1995). Conventionally, remediation of heavy-metal contaminated soils involves either onsite management or excavation and subsequent disposal to a landfill site. This method of disposal solely shifts the contamination problem elsewhere along with the hazards associated with transportation of contaminated soil and migration of contaminants from landfill into an adjacent environment. Soil washing for removing contaminated soil is an alternative way to excavation and disposal to landfill. This method is very costly and produces a residue rich in heavy metals, which will require further treatment. Moreover, these physio-chemical technologies used for soil remediation render the land usage as a medium for plant growth, as they remove all biological activities (Gaur and Adholeya 2004).

Recent concerns regarding the environmental contamination have initiated the development of appropriate technologies to assess the presence and mobility of metals in soil (Shtangeeva et al. 2004), water, and wastewater. Since, these heavy metals cannot be degraded from the nature, at great extent we can take the advantage of natural solar driven process of plants for translocating these contaminants to various parts from the soil or water. Hence, this process is called phytoremediation. Presently, phytoremediation has become an effective and affordable technological solution used to extract or remove inactive metals and metal pollutants from contaminated soil. The term phytoremediation ("phyto" meaning plant, and the Latin suffix "remedium" meaning to clean or restore) refers to a diverse collection of plant based technologies that use either naturally occurring, or genetically engineered, plants to clean up the contaminated soil or water environment (Flathman and Lanza 1998). Phytoremediation is natural, simple, cost effective, non environmentally disruptive green technology and most importantly, its by products can find a range of other economical uses (Truong 1999, 2003). Phytoremediation takes the advantage of the unique and selective uptake capabilities of plant root systems, together with the translocation, bioaccumulation, and contaminant degradation abilities of the entire plant body (Hinchman et al. 1995).

The main objective of this work is to highlight the potential of phyto-products from various plants applied in remediation of heavy metal contaminated soils as well as the usage of the polluted water for co-generation of biofuels using algae (Samardjieva et al. 2011).

Phytoremediation is the use of plants to accumulate, remove or render harmless toxic compounds contaminating the environment. Plants absorb/exclude, translocate, store or detoxify inorganic and organic. Thereby they contribute significantly to the fate of these contaminants from the biosphere. Thus, contaminants (inorganic and organic) can enter the food chain when bio-available, which would cause unwanted effects (Abhilash and Yunus 2011). Many think that phytoremediation is a temporary solution and often the following questions are posed:

(a) There has been a disbelief among scientists and regulators about phytoremediation
(b) How does phytoremediation works?
(c) How to select plants suitable for fostering remediation?
(d) Will phytoremediation work on every heavy metal contaminated/polluted site?
(e) How to dispose of the plants contaminated in this decontamination process?
(f) How to manage the risk based phytoremediation?

Often policy and decision makers, academia and civic governance think that phytoremediation is a temporary solution of transferring the pollutants and contaminants are transferred from one place to another. Scientists and academia are no exception to this feeling. Regulators have expressed apprehensions about phyto-remediation due to lack of knowledge of contemporary knowledge of environmental sustainability. Often, it is generally believed that this kind of environmental decontamination is a temporary solution. How to dispose of the contaminated phytomass is a puzzling question by environmental managers and regulators.

Soil, water and air are the important natural resources that must be cleaned. Unfortunately, natural resources are polluted globally. Rapid industrialization and extraction of large quantity of natural resources including indiscriminate extraction of ground water have resulted in environmental contamination and pollution. Large amounts of toxic waste have been dispersed in thousands of sites spread across the globe resulting in varying degrees of contamination and pollution. Thus, every one of us are being exposed to contamination from past and present industrial practices, emissions in natural resources (air, water, and soil) even in the most remote regions. The risk to human and environmental health is rising and there is evidence that this cocktail of pollutants is a contributor to the global epidemic of cancers, lungs and other degenerative diseases. These pollutants belong to two main classes: inorganic and organic. The challenge is to develop innovative and cost-effective solutions to decontaminate polluted environments.

Bioremediation includes variety of technologies using plants and microbes to remediate or contain contaminants in soil, groundwater, surface water, and/or sediments including air. These technologies have become attractive alternatives to conventional cleanup technologies due to relatively low capital costs and the inherently aesthetic nature.

Bioremediation is the use of biological interventions of biodiversity for mitigation (and wherever possible, complete elimination) of the noxious effects caused by environmental pollutants in a given site. It operates through the principles of biogeochemical cycling. If the process occurs in the same place affected by pollution then it is called *in-situ* bioremediation. In contrast, deliberate relocation of the contaminated material (soil and water) to a different place to accelerate biocatalysis is referred to as *ex-situ* bioremediation. Bioremediation has been successfully applied for cleanup of soil, surface water, groundwater, sediments and ecosystem restoration. It has been unequivocally demonstrated that a number of xenobiotics including nitro-glycerine (explosive) can be cleaned up through bioremediation. Bioremediation is generally considered to include natural attenuation (little or no human action), bio-stimulation or bio-augmentation, the deliberate addition of natural or engineered micro-organisms to accelerate the desired catalytic capabilities. Thus bioremediation, phytoremediation and rhizoremediation contribute significantly to the fate of hazardous waste and can be used to remove these unwanted compounds from the biosphere (Rajkumar et al. 2009, 2010, 2012).

Hibiscus cannabinus (Kenaf), *Brassica juncea* (Indian mustard), *Helianthus annus* (Sunflower), *Ricinus communis* (Castor), *Vetiveria zizanioides* (Khus Khus grass) and *Prosopis juliflora* (Velvet Mesquite) are considered as environmental/ phytoremediation crops. Therefore, phytoproducts (value additives) from the plants that are applied in phytoremediation and safe disposal of contaminated phytomass (risk based remediation) would propel bioeconomy.

7.3.1 Products from Phytoremediated Crops

Due to rapid globalization of the current century, the demand for energy is steeply increasing. Use of fossil fuels have an adverse environmental impact although economic viability and efficiency are in favour of fossil fuels. To reduce the negative impact of fossil fuels, use of contaminated substrates for bioenergy is being researched globally (Meers et al. 2010). This strategy avoids competition over land use for food crops. Further, vast resources that can be harvested from contaminated substrates, conversion through efficient biorefineries via research and development is being tapped in many countries. Technically, biomass and the conversion products can be used as supplemental sources for conventional fuels (fossil fuels) and chemical feedstock for various industries. This approach is promising in the area of alternate sources of energy and dependency on fossil fuels can be minimized to a considerable extent. The production of biomass involves the use of abundantly available and rapidly growing non-agricultural plants, preferably with good coppicing and nitrogen fixing capabilities to produce energy products by suitable conversion technologies (Table 7.2).

The annual photosynthetic storage of energy in biomass is eight times more that of energy use from all sources. This estimate clearly illustrates the immense potential of biomass resources, if harnessed and managed sustainably. Further, biomass derived constituents serve as analogues of fossil fuel derivatives. Sunflower and Indian mustard are popular energy and environmental crops (Fozia et al. 2008; Madejón et al. 2003).

Large tracts of contaminated sites are available in different agro-climatic zones of our country. A number of multiuse plant species with energy rich chemicals have been identified. It is therefore, necessary to integrate such energy plantations into a system of rotational cycle to suit socio-economic aspects of the people. Broad areas of biomass energy sources.

Energy crops include corn, sugar cane, sugar beet, cassava, soyabean and other sugar and starch producing crops. In Brazil, sugar cane cultivation was expanded to some extent at the expense of food crop. Brazil, USA, Philippines, Germany use blended mixture of gasoline and alcohol. Such a blended fuel was termed Gasohol or alcogas. Brazil's National alcohol and pure alcohol programmed "Proalcool" aims at running cars on gasohol and pure alcohol. Cassava (mandioca, tapioca), grown in many developing countries is rich in starch. It can be cultivated

Table 7.2 Metal accumulation reports with reference to *Brassica juncea* (Indian mustard)

As	Srivastava et al. (2013)
As, Cd, Cu, Fe, Mn, Pb, Zn,	Clemente et al. (2005, 2006)
Au	Bali et al. (2010)
B	Giansoldati et al. (2012)
Cd	Schneider et al. (1999), Lee and Leustek (1999), Qadir et al. (2004), Quartacci et al. (2005), Minglin et al. (2005), Nouairi et al. (2006), Manciulea and Ramsey (2006), Mobin and Khan (2007), Hayat et al. (2007), Seth et al. (2008), Szőllősi et al. (2009), Hong-Xia et al. (2009), Ahmad et al. (2011); Bauddh and Singh (2012a, b)
Cd, Zn	Sridhar et al. (2005)
Cd, Co, Cr, Cu, Fe, K, Mn, Na, Ni, Pb, Zn,	Gupta and Sinha (2007)
Cd, Cr, Cu, Fe, Mn, Ni, Pb, Zn,	Gupta and Sinha (2006)
Cd, Cr, Cu, Pb, Zn	Quartacci et al. (2006)
Cd, Cu, Ni, Zn,	do Nascimento et al. (2006)
Cd, Cu, Pb, Zn,	Wu et al. (2004)
Cd, Cu, Zn	Schafer et al. (1997)
Cd, Fe	Qureshi et al. (2010)
Cd, Ni	Cao et al. (2008)
Cr	Pandey et al. (2005), Pandey (2013), Rajkumar et al. (2006), Sinha et al. (2009)
Cr, Ni	Hsiao et al. (2007)
Cu	Fariduddin et al. (2009), Singh et al. (2010), Chigbo et al. (2013)
Cu, Fe, Mn, Pb, Zn	Walker et al. (2003)
Hg	Shiyab et al. (2009), Meng et al. (2011), Cassina et al. (2012)
Mo, Se	Schiavon et al. (2012)
Ni	Rajkumar and Freitas (2008a, b), Ali et al. (2008), Kanwar et al. (2012)
Pb	Liua et al. (2000), Lim et al. (2004), Meyers et al. (2008), Zaier et al. (2010), Ghnaya et al. (2013)
Pu	Lee et al. (2002)
Se	Grant et al. (2004), Kahakachchi et al. (2004), Yathavakilla et al. (2005), Hladun et al. (2011), Jaiswal et al. (2012)
Zn	Alia and Saradhi (1995), Prasad et al. (1999), Saxena et al. (2005)

in wide range of agroclimatic situations and does not require high dosage of fertilizer and pesticides to produce high yields. The notable examples of biomass conversion to biofuels are sweet sorghum and *Miscanthus* (Babu et al. 2014; Brosse et al. 2009; Chami et al. 2014; El Hage et al. 2010; Pavel et al. 2014; Płażek et al. 2014; Sharmin et al. 2012).

Bioenergy is renewable alternative fuel/energy produced from materials derived from biological sources such as plants, animal oils or fermentation. Biomass is any organic material which has stored sunlight in the form of chemical energy. As a fuel it may include wood, wood waste, straw, manure, sugarcane, and many other byproducts from a variety of agricultural processes. Energy is a key to socio-economic development (Figs. 7.14–7.16).

Sunflower crop can be grown in the pollution affected areas. The root system of the plan develops well in varying soil pH conditions. Moistening and acidification of the rhizosphere could enhance the nutrient uptake mechanism of the crop, inducing a 'fertilising' effect (Junkang Guo et al. 2014; Juwarkar et al. 2008; Khouja et al. 2013; Kötschau et al. 2014; Liphadzi and Kirkham 2006; Madrid et al. 2008; Rajkumar et al. 2009; Stanbrough et al. 2013). Sunflower is a multipurpose crop producing a variety of industrial feed stock.

In recent years environmental crops for e.g. Indian mustard, sunflower and sweet sorghum etc. have gained considerable attention. The primary goal of this exercise is to cultivate these crops using contaminated substrates with two objectives. (a) to combat environmental pollution, and (b) to produce beneficial and regenerable products.

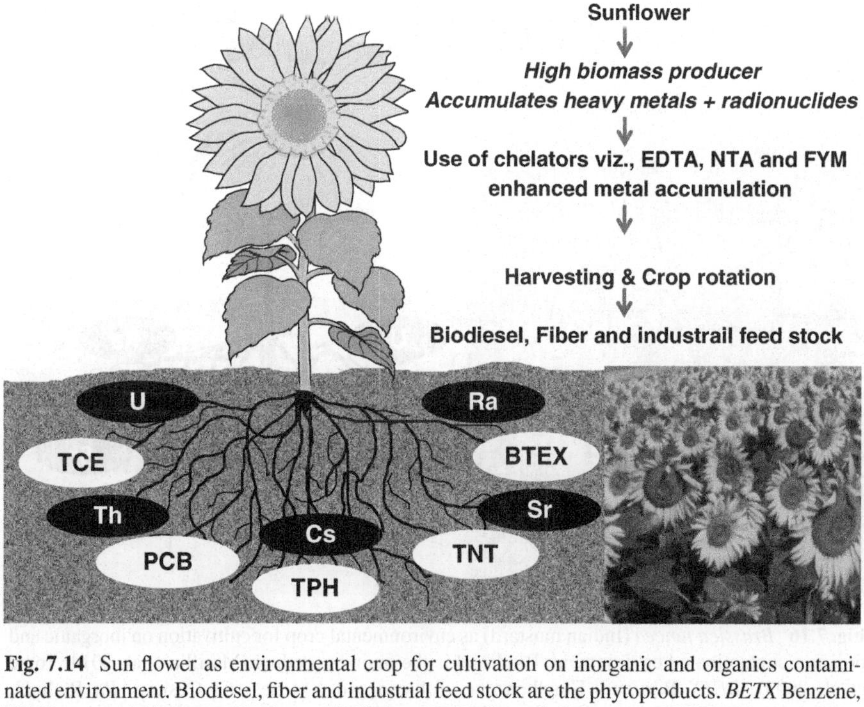

Fig. 7.14 Sun flower as environmental crop for cultivation on inorganic and organics contaminated environment. Biodiesel, fiber and industrial feed stock are the phytoproducts. *BETX* Benzene, Ethylbenzene, Toluene and Xylenes, *Cs* Cesium, *Ra* Radium, *Sr* Strontium, *TCE* Trichloroethylene, *Th* Thorium, *TNT* Trinitrotoluene, *TPH* Total petroleum hydrocarbons, *U* Uranium

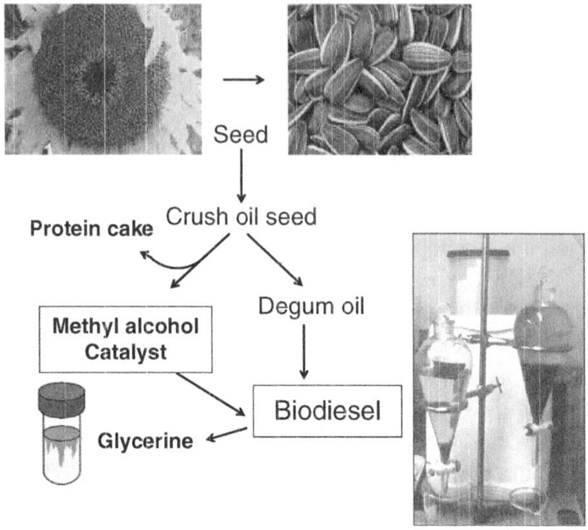

Fig. 7.15 Biodiesel and cogeneration of industrial feed stock from sun flower from phytoremediated crop produce

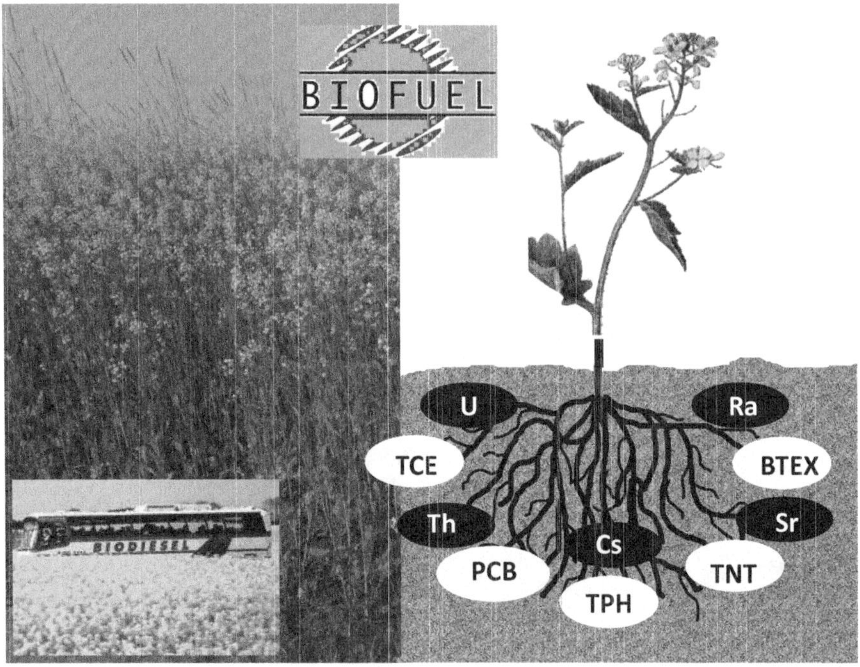

Fig. 7.16 *Brassica juncea* (Indian mustard) as environmental crop for cultivation on inorganic and organics contaminated environment. Biodiesel is the main byproduct (Marillia et al. 2014; Tomé et al. 2009). *BETX* Benzene, Ethylbenzene, Toluene and Xylenes, *Cs* Cesium, *Ra* Radium, *Sr* Strontium, *TCE* Trichloroethylene, *Th* Thorium, *TNT* Trinitrotoluene, *TPH* Total petroleum hydrocarbons, *U* Uranium

The biofuel generations are the following:
1st generation – Jatropha based
2nd generations – lignocellulosic ethanol
3rd generation – algal biofuels
4th generation – Bio-butanol, bio-hydrogen, biomethyl furan etc.

7.3.2 Medicinal and Aromatic Plants on Contaminated Soils

Aromatic crops, used for production of essential oils as opposed to food or feed, may be suitable alternative crops in heavy metal contaminated agricultural as well as non-agricultural soils. Essential oils are low-volume, high value products that are widely used as aromatic agents in various non-food industries, such as perfumery, cosmetics, and aromatherapy suggesting the possibility that such plants could be used in phytoremediation of contaminated soils. The use of aromatic plants for phytoremediation may have several advantages over other crop plants in that the harvested foliage is a source of essential oils, which are the marketable revenue-generating products of aromatic crops. Growing of these aromatic crops in metal contaminated areas may not introduce heavy metals into the food chain and may not result in an economic penalty compared to most other edible crops. In the process of oil extraction by distillation, heavy metals remain in the extracted plant residues, limiting the quantities of heavy metals in the commercial oil product (Zheljazkov and Nielsen 1996; Scora and Chang 1997; Zheljazkov and Warman 2004). Thus, significant amounts of heavy metals could be removed from the soil through proper disposal of the metal contaminated plant residues, while the metal-free, extracted oils could be safely marketed. High-value aromatic crops may be a better alternative for heavy metal contaminated agricultural soils than the suggested woody species such as *Salix* and *Betula* (Hammer et al. 2003; Rosselli et al. 2003), or other plants like *Sesbania drummondii* that have been shown to hyperaccumulate Pb (Sahi et al. 2002). For disposal of metal contaminated phytomass, several approaches such as composting, incineration, ashing, pyrolysis, direct disposal and liquid extraction have been proposed (Sas-Nowosielska et al. 2004; Keller et al. 2005).

Some aromatic plants appear capable of accumulating heavy metals from contaminated soil (Chand et al. 2012; Zheljazkov and Nielsen 1996; Zheljazkov and Warman 2004; Zheljazkov et al. 2008b; Chaiyarat et al. 2011; Aziz et al. 2011; Amirmoradi et al. 2012), and some might demonstrate significant phytoremediation potential if coupled to other means for increasing bioavailability and uptake of Cd, Pb, and Cu, such as chelates (Schmidt 2003) or biosurfactants (Mulligan et al. 2001) (Fig. 7.17).

The effects of these metals on growth, essential oil production, and metal accumulation of most commercially important essential oil producing aromatic crops, such as coriander (*Coriandrum sativum* L.), dill (*Anethum graveolens* L.), chamomile (*Chamomilla recutita* (L.)K.), peppermint (*Mentha x piperita* L.), basil (*Ocimum basilicum* L.), hyssop (*Hyssopus officinalis* L.), lemon balm (*Melissa officinalis* L.), and sage (*Salvia officinalis* L.) are, however, largely

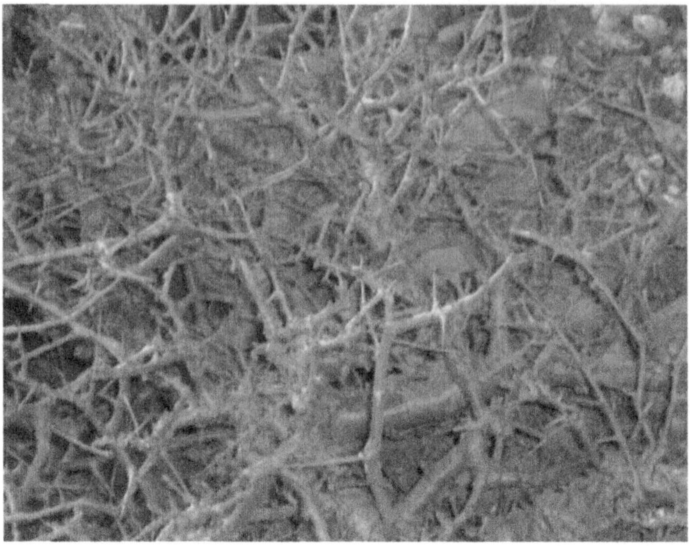

Fig. 7.17 *Commiphora wightii*, a source of guggul steroids capable of growing on lime stone quarry waste

unknown. Coriander, dill, chamomile, peppermint, basil, hyssop, lemon balm, and sage are aromatic crops that have been traditionally grown as cash crops in Europe (Topalov 1962), and North America. Coriander, dill, chamomile, peppermint, basil, hyssop, lemon balmand sage have very high phytoremediation potential (Brown et al. 1998) (Table 7.4).

Traditionally, peppermint, basil, and sage tissue wastes after distillation are used as feed for sheep (Topalov 1962). Results of experiments demonstrated that if peppermint, basil, and sage are grown in highly heavy metal contaminated medium, Cd, Pb, and Cu may accumulate in shoots and wastes from distillation above the maximum permissible concentrations for these elements in animal feed (NRC 1985), making these waste products unsuitable as animal feed. To reduce the amount of Cd, Pb, and Cu and produce usable final product, metal contaminated distillation wastes could be composted after mixing with wastes from other low metal feedstock.

Trace elements in the healing plants can act as remedies (Prasad 2008). Trace elements accumulated in medicinal plants have healing power in numerous ailments and disorders. Herbal preparation known as 'bhasmas' (ash of the polyherbals and specific plants and or their parts) are popular in Ayurveda, Indian traditional medicinal system. According to this medicinal system, metal based drugs known as 'bhasma' involve the conversion of a metal into its mixed oxides. Plants that accumulate metals and metalloids have gained considerable significance and are implicated in healing function. Plants that accumulate essential trace elements are implicated in propelling metabolic processes (metallomics) (Prasad 2008).

7.3.3 *Ornamentals for Environmental Moderation and Toxic Trace Metal Cleanup*

Several ornamental plants have been successfully applied in environmental toxic cleanup (Figs. 7.17–7.25) (Abad et al. 2001; Bosiacki 2009a, b; Ding and Hu 2012). For e.g. Lemon-scented geraniums (*Pelargonium* sp. 'Frensham', or scented geranium) accumulated large amounts of Cd, Pb, Ni and Cu from soil in greenhouse experiments (Bosiacki 2008; Dan et al. 2000; Saxena et al. 1999). Biotechnological interventions through hairy root regenerants are useful in floriculture (Giri and Narasu 2000; Giovanni et al. 1997). Pellegrineschi et al. (1994) improved the ornamental quality of scented *Pelargonium* spp. This plant has pleasant odor that adds scent to the toxic metal contaminated soil.

Vetiveria zizanioides (Vetiver grass): It is known to have multiple uses. This plant had several popular names such as 'the miracle grass', 'a wonder grass', 'a magic grass', 'an unique plant', 'an essential grass', 'an amazing plant', 'an amazing grass', 'a versatile plant', 'a living barrier', 'a living dam', 'a living nail', 'a living

Fig. 7.18 Cultivation of ornamentals on contaminated soils in peri urban Greater Hyderabad. (**a**) *Chrysanthemum* sp. (*pink* flowers) (**b**) *Chrysanthemum* (*Yellow* flowers), (**c**) *Gaillardia* sp. and (**d**) *Rosa* sp. (Colour figures online)

Fig. 7.19 Cultivation of ornamentals on contaminated soil in perdi urban Greater Hyderabad (**a**) and (**b**) Jasminum *officinale* L. with Musi river contaminated water. (**c**) and (**d**) *Tagetes* sp. (Patel and Patra 2014)

Fig. 7.20 Cultivation of (**a**) *Crossandra infundibuliformis* (L.) Nees with Musi river contaminated water (**b**) flowers of *C. infundibuliform* (**c**) *Asparagus setaceus* (Kunth) Jessop and (**d**) *Origanum majorina*

Fig. 7.21 Ornamentals in commercial market (**a**) *Chrysanthemum* sp. (**b**) *Alternanthera* sp.

wall', 'an eco-friendly grass'. This extraordinary grass is adaptable to multiple environmental conditions and it is globally recognized as an easy and economical alternative to control soil erosion and to solve a variety of environmental problems. It has been used for restoration, conservation and protection of land disrupted by man activities like agriculture, mining, construction sites, oil exploration and extraction, infrastructure corridors, as well as used for water conservation in watershed management, disaster mitigation and treatment of contaminated water and soil. Research at the global level has proved the relevance of vetiver in multiple applications. *Vetiveria zizanioides* exhibited high metal tolerance (Andra et al. 2009, 2010, 2011; Chintakovid et al. 2008; Chen et al. 2012; Chiu et al. 2006; Chong and Chu 2007; Danh et al. 2011; Rotkittikhun et al. 2007; Makris et al. 2007; Pang et al. 2003; Truong 2000; Wilde et al. 2005) and is used in stabilization of mine tailings in different parts of the world (Frérot et al. 2006; Pérez-López et al. 2014; Rocio et al. 2013).

The debris generated from the ornamentals containing the toxic metal residues can be treated easily as the biomass would be relatively less in view of high water content. Appropriate techno-economic feasible options based on integrated model systems was recently suggested for the appropriate use of *Eichhorinia crassipes*

Fig. 7.22 Ornamentals in commercial market (**a**) *Rosa* sp. (**b**) *Lilium* sp. (**c**) *Jasminum officinale* L.

Fig. 7.23 Water ornamentals. (**a**) *Pistia stratiotes* L. dry phytomass. (**b**) *Nymphaea odorata* Aiton. (**c**) *N. nouchali* Burm. f. (**d**) *N. pubescens* Willd.

Fig. 7.24 (**a** and **b**) *Vetiveria zizanioides* and (**c** and **d**) Water lilies (*Nymphaea*) in commercial market

Fig. 7.25 Ornamental *Canna indica* in soil scape filtration for removal of nutrients and colouring material from waste water (Huiping Xiao et al. 2010)

(water hyacinth) (Malik 2007). Similar solutions need to be worked out for the ornamentals proposed for toxic metal celanup.

Ornamental plants have an added advantage of enhancing environmental aesthetics besides cleaning the environment. This approach has several advantages for environmental moderation, cleanup and generation of revenue. Therefore, this approach will add new dimension to the field of phytoremediation of contaminated environment. The compost generated from the phytoremediated ornamental plants serves as a medium for use as growing media for production ornamentals (Page et al. 2014).

"Scent the soil" with aromatic plants is a fascinating proposition. Several potential and promising options exist for environmental management of contaminated soils using medicinal and aromatic plants. Yet certain bottlenecks are to be investigated for wider applications. The regulatory bodies and biosafety issues might complicate this approach. The compost generated from the ornamentals plants used in remediation serves as a compost and can be reused as growing media for production ornamentals. The quantum of biomass generated would be relatively less in view of high water content in several of these ornamentals. Integrated model and appropriate techno-economic feasible options were suggested for the use of *Eichhorinia crassipes* (water hyacinth) (Malik 2007). Similar solutions need to be worked out for the ornamental proposed for toxic metal celanup.

7.3.4 Non-edible Oil Plants for Biofuels on Contaminated Soil

Non-edible vegetable oils or the second generation feed stocks have become more attractive for biodiesel production. These feed stocks are very promising for the sustainable production of biodiesel (Agbogidi and Eruotor 2012; Agbogidi et al. 2013; Ahmadpour et al. 2010; Jamil et al. 2009; Liang et al. 2012; Luhach and Chaudhry 2012; Majid et al. 2012; Mangkoedihardjo and Surahmaida 2008).

Taking the advantage of second generation biodiesel plants and obtaining the biodiesels from these plants as most of these plants naturally grown in highly contaminated with variety of inorganic and organic pollutants. Most of the these plants grow widely in contaminated substrates. Our field work survey reveled that some of the potential plants are *Ricinus communis, Jatropha curcas, Azadiracta indica*. The world's dependence on fossil fuels is a perfect example. Therefore, biodiesel feedstock should be as diversified as possible, depending on geographical locations in the world. Therefore, exploring alternative biodiesel feedstocks like non-edible vegetable oils should be an important objective for biodiesel industries in near future.

In developing countries like India, biofuel program of Government of India has launched alcohol production from sugar/starch based feedstock the *Jatropha curcas* plant has been planted on massive scale on waste/degraded or other lands to open new avenues for producing biodiesel from *Jatropha curcas* oil thereby reducing/saving foreign exchange needed to import the petroleum fuel (Abioye et al. 2010; Acharya et al. 2012; Gao et al. 2009; Ghavri and Singh 2012; Kumar et al. 2008, 2013; Singh et al. 2013, 2014; Wu et al. 2011; Yadav et al. 2009). In biodiesel production, India is on 8th position with 0.03 billion liters of biodiesel production per year.

The above problem can be solved by using cheapest, low cost several potential tree borne oil seeds (TBOs) and non-edible crop source have been identified as suitable feed stock for biodiesel. However, it must be pointed out that global biodiesel feed stocks should not rely on certain sources as it could bring harmful influence in the long run. The worlds' dependence on fossil fuels is a perfect example.

Therefore, biodiesel feed stock should be as diversified as possible, depending on geographical locations in the world. The determined properties beside the engine performance and emission characteristics of non-edible biodiesel covered in this review indicated that there is a huge chance to produce biodiesel from non-edible sources in the future.

Since, India is net importer of edible vegetable oils, these oils are therefore not available for conversion to biodiesel. India has the potential of becoming the world's leading producer of biodiesel, as it can be "harvested," and produced from non-edible oils like *Jatropha curcas, Pongamia pinnata, Madhuca indica* plants etc (Awwokunmi et al. 2012; Bauddh and Singh 2012a, b).

The increasing use of biodiesel in India will provide green cover to wasteland, support the agricultural and rural economy, reduce the dependency on imported crude oil and improve the environmental emissions. Even though, 12 *Jatropha* species are reported by several Indian floras, research has been confined to nine species only. Among all the *Jatropha* species, *J. curcas* is the most primitive form and has the potential for biodiesel and for medicinal (Abd El-Kader et al. 2012).

There are several tree species that are suitable for Biodiesel production like *Jatropha sps., Pongamia pinnata (Millettia pinnata), Moringa, Simarouba glauca, Calophyllum inophyllum, Citrullus colocynthis, Simmondsia chinesis, Helianthus tuberosus, Garcinia indica, Madhuca indica, Azadirachta indica, Linum usitatissimum* etc. (Fig. 7.26).

Advantages of non-edible oil as Biodiesel has the following advantages:

- Mitigate Environmental Global Warming
- Energy Security
- Rural and Urban Economic Development
- Reduce Sulfur/Aromatics
- Clean Air
- Health Effects
- Biodiesel Exhausts reduce Carcinogen Emissions
- High Cetane and Lubricity
- Biodegradable
- Non-Toxic
- The cancer causing potential of diesel exhaust largely a function of:

  Amount, size, and composition particulates
  Mutagenicity of exhaust gases
  High exhaust mutagenicity can cause:
  shortened life, birth defects
  asthma and respiratory problems
  specially in infants or the elders

Fig. 7.26 Cultivation of biodiesel and oil producing crops contaminated soil (**a**) *Ricinus communis* (**b**) *Vetiveria zizanioides* (**c**) *Jatropha curcas* and (**d**) *Moringa oleifera*

7.3.5 Tree Crops

The following are some examples *Casurina equisetifolia, Leucaena leucocephala* (Subabul, koobabul, ipil-ipil, lead tree, tan tan, white popinac etc.) *Parkinsonia aculeata* (Jerusalem thorn), *Pithecellobium dulce* (Jangil jailebi), *Bauhinia variegata, Cassia siamea, Prosopis juliflora* (Mesquite), *P. chiliensis* (Paradeshi Babul), *P. cineraria* (Khejri), *Peltophorum pterocarpum* (Copper pod), *Sesbania gradiflora* (Avisa), and *S.bispinosa* (Jeelgu) (Yang et al. 2003; Zalesny et al. 2009). However, only a few like *Eucalyptus, Leucaena* and *Casuarina* are mostly grown (monoculture). In such plantations, wood production is about 20 cu m/ha/year/ to 60 cu m/ha/year. *Eucalyptus* wood has industrial value in pulp, rayon charcol, methanol distilleries (Fig. 7.27).

In the contemporary time energy and environment security are at stake in our situation. Land resources are under immense pressure glocally (globally and locally). The pressure on available land resources is increasing due to land degradation, population explosion, global economic development and urbanization. Land degradation is mainly due to erosion, nutrient depletion, loss of organic matter, structural losses, induced land degradation, faulty irrigation, atmospheric deposition of pollutants, poor agricultural practices. Climate changes increases the inten-

Fig. 7.27 Stabilization of mine tailings with *Cymbopogon citratus* (lemon grass) miracle grass, *Vetiveria zizanioides* and Jatropha curcas (**a**) Nyveli lignite mine (**b**) Mine over burden (**c**) *Jatropha curcas* plantation in mine over burden and (**d**) Jatropha biodiesel

sity of land degradation. Therefore, utilization of contaminated environment (terrestrial and aquatic) following the principles of phytoremediation for biofuel production would be one of the sustainable options (Núñez-López et al. 2008). Although this strategy has immense scope and limitations (Tables 7.3 and 7.4) considerable progress and successful demonstration projects have provided convincing evidence for environmental moderation, cleanup and co-generation of revenue. Therefore, this approach will add new dimension to the field of sustainable development.

Natural vegetation of *Prosopis juliflora* in Patancheru Industrial Development Area contaminated with highly above permissible limits of heavy metals in soil and surface and ground water. *Prosopis juliflora* is an evergreen, fast growing, drought resistant, widely distributed phreatophyte not only in India but also in other arid and semi-arid tropical countries. A valued tree for shade, timber and forage It is a thorny, deciduous, large crowned and deep rooted bush or tree which grows up to 10 m height or more, depending on the variety and climatic conditions. It is widely distributed in the dry tropical and sub-tropical regions of Central America and Northern South America. Also, it is widely propagated in Africa and Asia. It is the only exotic species capable of growing on a wide variety of soils and climatic conditions. It is an ideal species for afforestation and helps in the reclamation of waste lands, stabi-

Table 7.3 Metal accumulation reports with reference to *Helianthus annus* (sun flower) (Ruiz et al. 2007)

Al, Cd, Cr	Gallego et al. (2002)
Al, Cd, Cu, Fe, Mn, Zn	Fauziah et al. (2011)
Al, Cd, Cu, Ni, Pb, Zn	Chakravarty and Srivastava (1992)
As	Baroni et al. (2004)
As, Ca, Cd, Co, Cr, Cu, Mn, Na, Ni, Pb, U, V, Zn	Paun et al. (2012)
As, Cd, Co, Cu, Pb, Zn	Marchiol et al. (2007)
As, Cd, Cr, Fe, Ni,	January et al. (2008), Cutright et al. (2010)
Ca, Cd, Co, Cu, K, La, Li, Mg, Mn, Na, Ni, Zn	Enache et al. (2003)
Cd	Yurekli and Kucukbay (2003), Gallego et al. (2005)
Cd, Cr, Cu, Pb, Zn	Soudek et al. (2010)
Cd, Cr, Ni	Chen and Cutright (2001)
Cd, Cr, Pb,	Ullah et al. (2011)
Cd, Cu, Fe, Mn, Ni, Pb, Zn	Liphadzi et al. (2003)
Cd, Cu, Ni, Pb, Zn	Meers et al. (2005), Jadia and Fulekar (2008)
Cd, Cu, Ni, V, Zn	Lombi et al. (1998)
Cd, Cu, Pb,	Adewole et al. (2010)
Cd, Cu, Pb, Zn	Garcia et al. (2006, 2009), Nathan et al. (2012), Mayora et al. (2012), Angelova et al. (2012)
Cd, Hg	Rai and Kumar (2010)
Cd, Pb	Lotfy et al. (2009) Awotoye et al. (2009)
Cd, Pb, Zn	Nehnevajova et al. (2005, 2007)
Cd, Zn	Xiu-Zhen et al. (2012)
Cr	Fozia et al. (2008)
Cr, Fe, Mn, Zn,	Singh et al. (2004)
Cr, Pb	Araiza-Arvilla et al. (2006)
Cs	Soudek et al. (2004)
Cs, Sr, U	Prasad (2007b)
Cu	Zengin and Kirbag (2007), Herrera-Rodríguez et al. (2007)
Cu, Cd, Pb, Zn,	Lesage et al. (2005)
Cu, Fe, Mn, Si, Sr, Ti, Zn,	Busuioc et al. (2009)
Cu, Fe, Mn, Zn	Sabudak et al. (2007)
Cu, Ni, Zn	Rajkumar et al. (2008)
Cu, Pb, Zn,	Tandy et al. (2006)
Hg	Pedron et al. (2013)
Ni	Szymaska and Matraszek (2005), Najafi et al. (2011)
Pb	Azhar et al. (2006), Sinegani and Khalilikhah (2008), Melo et al. (2009), Azhar et al. (2009), Sinha et al. (2011), Usha et al. (2011), Azad et al. (2011), Chandra et al. (2011), Seth et al. (2012)
Pb, Zn	Solhi et al. (2005), Adesodun et al. (2010)
Ra, U	Tomé et al. (2008)
Zn	Fauziah et al. (2011)

Table 7.4 Medicinal and aromatic plants capable of growing on contaminated soils

Plant name	Common name	Family	Parts used
Abies densa	Silver fir	Pinaceae	Wood
Acorus calamus	Sweet flag	Araceae	Rhizome
Acorus gramineus	Sweetflag	Araceae	Aerial parts
Acorus gramineus	–	Araceae	Rhizome
Adenosma indicum	–	Scrophulariaceae	Aerial parts
Amomum sp.	Cardamom	Zingiberaceae	Fruit
Aquilaria crassna	Eaglewood	Thymelaeaceae	Wood
Artemisia vulgaris	Mugwort	Asteriaceae	Aerial parts
Blumea balsamifera	–	Asteriaceae	Leaf
Chenopodium ambrosioides		Chenopodiaceae	Aerial parts
Cinnamomum camphora	Camphor	Lauraceae	Wood
Cinnamomum cassia	Cassia	Lauraceae	Bark, leaf
Cinnamomum glaucescens	–	Lauraceae	Berry
Cinnamomum iners	Thai cinnamon	Lauraceae	Bark
Cinnamomum loureirii	Vietnamese cassia	Lauraceae	Bark
Cinnamomum obtusifolium	–	Lauraceae	Bark, leaf
Cinnamomum tamala	Indian cinnamon	Lauraceae	Leaf
Citrus hystrix	Leech lime	Rutaceae	Fruit peel
Cunninghamia sinensis	–	Pinaceae	Saw dust
Cymbopogon distans	–	Poaceae	Aerial parts
Elscholtzia cristata	–	Lamiaceae	Aerial parts
Elsholtzia blanda	–	Lamiaceae	Aerial parts
Eucalyptus globulus	Eucalypt	Myrtaceae	Leaf
Gaulteria fragrantissima	Wintergreen	Ericaceae	Leaf
Homalomena aromatica	–	Araceae	Rhizome
Homalomena occulta	–	Araceae	Rhizome
Hyptis suaveolens	–	Lamiaceae	Herb
Jasminum sambac	Arabian jasmine	Oleaceae	Flower
Juniperus indicus	Juniper	Cupressaceae	Berry
Lavandula angustifolia	Lavender	Lamiaceae	Flower
Lavandula officinale	Lavender	Lamiaceae	Flower
Litsea cubeba	Cubeb	Araceae	Fruit
Lonicera japonica	Honeysuckle	Caprifoliaceae	Flower
Michelia alba	Champi	Annonaceae	Flower
Nardostachys grandiflora	Spikenard	Valerianaceae	Rhizome
Ocimum gratissimum	Lemon basil	Lamiaceae	Aerial parts
Ocimum tenuiflorum	Holy basil	Lamiaceae	Aerial parts
Parmelia nepalensis,	Lichens/ Tree moss	Parmeliaceae	Whole plant
Pelargonium fragrans	Nutmeg-scented geranium	Geraniaceae	Leaf
Pelargonium capitatum	Alta of rose geranium	Geraniaceae	Leaf
Pelargonium crispum	Curly-leaved geranium	Geraniaceae	Leaf
Pelargonium graveolens	Pot geranium	Geraniaceae	Leaf
Pelargonium macrorrhizum	Scented geranium	Geraniaceae	Twig
Pelargonium pratense	Scented geranium	Geraniaceae	Twig
Pinus roxburghii	Pine	Pinaceae	Resin
Pinus khasya	Pine	Pinaceae	Resin
Pinus merkusii	Pine	Pinaceae	Resin

(continued)

Table 7.4 (continued)

Plant name	Common name	Family	Parts used
Rhododendron anthopagon	Rhododendron	Ericaceae	Twig
Rosa damascena	Damask rose	Rosaceae	Flowers
Usnea sp. *Ramaliana* spp.	Lichens/Tree moss	Usneaceae	Whole plant
Vetiveria zizanioides	Vetiver	Poaceae	Root
Zanthoxylum armatum	Zanthozylum	Rutaceae	Fruit
Zingiber purpureum	Phlai'	Zingiberaceae	Rhizomes

Table 7.5 Metal accumulation in selected Medicinal and aromatic plants

Plant name	Metals	References
Anethum graveolens	Cd, Pb, Cu, Mn, Zn	Zheljazkov et al. (2008a, b)
Chamomilla recutita		
Coriandrum sativum		
Hyssopus officinalis		
Melissa officinalis		
Mentha x piperita .		
Ocimum basilicum		
Salvia officinalis		
Acorus calamus	Al, Ca, Fe, K, Mg, Na, P, Zn	Ozcan and Akbulut (2007)
Artemisia absinthium		
Brassica alba		
Capparis ovata		
Capsicum frutescens		
Carum copticum		
Cinnamomum zeylanicum		
Cuminum cyminum		
Echinophora tenuifolia		
Foeniculum vulgare ssp. piperitum		
Glycyrrhiza glabra		
Laurusnobilis		
Matricaria chamomilla		
Melissa officinalis		
Mentha piperita		
Myrtus communis		
Nigella sativa		
Ocimum minumum		
Pimpirella anisum.		
Piper nigrum		
Rhus coriaria		
Rosmarinus officinalis		
Salvia aucheri		
Salvia fruticase		
Satureja hortensis		
Sesamum indicum		
Syzygium aromaticum		
Thymbra spicata		
Tilia cordata		

(continued)

Table 7.5 (continued)

Plant name	Metals	References
Ocimum basilicum	Cd, Pb, Zn	Galeş et al. (2009)
Salvia officinalis		
Vetiveria zizanioides	Cd, Pb	Minh and Khoa (2009)
Catharanthus roseus *Hyptis suaveolens*	Cu, Fe, Al, Cr	Deo et al. (2011)
Woodfordia fruiticosa		
Taraxacum officinale	Pb, Cd, Cu, Zn, Hg, Fe, Co, Cr, Mo	Malawska and Wiłkomirski (2001)
Artemisia herba alba,	Al, Ca, Fe, K, Mg, Mn, P, Ba, Cd, Pb, Zn	Imelouane et al. (2011)
Lavandula dentata,		
Rosmarinus tournefortii,		
Thymus vulgaris		
Salvia officinalis	Pb, Cd, Cr, Co, Ni, Zn, Fe, Cu, Mn	Abu-Darwish et al. (2011)
Salvia officinalis	Pb, Cd, Cu, Zn	Blagojević et al. (2009)
Thymus *vulgaris*	Cd, Cr, Pb, Ni, Cu, Mn, Zn, Fe, Co	Abu-Darwish (2009)
Thymus serpyllum		
Salvia officinallis		
Tribulus terrestris	Cd, Pb, Zn	Stancheva et al. (2011)
Salvia officinalis	Cd, Cu, Pb, Zn	Stancheva et al. (2009)
Hyssopus officinalis	Cd, Pb, Cu, Mn, Zn, Fe	Roodi et al. (2012)
Satureja montana		
Hypericum perforatum		
Achillea millefolium		
Mentha piperita		
Mentha arvensis var *piperascens* Malinv.		
Majorana hortensis	Ca, Na, Mg, Fe, Mn, Zn, Cu, Cd, Ni, Pb	Khalifa et al. (2011)
Mentha pipperita		
Pelargonium graveolens		
Foeniculum vulgaris		
Matricaria chamomilla		
Abutilon indicum	V, Cr, Mn, Ni, Co, Cu, Zn, As, Se, Rb, Sr, Mo, Ag, Cd, Sb, Ba, Ti, Pb, U, Na, Mg, Al, Si, K, Ca, Fe	Kumari et al. (2012), Bauddh and Singh (2012a, b)
Achyranthes aspera var.		
Adhatoda vasica Nees,		
Alternanthera sessilis		
Azadirachta indica		
Eucalyptus globulus		
Hyptis pectinata		
Ocimum sanctum,		
perphyristachya		
Ricinus communis		
Tinospora cordifolia		
Vetiveria zizanioides	As, Cd, Cr, Cu, Hg, Ni, Pb, Se, Zn	Danh et al. (2009)
Vetiveria zizanioides	Zn, Cu, Ni, and Cr	Roongtanakiat et al. (2003)

(continued)

Table 7.5 (continued)

Plant name	Metals	References
Rosmarinus officinalis *Origanum majorana*	Fe, Mn, Zn, Ni, Cu, Pb, Cd, Cr, Co	Koc and Sari (2009)
Sideritis congesta		
Mytrus communis		
Hypericum perforatum		
Capsicum annum		
Thymus vulgaris		
Saturage hortensis		
Melisa officinalis		
Lavandula officinalis		
Vitex agnis		
Mentha piperita		
Laurus nobilis		
Chrysopogon zizanioides	Fe, Zn, Mn, Cu	Roongtanakiat et al. (2009)
Chrysopogon nemoralis	Zn, Cd, Pb	Roongtanakiat and Sanoh (2011)
Chrysopogon zizanioides		
Acorus calamus	Pb, Cd, Ni, Cu, Fe, Mn, Co, Zn	Malenčić et al. (2005)
Thymus montanus		
Salvia verticillata		
Ocimum gratissimum	Cd, Zn	Chaiyarat et al. (2011)
Mentha piperita	Co	Aziz et al. (2011)
Mentha piperita	Cd, Pb	Amirmoradi et al. (2012)
Majorana hortensis,	Ca, Mg, Na, Fe, Mn, Zn, Cu, Cd, Ni, Pb	Hussein et al. (2006)
Mentha pipperita		
Pelargonium graveolens,		
Foeniculum vulgaris,		
Matricaria chamomilla,		
Salvia officinalis	Cd, Cu, Pb, Zn	Stancheva et al. (2009)
Azadiractha indica	Zn, Pb, Cd	Princewill-Ogbonna and Ogbonna (2011)
Occimum gratissimum		
Vernonia amygdaline		
Mentha arvensis	Pb, Ni, Cr, Zn, Cu, Mn	Chand et al. (2012)
Catharanthus roseus	Cd, Mn, Ni, Pb	Srivastava and Srivastava (2010)
Rosmarinus officinalis	Zn, Pb, Cd	Gaïda et al. (2013)
Vetiveria nigritana,	Cd, As, Pb	Oshunsanya et al. (2012)
Vetiveria zizanioides		
Matricaria recutita,	Cd, Pb, Ni	Lydakis-simantiris et al. (2012)
Salvia officinalis		
Thymus vulgaris		
Phyllanthus amarus,	Al, Cd, Pb, As	Jadhav et al. (2012), Agamuthu et al. (2010)
Jatropha gossypifolia		
Ruta graveolens		

lization and prevents soil erosion. It has tremendous field potential for reclamation of a variety of mine soils, fly ash land fills and disturbed ecosystems.

P. juliflora is an ideal species for stabilizing the pegmatitic tailings of mica mines in Nellore district of Andhra Pradesh. Research findings revealed that it is helpful for reclamation of copper, tungsten, marble, dolomite mine tailings and is a green solution to heavy metal contaminated soils. This is an appropriate species for rehabilitation of gypsum mine spoil in arid zone restoration of sodic soils. It outperformed all other tree species in sand dune stabilization. Arbuscular mycorrhizal inocula have been isolated from its rhizosphere (low cost agrotechnology) were found to accelerate the growth of their agroforestry and social forestry legumes in perturbed ecosystems (Patrick Audet 2014).

Prosopis juliflora colonized the industrial effluent produced by textile, paper products, tannery, chemical products, basic metal products, machinery parts and transport equipments industry. *P. juliflora* and *Leptochloa fusca* association was successful for revegetating salt laden lands (Singh 1995; Shelef et al. 2012) (Figs. 7.28, 7.29, 7.30, 7.31, and 7.32).

Grass-legume-tree association need to be tested on different sites for remediation, if necessary with biotic and abiotic amendments. Restoration of fly ash landfills with *P. juliflora* following different amendments and *Rhizobium* inoculation

Fig. 7.28 *Prosopis juliflora* in industrial areas contaminated with heavy metals. is the feed stock in biomass based power plants (**a**) *Prosopis juliflora* (**b**) Prosopis wood (**c**) Biomass fired power plant

Fig. 7.29 *Prosopis juliflora* – charcoal production (**a**) Prosopis wood (**b**) Controlled burning to produce charcoal (**c** and **d**) Charcoal

yielded promising results (Dwivedi et al. 2008). Mycorrhizae improved the growth of *P. juliflora* on high pH soils. *P. juliflora* seedlings growing in gypsum mine had high frequency arbuscular mycorrhizal fungal infection. However, *P. juliflora* has some biological characters that foster invasion, hence appropriate management practices needed to be developed for recommending it for phytoremediation.

Industrial and urban activities impact our environment, especially in terms of soil pollution. Large numbers of sites are nowadays contaminated by pollutions of chemical or organic origin. Treating these pollutions represents an economic need, which often remains unanswered by conventional civil engineering methods, due to their inappropriateness, their environmental impact and costs (notably for large sites). During the last two decades, we have witnessed the emergence of gentle soil remediation techniques using various plant species and the combination of microbial biotechnologies. Several phytotechnologies can be considered and applied to polluted soils Conesa et al. (2012): (1) phytostabilisation, which uses perennials able to sorb and immobilize potentially toxic trace elements (PTTE) in the root zone, avoiding their transfer toward groundwater and aerial parts and preventing their bioaccumulation in the food chain as well as dispersion by natural agents (wind erosion, water, etc.) (2) phytoextraction, based on root-to-shoot transfer and storage of PTTE in harvestable plant parts. *P. juliflora* produced phytoproducts are shown in Figs. 7.33–7.40).

Fig. 7.30 *Prosopis juliflora* wood is used as fuel and also for production of charcoal which is also used as a carrier of biofertilizing microbes (**a**) Charcoal production (**b**) Charcoal (**c**) Charcoal is used as carrier for plant growth promoting bacteria (**d**) Application of biofertilizer using charcoal as carrier

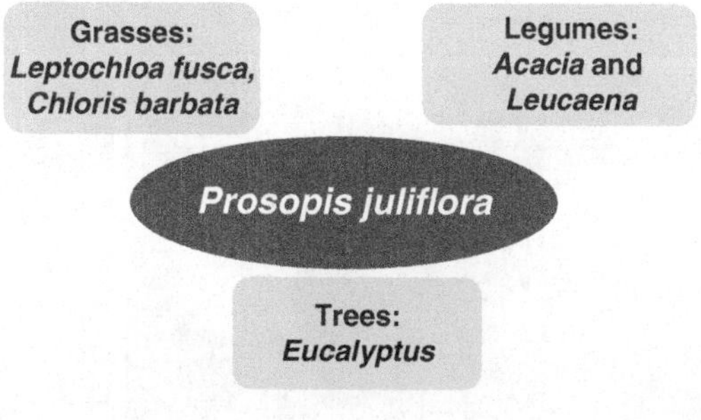

Fig. 7.31 Tree – grass – legume association was found to be the best combination for restoration of mica, copper, tungsten, marble, dolomite, limestone, and mine spoils of Rajasthan State and else where in India

7.3.6 Products from Phytoremediated Fiber Crops

Since ancient times, plants were of considerable help in satisfying man's necessities in respect of food, clothing and shelter. In those days, man also required some form of cordage for his snares, bow-strings, nets, etc. and also for better types of covering

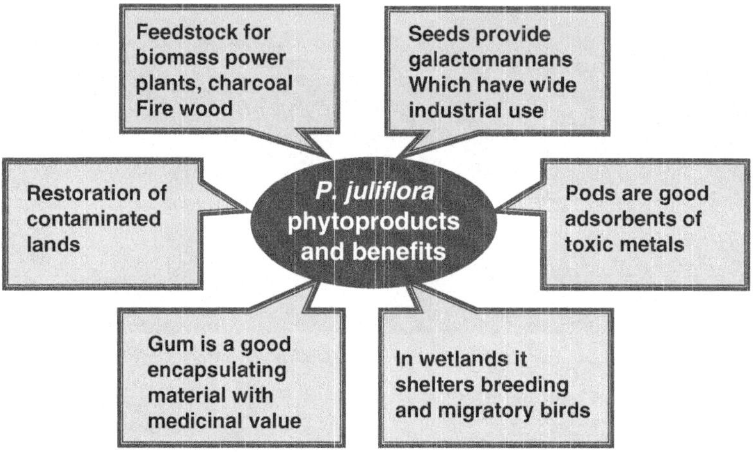

Fig. 7.32 *Prosopis juliflora* benefits and products

Fig. 7.33 Cultivation of *Gossypium arboreum* in polluted (**a** and **b**) (**c**) Cotton ready for hearvest (**d**) Cotton godown

Fig. 7.34 (**a**) Cultivation of *Vetiveria*, (**b**) *Hibiscus cannabinus*, (**c**) *H. sabdariffolia* and (**d**) *Jatropha gossipyfolia* on contaminated lands. Variety of non-edible products are produced from these plants

Fig. 7.35 Novel composites containing polyaniline coated short kenaf (*Hibiscus cannabinus*) bast fibers and polyaniline nanowires

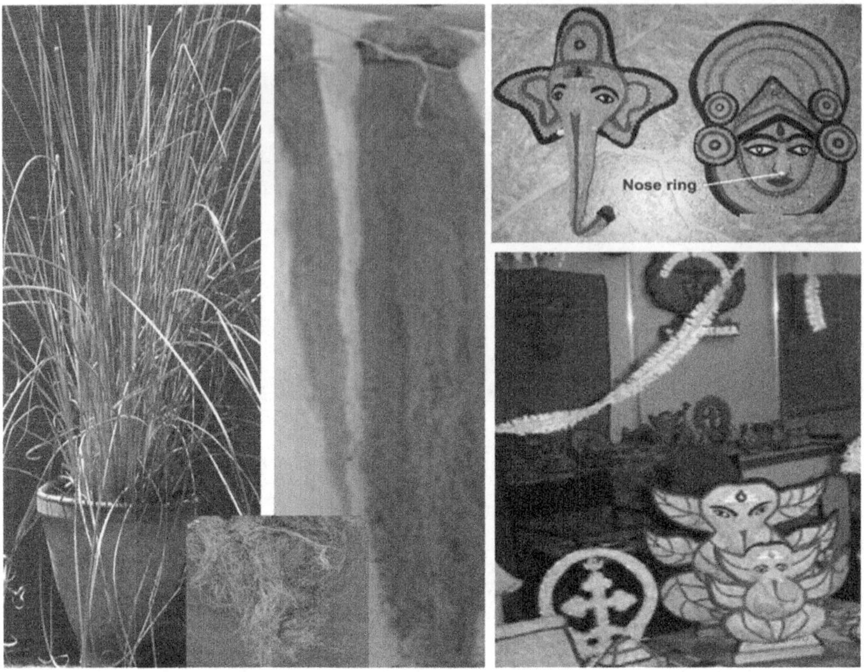

Fig. 7.36 Variety of handi crafts made out of vetiver roots (Source: www.vetiver.org/IND_handicrafts.pdf; www.vetiver.org and Dr Paul Truong)

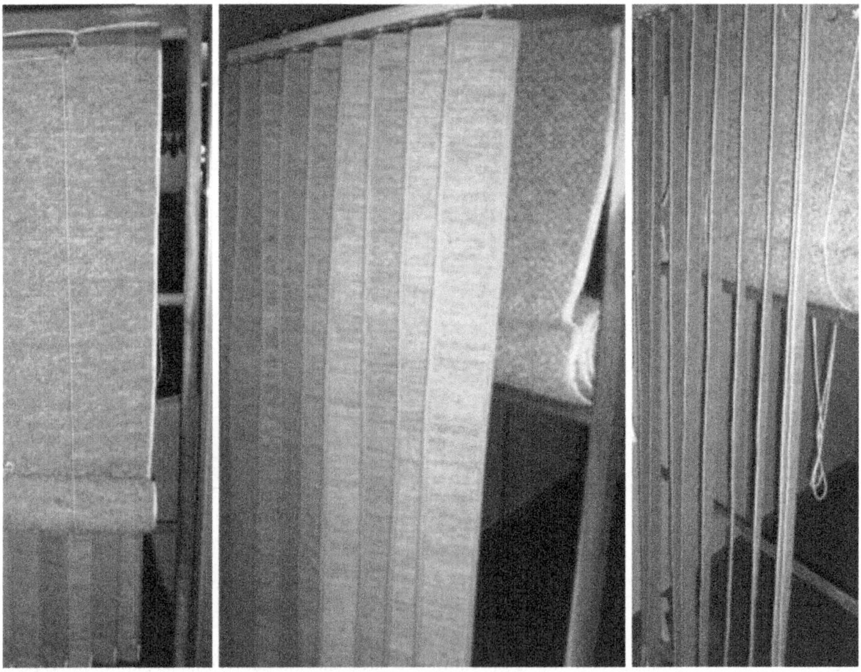

Fig. 7.37 Venetian blinds and door curtains made out of vetiver roots (Source: www.vetiver.org/IND_handicrafts.pdf; www.vetiver.org and Dr Paul Truong)

Fig. 7.38 Variety of household items made out of vetiver roots (Source: www.vetiver.org/IND_handicrafts.pdf; www.vetiver.org and Dr Paul Truong)

Fig. 7.39 Garlands and a wide variety of household items made out of vetiver roots (Source: www.vetiver.org/IND_handicrafts.pdf; www.vetiver.org and Dr Paul Truong)

Fig. 7.40 Soft sandals and wide variety of household items made out of vetiver roots (Source: www.vetiver.org/IND_handicrafts.pdf; www.vetiver.org and Dr Paul Truong)

for his shelter. Tough, flexible fibres obtained from stems, leaves, roots, etc., of various plants served the above purposes very well. With the advancement of civilization, the use of plant fibres has gradually increased and their importance today is very great. Fibre-yielding plants have been of great importance to humans and they rank second only to food plants in their usefulness. Although many different species of plants, roughly about two thousand or more, are now known to yield fibres, commercially important ones are quite small in number (Tables 7.2 and 7.3). Fibers are long narrow tapering cell, dead and hollow at maturity, thick cell wall composed mostly of cellulose and lignin. They are rigid for support, found mainly in vascular tissue. Most importantly natural fibers are biodegradable (Figs. 7.33–7.40).

7.3.7 Dye Yielding Plants

Dye yielding plants are shown is Table 7.12.

7.3.8 Phytoremediation Crops for Carbon Sequestration

Biofuel production, carbon sequestration are the two sides of the coin of the "Phytoremediation crops" (Aggarwal and Goyal 2007). Before the industrialization revolution (1760–1830 A.D), the amount of greenhouse gases (GHGs) such as Carbon dioxide (CO_2), Carbon monoxide (CO), Nitrous oxide (N_2O), Methane (CH_4), Ozone (O_3), Chlorofluorocarbon (CFC) i.e., Hydrofluoro-carbons (HFCs), Perfluorocarbons (PFCs), Sulfur hexafluoride (SF_6) in the atmosphere remained relatively constant. Except for slow changes on a geological time scale, the absorption through photosynthetic process by green plants and release of carbon from various sources was at equilibrium.

The worldwide economic growth and development basically requires energy. Among the many human activities that produce GHGs, the use of energy represents by far the largest source of emissions of GHGs, direct combustion of fossil fuels dominates the GHGs emissions from the energy sector. CO_2 results from the oxidation of carbon in fuels (www.iea.org). According to available data 2011 A.D, total CO_2 emissions from the consumption of energy is 32,578.645 Million metric tonnes (http://www.eia.gov). The world's current data for atmospheric CO_2 is measured at the Mauna Loa Observatory in Hawaii. Measurements are made and reported independently by two scientific institutions: Scripps Institution of Oceanography and the National Oceanic and Atmospheric Administration (NOAA). According to the recent data, the upper safety limit for atmospheric CO_2 is 350 parts per million (ppm) has crossed in 1988 itself and in July 2013 the atmospheric CO_2 raised to 398.58 ppm and projected to cross the 400 ppm mark (http://co2now.org) (Table 7.6).

Although some of the effects of increased CO_2 levels on the global climate are uncertain, most scientists agree that doubling atmospheric CO_2 concentrations may cause serious environmental consequences. The ten indicators of warming world are increasing the humidity, air temperature near surface (Troposphere), temperature over oceans, sea surface temperature, sea levels, temperature over land, ocean heat content, decreasing the glaciers, snow cover, and sea ice. Increasing global temperatures could raise sea levels, change precipitation patterns and affect both weather and climatic conditions. Currently, we are experiencing all the effects of global-warming and in future our next generations have to face.

Anything that removes carbon from the atmosphere is a 'sink'. In order to be effective in combating climate change, the sink must be large and the carbon must stay in the sink. Carbon is continuously exchanged between atmosphere, soil, ocean and life, which is predominately, plants. Since, CO_2 is important GHGs, one strategy that can partially combat global warming and climate change is to increase the amount of carbon stored in plants. By increasing the amount of plant life on earth, or altering it to plant types that store the most carbon, more carbon dioxide may be pulled out of the atmosphere and stored for a period of time through photosynthetic process. It is presumed that only 26 % (9.5 million metric tonnes) of CO_2 is trapped again in the photosynthetic process by green plants. However, real time data reveals the rapid land changing pattern i.e., clearing the forest lands, increasing the

Table 7.6 Ornamental plants capable of accumulating metals

Metal	Metal accumulation values	References
As	0.06–0.58 mg/g	Meera and Agamuthu (2011)
B		Bañuelos et al. (1993)
Cd		Carlson et al. (1982)
		Hiroyuki et al. (2005)
		Hattori et al. (2006)
	1.26 mg/Kg	Subramanian et al. (2012)
		Jyothi et al. (2003)
Cr		Carlson et al. (1982)
Cu		Carlson et al. (1982)
	87.5 mg/Kg	Jyothi et al. (2003)
	65.39 mg/Kg	Subramanian et al. (2012)
Fe		Carlson et al. (1982)
		Meera and Agamuthu (2011)
		Abioye et al. (2012)
		Subramanian et al. (2012)
Hg		Carlson et al. (1982)
Mg	410 mg/Kg	Subramanian et al. (2012)
Mn		Carlson et al. (1982)
		Subramanian et al. (2012)
		Kanchi et al. (2012)
Na	782.42 mg/Kg	Subramanian et al. (2012)
Pb		Ho et al. (2008), Kiliç et al. (2008), Rahi et al. (2013)
		Bada and Kalejaiye (2010)
		Jyothi et al. (2003)

agricultural lands (http://www.iaees.org) and burning fossil fuels more rapidly. Due to shrinking of plant life in due course of time the present percentage of carbon sequestration is still further reduces. This phenomenon forced to increasing the global temperature and rise of sea levels, which are significant effects of GHGs. In the process of human development further, industries are necessary and release of pollutants is obvious end products. Hence to combat and control the GHGs and other pollutants (inorganic and organic), huge plantation programme should be conducted and see that its survival in industrial zones as marginal lands. This helps in carbon sequestration as well as pollutants remediation from those industrial zones. The latest carbon sequestration method is digging the deep wells and release into the injection zone (i.e., 7,000 ft below the surface) (http://www.epa.gov). Taking the advantage of latest carbon sequestration method integrated approach of phytoremediation techniques in the vicinity of industrial zones is both way getting benefited i.e., release of GHGs are significantly reduced and effectively implement the phytoremediation.

Selection and plantation of right choice of plants for effective phytoremediation as well as carbon sequestration is important for obtaining the phyto products in industrial zones and contaminated abandoned lands. Revegetating with right choice of plants on former mining sites can provide phytoremediation services as well as carbon sequestration. Though, the process is cost effective it requires considerable time and should be employed at sites where remediation can occur over a long period of time. Generally, long term carbon sequestration can be achieved when carbon from above ground biomass transfers to the roots and enters the pool of Soil Organic carbon (SOC) or Soil Inorganic Carbon (SIC) which is possible for perennial trees and herbaceous plants with extensive root systems (Jansson et al. 2010) (Fig. 7.41).

Photosynthesis

$$6CO_2 + 6H_2O + 674\ \text{Kcal} \underset{\text{Solar energy}}{\overset{\text{Chlorophyll}}{\rightarrow}} \underset{\text{Carbohydrate 1 mol}}{C_6H_{12}O_6} + 6O_2$$

One of the important limiting factor of solar energy conversion by plants is their photosynthetic efficiency

$$IS \rightarrow P \rightarrow H \rightarrow C \rightarrow TC$$

IS = Insolation (in coming solar radiation)
P = Producer, H = Herbivore, C = Carnivore
TC = Top carnivore
PN = Q.β.ε-R (PN= Net Productivity)
Q = PAR, the quantity of incident light

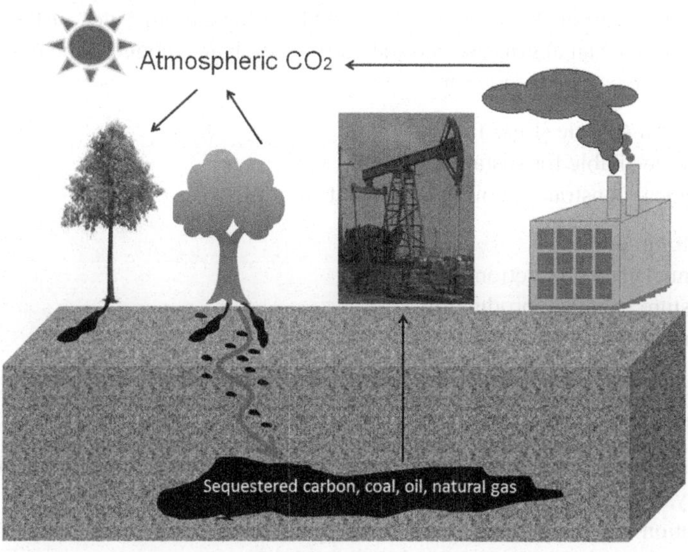

Fig. 7.41 Simplified carbon cycle. Unlike fossil fuels, biomass does not increase atmospheric green house gases when burned. Closed carbon cycle

β = the proportion of that light intercepted by green plant organs (canopy size, structure, pigments)
ε = the efficiency of photosynthetic conversion of the intercepted light into biomass
R = respiration

$$CO_2 + 2H_2O + \text{light}(8\,\text{quata}) \rightarrow (CH_2O) + H_2O + O_2 \text{where} CH_2O = 1/6\,\text{glucose molecule}$$

Gibbs free energy stored per glucose is 477 KJ and 8Q of light (400–700 nm monochromatic) of 575 nm is required.

8Q of light have energy content of 1,665 KJ, which gives maximum photosynthetic efficiency of 477/1,665 = 0.286

Maximum photosynthetic efficiency of PAR constitute about 43 %

Canopy absorb only 80 % available PAR

Respiratory losses and energy requirement for life accounts for 1/3 of stored energy = 66.7 %

Hence, over all photosynthetic efficiency of conversion of solar energy into stored chemical energy by terrestrial plants is $0.43 \times 0.8 \times 0.286 \times 0.667 = 6.6$ %

Therefore, in view of the above explained limitation, the only way is to use contaminated land for phytoremediation crops and thereby enhance carbon capture.

1. Strategies of wasteland development (NWDB)

   Remediation have been frequently used in the literature.
   Restoration: replication of site conditions prior to disturbance
   Reclamation: rendering a site habitable to indigenous organisms
   Rehabilitation: disturbed land will be returned to a form and productivity in conformity with a prior to land use plan
   Full restoration: Restoration of a site to its pre-damaged condition
   Partial restoration: Restoration of selected ecological attributes of the site and Creation of an alternative ecosystem type the latter though often desirable, is not to be called restoration.

2. Biotic and Abiotic stress factors
3. Species available for sustained yields
4. Ecological constraints/Soil management for

   - nutrient depletion
   - annual crop production
   - pasture livestock production
   - perennial and tree crops

5. Role of biofertilizers in restoring soil fertility, *Azobacter, Azospirillum Rhizobium*, Legume green manure

Also stated that, bioenergy crops (bioethanol, biogas, biodiesel, feedstock for electricity, charcoal) occupy a distinctive position in future terrestrial carbon sequestration and the vast areas of bioenergy cultivation envisioned for sustainable biofuel production (Smith et al. 2013), especially from perennial grasses and woody species, offer the potential for substantial mitigation of GHGs emissions both by displacing fossil fuels and through phyto-sequestration through extensive root systems. It is well known that most of these plants can accumulate pollutants by any of the phytoremediation process (Fig. 7.42).

Promising short rotation woody energy crops (SRWEC) are *Populus* ssp. (Imada et al. 2009; Sebastiani et al. 2004; Wu et al. 2010), *Salix* spp. (Lewandowski et al. 2006), *Liquidambar styraciflua, Platanus occidentalis, Robinia pseudoacacia, Acer saccharinum* L. and *Eucalyptus* can be grown for other uses also such as paper production and the waste can be utilized for energy (Capuana 2011; Claudia et al. 2012; Delplanque et al. 2013). *Hibiscus cannabinus* (Kenaf), *Brassica juncea* (Indian mustard), *Helianthus annus* (Sunflower), *Ricinus communis* (Castor), *Vetiveria zizanioides* (Khus Khus grass) *Prosopis juliflora* (Velvet Mesquite), are potential candidates for phytoremediation and carbon sequestration. Conversion of phytoremediation- borne biomass for (1) green-fine chemistry (catalyst production from

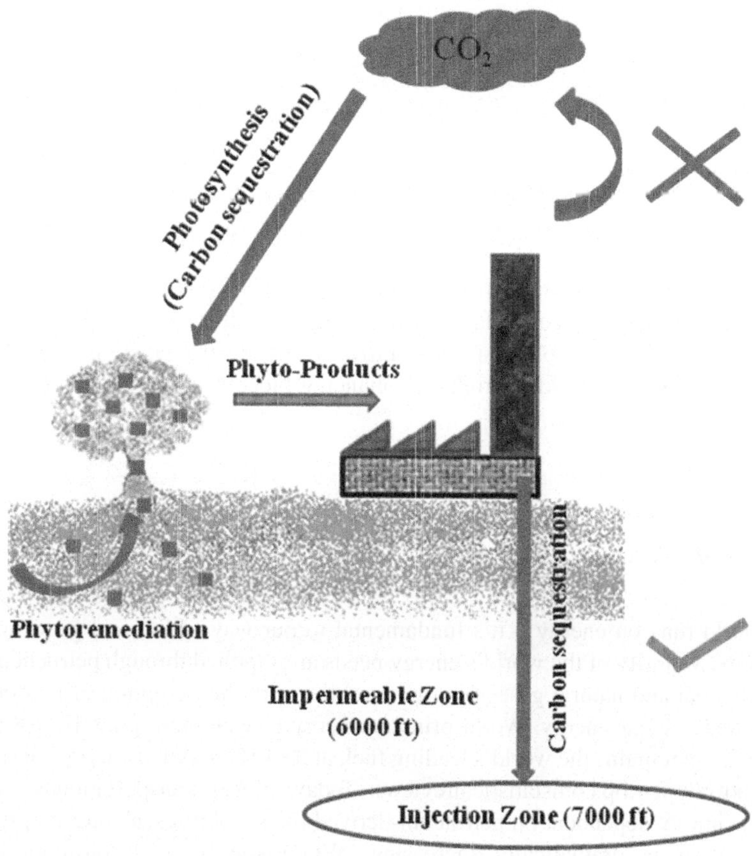

Fig. 7.42 Scheme showing carbon sequestration by phytoremediation crops

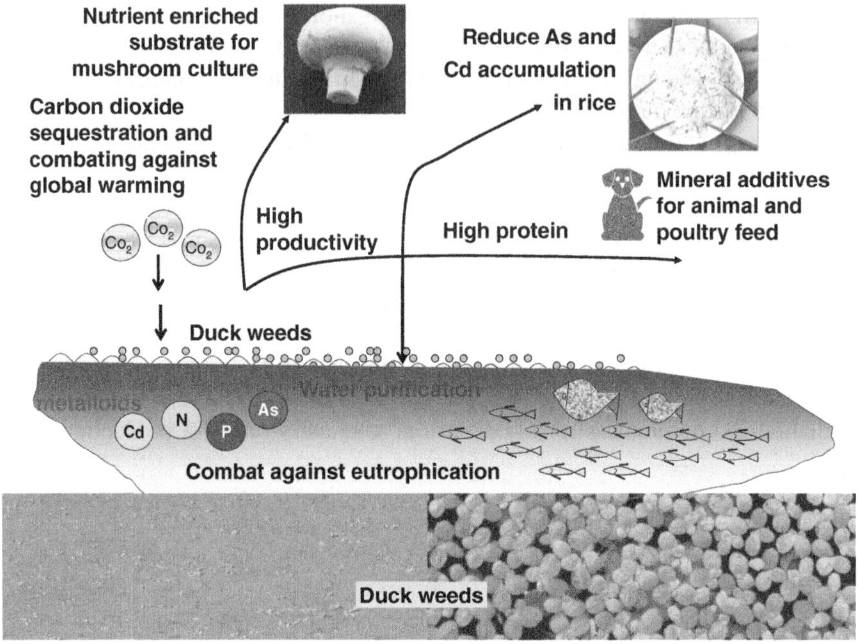

Fig. 7.43 Carbon sequestration by aquatic macrophytes and byproduct generation (Prasad 2012, © Springer Science+Business Media, LLC 2012)

metal accumulating biomass) (2) biorefinery (prehydrolysis and organosolv pretreatment from metal accumulating woody lignocellulosic biomass) and (3) by increasing the panel of plant species cultivated on metal contaminated soils for value chain and value added products enhancing bioeconomy are emerging fields (Fig. 7.43).

7.3.9 Non-edible Oil Plants for Bio-Fuels on Contaminated Soils

Our world runs on energy – it's fundamental to our way of life and growing our economy. Majority of the world's energy needs are supplied through petrochemical sources, coal and natural gases (non-renewable), with the exception of hydroelectricity and nuclear energy. World primary energy consumption grew by 1.8 % in 2012 and oil remains the world's leading fuel, at 33.1 % of global energy consumption (http://www.bp.comstatisticalreview). Today, global transportation sector is almost entirely dependent on petroleum-derived fuels and its escalating supply and consumption is dramatically increasing. While supply and consumption is increasing, its associated problems such as greenhouse gases especially CO_2 emission is also increasing (http://www.eia.gov/). Petroleum-based products are one of the main causes of anthropogenic carbon dioxide (CO_2) emissions to the atmosphere. Moreover, the increase in pollutants emissions from the use of petroleum fuel will affect human health, such as respiratory system, nervous system and skin diseases etc. The conventional source of energy is non-renewable resources and in the recent times the world has been confronted with an energy crisis due to rapid depletion of natural resources, increased social and environmental problems. It is important to develop suitable long-term strategies based on utilization of renewable fuel that would gradually substitute the declining fossil fuel production. Various groups of biomass generated and its uses are shown in Table 7.7.

Biodiesel is considered to be a possible substitute for conventional diesel getting the interest of scientists and workers all over the world. Biodiesel, mixture of fatty acid methyl esters (FAME), is generally produced from a varied range of edible (First generation feedstock) and non-edible (Second generation feedstock) vegetable oils, algae (Third generation feedstock), animal fats, used frying oils, and waste cooking oils and waste soap stocks from the oleo-chemical industries have been identified as a source of biodiesel feed-stock.

The biodiesel is quite similar to conventional diesel fuel in its physical characteristics and can be used as a direct substitute for petrodiesel and is technically called B100. The preferred ratio of mixture ranges between 5 % (B5) and 20 % (B20). Up to 20 % blending of biodiesel with diesel has shown no problems.

There are a great number of advantages as the primary feedstock can grow season after season and it provides a market for excess production of vegetable oils, and animal fats, thus enhancing the rural economies. Also for using biodiesel apart from reduces the country's dependence on imported petroleum and specifically it is biodegradable, non-toxic, renewable and closed carbon cycle. Also it reduces emission

Table 7.7 Metal accumulation capacity in selected high biomass producing ornamentals (Kaiser et al. 2009)

	References
Canna indica	
Cd, Cr, Cu, Fe, Mn, Ni, Pb, Zn	Bose et al. (2008)
Co, Cr, Cu, Ni, Zn	Yadav et al. (2012)
Cr, Ni	Yadav et al. (2010)
Ipomoea carnea	
Cd	Ghosh and Singh (2005)
Cd, Cr, Cu, Mn, Ni, Pb,	Pandey (2012a, b)
Alternanthera sessilis	
Cd, Cr, Cu, Fe, Mn, Pb	Rai et al. (1995)
Cd, Cr, Cu, Fe, Mn, Pb	Marchand et al. (2010)
Cd, Cr, Pb	Chandran et al. (2012)
Cd, Cu, Ni, Pb, Zn	Premarathna et al. (2011)
Cr	Sinha et al. (2002)
Cr, Na	Bareen and Tahira (2011)
Cr, Ni, Pb	Moodley et al. (2007)

of CO, CO_2, SO_2, particulate matter, volatile organic compounds and unburned hydrocarbons as compared to conventional diesel. In addition it has higher cetane number and flash point greater than 423 K as compared to 350 K for petroleum based diesel fuel. World production of various sources of vegetable oils in 2012–2013 was 160.22 Million metric tonnes and mostly used for consumption purposes (www.fas.usda.gov). However, in different countries of the world, same vegetable oils resources are used for biodiesel production. The source for biodiesel production is chosen according to physico-chemical properties, production cost, transportation and policy. In USA, Europe, Brazil other parts, 95 % used type of biodiesel fuel is first generation feedstock i.e., edible oils like soybean, rapeseed, sunflower, safflower, canola, palm, Coconut and fish oils are used to reduce air pollution and dependency on fossil fuel, which are limited and localized to specific regions (Jeonng and Park 2008; Darnoko and Cheryman 2000; Vicente et al. 2004; Ramadhs et al. 2004; Cheng et al. 2004; de Oliveira et al. 2005; El Mashad et al. 2008; Sarin and Sharma 2007; Meka et al. 2007).

Although, biodiesel are mainly produced in many regions recently, environmentalists have started to debate on the negative impact of biodiesel production from edible oils (Butler 2006). Recently, the use of edible vegetable oils or the first generation feed stocks has been of great concern and the major obstacle for commercialization of biodiesel is its cost from the feed stocks (Canakci and Van Gerpen 2001). Cost of edible oils is very higher than petroleum diesel and we use edible oils for biodiesel production leads food oil crisis. This is because they raise many concerns such as food versus fuel debate that might cause starvation especially in the developing countries and other environmental problems caused by utilizing much of the available arable land. This problem can create serious ecological imbalances as countries around the world began cutting down forests for plantation purposes and perfect examples is in countries like Malaysia, Indonesia and Brazil. In fact, the real time world land alterations can be glanced at on http://www.iaees.org.

Taking these factors into consideration, the above problem can be solved by using cheapest, low cost non edible oils which are not suitable for human consumption because of the presence of some toxic secondary metabolites in their oils. Non edible oil crops can be grown in waste lands and the cost of cultivation is much lower because these crops can still sustain reasonably high yield without intensive care.

Exploring the availability of non-edible oil seed as alternative biodiesel feedstock in the transportation sector is critical towards achieving higher self-reliance energy security. This situation offers a challenge as well as an opportunity to look for replacement of fossil fuels as well as first generation feedstock of biodiesels for both economic (Food *verses* Fuel) and environmental benefits. The following are the advantages of non-edible oil:

Non-edible vegetable oils are not suitable for human consumption due to the presence of some toxic compounds in their oils. Non-edible oil plants are well adapted to arid, semi-arid, high rain fall zone conditions and require low fertility and moisture and to grow. Moreover, Non-edible biodiesel crops are expected to use lands that are largely unproductive and those that are located in poverty stricken areas and in degraded forests which can fix upto 10 t/ha/year CO_2 emissions. They can also be planted on cultivators' field boundaries, fallow lands, and in public land such as along railways, roads and irrigation canals. They do not compete with existing less farm land and agricultural resources. Hence they eliminate competition for food and feed. Non-edible biodiesel plant cultivation as mentioned above could become a major poverty alleviation program for the rural poor apart from providing energy security. Moreover, they can be propagated through seed or cuttings. Most of the non-edible oils are highly pest and disease resistant. Non-edible feed stock can produce useful by-products during the conversion process, which can be used in other chemical processes or burned for heat and power generation.

7.3.10 Bioenergy from Phytoremediated Phytomass via Gasification

Phytomass gasification is basically the conversion of wood chips and wood waste into a combustible gas mixture called as producer gas (Low Btu gas). This process involves partial combustion of phytomass. Given that phytomass contains carbon, hydrogen and oxygen, complete combustion would produce carbon dioxide and water vapour. Partial combustion produces carbon monoxide and hydrogen both of which are combustible gases. The gas thus generated in a "Gasifier" could be used for captive power generation (a few KWs to several hundred KWs). Fuelwood direct combustion has several disadvantages owing to its limited applications and is of low efficiency. Direct thermal applications for power generation require steam boilers with steam engine or turbine along with necessary equipments. Such a conversion technology is not only capital intensive but is of low conversion efficiency. Conversion of the same phytomass (wood chips/ wood waste) to combustible gases in gasifier and its utilisation for power generation has innumerable advantages. Phytomass gasification (using gasifiers of various capacities) generates electricity, and the gas can be utilised for direct thermal energy and for shaft power (Fig. 7.44).

7.3.10.1 Aquatic Biomass

Biomass of *Eichhornia crassipes* (water hyacinth), *Pistia statoides* (water lettuce) and other aquatic weeds can be used for gasification (Figs. 7.45–7.49). Macrophytes spread rapidly and clog the aquatic systems (Bi et al. 2011). Therefore, utilization of such freely available biomass for bioenergy would be a happy solution. Aquatic weed biomass when subjected to anaerobic digestion optimum yield of gas was recovered when the C/N ration was between 20 and 30. One hectare of water hyacinth grown on sewage can purify the wastewater and produces 0.8 ton of dry matter per day which can be converted to 200 cu m gas that is enough for generating 250 kw power.

It is possible for generating 100 L of biogas from 1 kg of dry matter under controlled conditions. In China and Japan, large scale sea weed farms are under operation. *Macrocystis* (Giant kelp) is grown for fuel production. The kelp yields are represented to be of the tune of 90 tons/ha/year. In USA, California expects an area

Fig. 7.44 (**a–d**) Harvested phytoremediated aquatic weed biomass is used as feed stock for biogas in anaerobic digester

Fig. 7.45 (**a–c**) Profuse growth of water hyacinth (*Eichhornia crassipes*) in polluted water. Harvested biomass is used as feed stock for biogas (please see also Fig. 7.22)

Fig. 7.46 Phytofilteration of toxic metals in aquaculture ponds using *Pistia stratiotes* (Prasad and Prasad 2012, © Walter de Gruyter, Berlin, Boston)

Fig. 7.47 After harvesting *Pistia stratoites* the water body is used for fish culture – *white arrows* show feed bags suspended with the help of rope across the water body

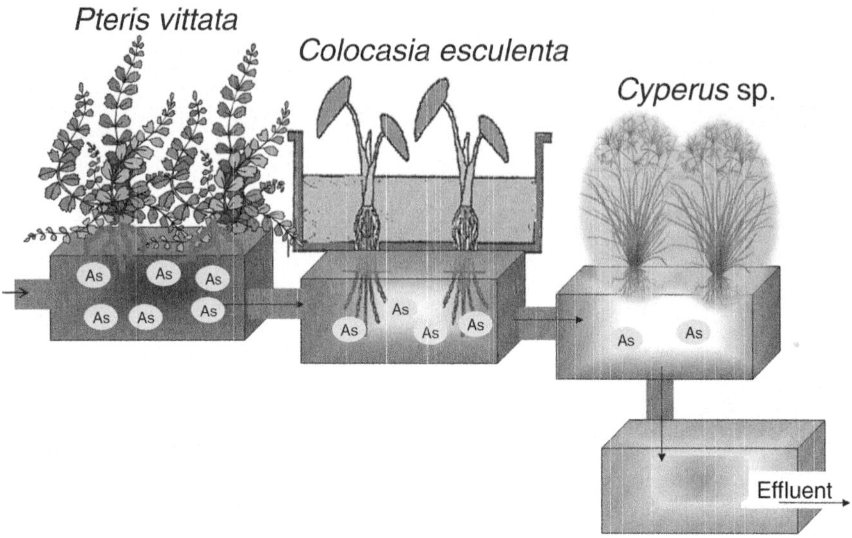

Fig. 7.48 Arsenic accumulating ornamentals are being used for treating arsenic contaminated water e.g. *Pteris vittata, Colocasia esculenta* and *Cyperus* sp.) (Kurosawa et al. 2008; Nakwanit et al. 2011; Prasad and Nakbanpote 2015, © Springer India 2015)

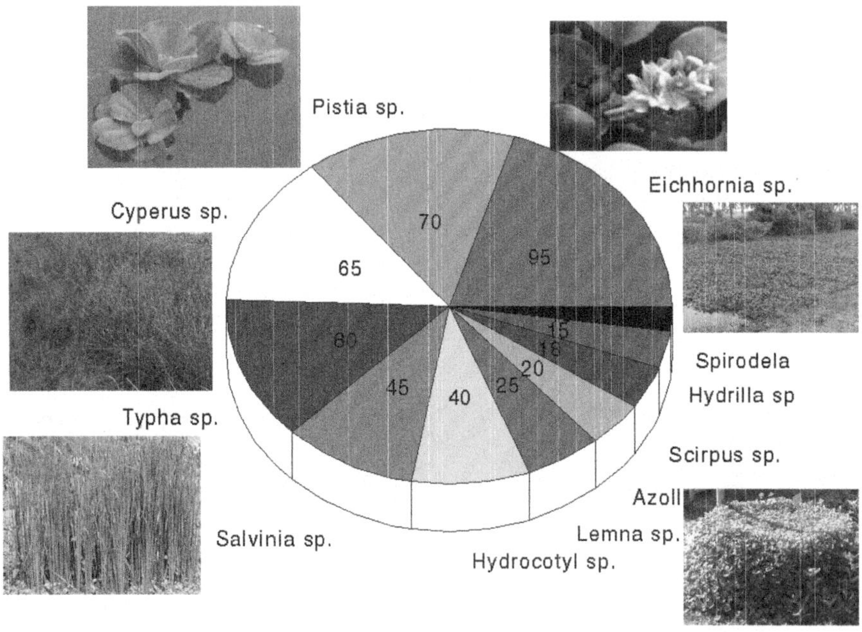

Fig. 7.49 Productivity and relative proximate dry matter production of potential macrophytes being applied for phytoremediation (Prasad et al. 2001)

of 40,000 ha of ocean energy farms by the turn of the century. Each hectare of cultivated kelp would yield about 10 million kilo calories of oil and about a 100 kilo calories of methane energy per year.

Plants endophytes relationships and the microbial communities play a key role in degrading the hazardous contaminants in rhizosphere to varying extents (Reichenauer and Germida 2008; Husain et al. 2009; Scow and Hicks 2005).

Arsenic contamination in Bengal Delta region is prevalent. The first known arsenic hyper accumulating plant is *Pteris vittata.* It is also known as Chinese brake fern, was discovered from an arsenic-contaminated site that was contaminated from pressure-treating lumber using chromated-copper-arsenate (CCA). *P. vittata* is reported to accumulate 22 g kg^{-1} of arsenic in its fronds (Carrier et al. 2011, 2012; Shelmerdine et al. 2009; Ye et al. 2011).

In addition to *P. vittata* and *P. cretica*, several other arsenic hyperaccumulating plants have been reported recently including *Pityrogramma calomelanos, Pteris longifolia* and *Pteris umbrosa.* Pilot-scale demonstration of phytofiltration for treatment of arsenic has been demonstrated in New Mexico for production of drinking water (McCutcheon and Schnoor 2003) (Fig. 7.24).

Bioconcentration factor: Bioconcentration of heavy metal by water weeds is described as the bioconcentration factor (BCF), which is the ratio of heavy metal accumulated by plants to that dissolved in the surrounding medium. For this, two bioconcentration factors were computed from the plant compartment concentrations as

$$BCF_R = CR_{root} / C_{water} \tag{7.1}$$

$$BCF_L = C_{leaf} / C_{water} \tag{7.2}$$

The translocation of heavy metal from the roots to harvestable aerial part is generally expressed as the translocation factor (TF) (Li et al. 2009, 2011, 2014). It was calculated on a dry weight basis by dividing the heavy metal concentration in aerial parts by the heavy metal concentration in root. Based on the above two equations (7.1) and (7.2), the translocation factor can be expressed as:

$$\text{Translocation factor}\left[TF\right] = BCF_{leaf} / BCF_{root} \tag{7.3}$$

Plants with efficient phytofiltration of toxic metals are beneficial for cleanup of contaminated fish ponds (*Pistia stratiotes* = Water lettuce/Water cabbage). For metal contaminants, plants with high potential for phytoextraction (uptake and recovery of contaminants into above-ground biomass) are desirable (Cao et al. 2007; Prasad 2007a, b) (Fig. 7.50)

Hazardous wastewater utilization for high rate algal ponds for production of biodiesel is being considered. Algae growing in wastewater treatment, high rate algal ponds [HRAPs] assimilate nutrients and thus subsequent harvest of the algal biomass recovers the nutrients from the wastewater (Figs. 7.51–7.54).

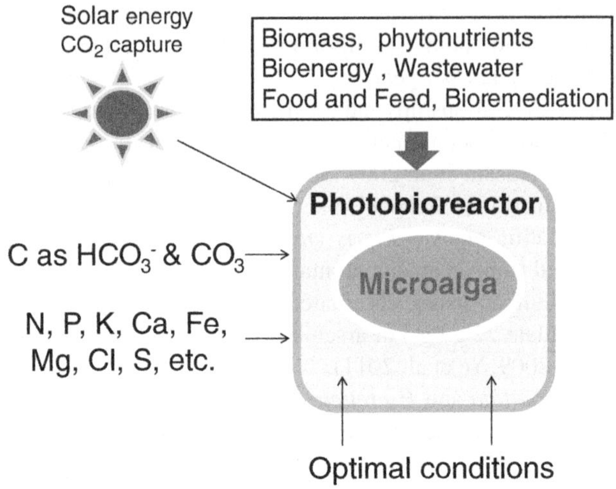

Fig. 7.50 Schematic view of photobioreactor for cultivation of algae in hazardous waste water (Chisti 2008)

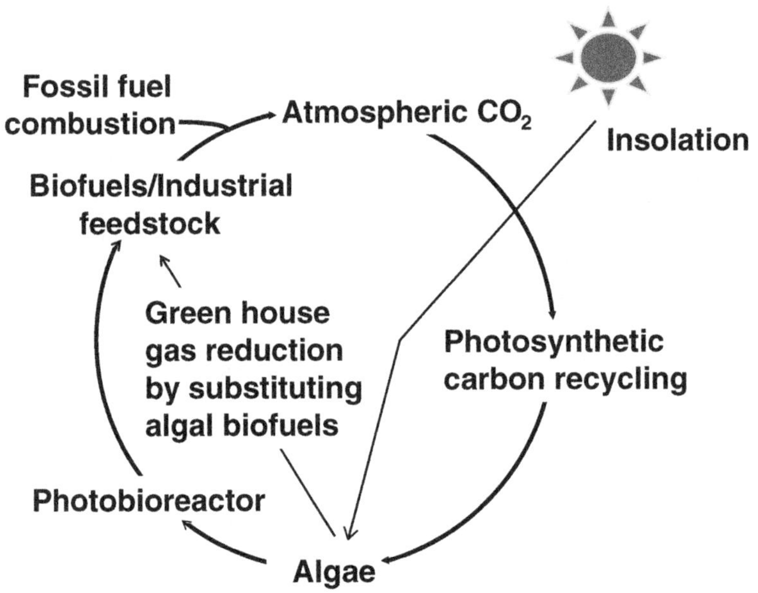

Fig. 7.51 Algae based photobioreactor for cultivation of algae in hazardous waste water for production of bio fuels, industrial feedstock and carbon dioxide sequestration. Leachate from hazardous waste dump sites are being used to establish **High Rate Algal Ponds** [HRAPs] for production of biodiesel (Park et al. 2011). The harvested phytomass of aquatic plants used for bioremediation serves as a valuable feedstock for biogas production (Abbasi et al. 1991) (Fig. 7.53)

Fig. 7.52 Nutrient rich and sewage water treatment in constructed engineered wetland using high biomass produced macrophytes viz. *Typha* and *Phragmites*. Production of low cost roofing material and compost from macrophytes used in nutrient rich and sewage water treatment in constructed engineered wetland (Zhang et al. 2014; Zheng et al. 2013)

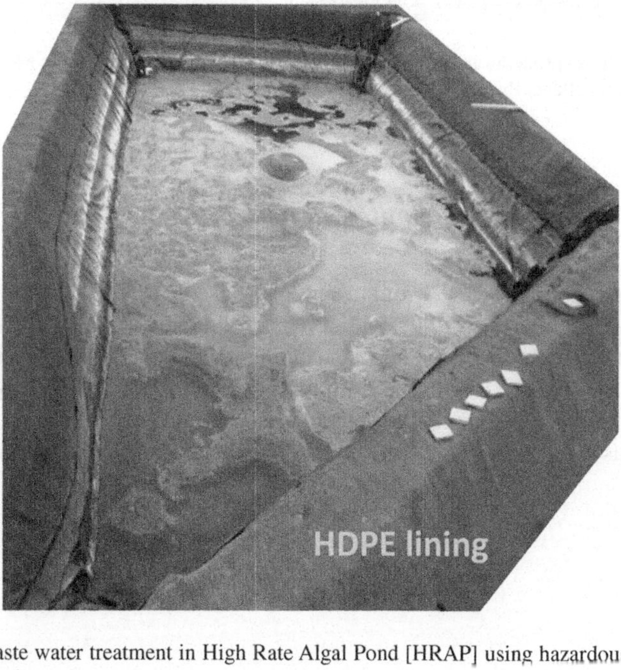

Fig. 7.53 Waste water treatment in High Rate Algal Pond [HRAP] using hazardous waste water (Lim et al. 2013)

Fig. 7.54 Biodiesel production using algae and waste water at lab scale (Source: Mr M.V. Bhaskar "Kadambari Consultants Pvt Ltd - Nualgi")

7.3.11 Stabilization of Contaminated Lands by Sodding

Mulberry is an economically important tree, used for feeding the silkworm, *Bombyx mori*. Mulberry exhibits spatial localization and serves as sink for Sr, Zn and Cd in leaves along with Ca and Si accumulation. It is the sole food source for the silk worm (Ashfaq et al. 2009; Hegde and Fletcher 1996; Katayama et al. 2013). The productivity of leaf biomass on coal mine over burden is quite high and serves as a resource. An estimate of 40 tons/ha/year (approximately 10 tons of dry matter) has been reported. Based on the yield data, the amount of Sr deposition in leaves is roughly estimated to be 300 kg/ha/year. Mulberry leaves posses not only a high Sr sink but also yield high biomass, indicating that mulberry is one of the most efficient bioaccumulator plants for Sr and can be used in cleaning or rehabilitating soil contaminated by radiostrontium (Fig. 7.55).

Sodding is a bioengineering technique that uses vegetation mats for mine soil stabilization and erosion control. This technique has been used with success with a variety of grasses (Bonanno 2012; Bonanno et al. 2013), Perennial grasses with sturdy adventitious root mat anchor and stabilize mine soils. The following is a list of recommendations for using vegetation mats as bioengineering materials (Fig. 7.56)

- To anchor perennial solid binding grass mats to a slope, mats can be cut to form any shape desired. A shallow, narrow trench built along the contour of a slope and planted with a vegetation mat may become an effective terrace.
- The mat should remain attached to stable vegetation and thus be held in place from the top. The mat can be pegged to prevent ripping and sliding. This technique could be used to stabilize the contaminated soil.
- Vegetation mats can be used as building bricks. Slice the mats into rectangular pieces and use them to construct a very steep, living wall. The bricks can be pegged to each other and to the underlying substrate. This technique may be useful around culverts or sunken walkways and controls erosion.

Tables 7.8–7.17 show various examples of biomass generation in phytoremediation and their possible economic uses.

7.4 Conclusions

Although phytoremediation is environmentally friendly, powerful and low-cost technology, one general belief is that it might take long time to clean-up inorganics in soil. However, it is relatively convenient to clean small water bodies and waste waster. Most of the previous researches have focused only on wild selected species (non-economic crops) that can tolerate and takeup large amounts of inorganic contaminants to increase the efficiency of phytoremediation. Much progress has been

Fig. 7.55 Luxurient growth of *Morus alba* (mulberry) on coal mine over burden of west bokaro coal field

made in countries like UK, USA, Canada, Australia, Japan and many European countries. In developing country like India, it is difficult to convince environmental regulators and local agencies to grow metal accumulators (the data on hyperaccumulators is scanty) in the contaminated areas for the sole purpose of removing pollutants from their environment unless financial renumeration or expenses are subsidized.

Fig. 7.56 Sodding, a bioengineering technique that uses grass mats for soil stabilization and erosion control. Perennial grasses are preferred. Grasses with sturdy adventitious root mat stabilize metalliferous soil. To anchor perennial solid binding grass mats to a slope, mats are cut to desirable sizes (1 × 1 foot). This technique would beneficial to stabilize the contaminated soil (Thangavel and Sridevi 2015, © Springer India 2015) (**a**) Cultivation of Zoysia grass with contaminated substrated (soil and water) (**b**) Harvested grass mats are ready for sale (**c**) and (**d**) road cutting are stabilized with grass mats

Table 7.8 Metal accumulation in *Ricinus communis* (Castor bean)

Ag, Cd, Cu, Pb	Figueroa et al. (2008), de Souza Costa et al. (2012)
As	Santos-Jallath et al. (2012)
As, Cd, Co, Cr, Cu, Mn, Ni, Pb	Varun et al. (2012)
B, Cu, Fe, Mn, Zn	de Abreu et al. (2012)
Ba	Coscione and Berton (2009)
Cd	Shi and Cai (2009), Huang et al. (2011), Prabavathi et al. (2011), Bauddh and Singh (2012a, b)
Cd, Cr, Cu, Fe, Mn, Ni, Pb, Zn	Singh et al. (2010)
Cd, Cu, Mn, Ni, Pb, Zn	Olivares et al. (2013)
Cd, Cu, Mn, Pb, Zn	Olivares et al. (2013)
Cd, Cu, Pb, Zn	Chaudhry et al. (1998), Olivares et al. (2013)
Cd, Pb	Zhi-xin et al. (2007), Niu et al. (2009), de Souza Costa et al. (2012)
Cd, Pb, Zn	Sas-Nowosielska et al. (2008)
Cu, Fe, Mn, Zn	Stephan et al. (1994), Schmidke and Stephan (1995), Stephan et al. (1995)
Cu, Pb, Zn	Xiaohai et al. (2008), Nazir et al. (2011)
Cu, Zn	Chaves et al. (2010)
Cu, Zn, Fe	Khanam and Singh (2012)
Hg	Siegel et al. (1984)
Mn, Ni, Pb, V	Vwioko et al. (2006)
Ni	Giordani et al. (2005)
Pb	Romeiro et al. (2006)

Table 7.9 *Prosopis juliflora*, a phreatophyte and its capability for accumulation of heavy metals

Al, As, Au, Ba, Br, Cl, Ce, Cs, Cu, Fe, Hf, In, K, La, Mg, Mn, Na, Sb, Sc, Sm, Th, Ti, U, V, W, Yb, Zn,	Gabriel and Patten (1994)
Al, B, Ba, Ca, Cr, Cu, Fe, K, Li, Mg, Mn, Na, Ni, Sr, V, Zn,	Nagaraju and Prasad (1998)
Al, Cr	Jamal et al. (2006)
As	Mokgalaka-Matlala et al. (2008, 2009)
As, Cd, Cr, Cu, Mn, Pb, Zn	Solís-Domínguez et al. (2011)
As, Cd, Cu, Pb, Zn	Al-Farraj and Al-Wabel (2007)
As, Cr, Cu, Mo, Zn	Haque et al. (2009)
B, Ba, Co, Cr, Cu, Fe, Mn, Mo, Ni, Sr, V, Zn	Chaudhary et al. (2009)
Cd	Khan (2007)
Cd, Cr, Cu, Fe, Mn, Ni, Pb, Zn	Shukla et al. (2011)
Cd, Cu, Zn	Usha et al. (2009)
Cd, Pb	Varun et al. (2012)
Ce, Zn,	Viezcas (2009)
Co, Fe, Mn, Ni	Naveed et al. (2012)
Cr	Arias et al. (2010)
Cr, Cu, Fe, Mn, Zn	Rai et al. (2004), Sinha and Gupta (2005)
Cr, Cu, Fe, Pb, Zn	Atiq-Ur-Rehman and Iqbal (2008)
Cu, Cd	Senthilkumar et al. (2005)
Cu, Pb, Zn	Iqbal et al. (1999)
F	Saini et al. (2012), Baunthiyal and Sharma (2012)
Pb	Naveed et al. (2010)
Zn	Hernandez-Viezcas et al. (2011)

Table 7.10 Fiber yielding plants that can be grown on contaminated sites (Bjelková et al. 2011; Griga and Bjelkova 2013; Linger et al. 2002; Smykalova et al. 2010)

Plant Source	Description
Seed fiber	Collected from seeds or seed cases (Eg: *Gossypium Sps., Ceiba pentandra.*
Leaf fiber	Collected from leaves (Eg: *Furcraea andina* (introduced in India), *Agave* Sps. *Sansevieria roxburghiana*, *Sansevieria hyacinthoides* (Bowstring Hemp).
Bast fiber (or) Skin fiber	Collected from the skin or bast surrounding the stem of their respective plant. These fibers have higher tensile strength than other fibers. Therefore, these fibers are used for durable yarn, fabric, packaging, and paper. Some examples are Flax (produces linen) *Linum usitatissimum*, Jute (widely used, cheapest fiber after cotton) *Corchorus capsularis, C. olitorius* Kenaf (The interior of the plant stem is also used for fiber.) *Hibiscus cannabinus,* Hemp (A soft, strong fiber, edible seeds.) *Cannabis sativa, Boehmeria nivea* 600 species of palms in the tribe Calameae (rattan) and vine fibers.
Fruit fiber	Collected from the fruit of the plant (Eg: *Cocos nucifera*).
Stalk fiber	Stalks of the plant (Eg: straws of *Triticum* sps. *Oryza sativa Hordeum vulgare* and other crops including bamboo and grass.
Other fibers	Bamboo fiber, Grewia optiva, Himalayan Nettle (*Urtica dioica* L.) Bhabar (*Eulaliolopsis binata*).

Table 7.11 Metal accumulation in *Vetiver zizanioides*

Metals	References
Al, As, B, Ba, Ca, Cd, Co, Cu, Fe, Hg, K, Mg, Mn, Mo, Ni, Pb, Rb, Sr, TI, Zn	Martin et al. (2010)
As	Datta et al. (2011)
As, Ca, Co, Cr, Cu, Fe, Hg, Mg, Mn, Na, Ni, Zn	Bhat et al. (2010)
As, Cd, Cr, Cu, Hg, Ni, Pb, Se, Zn	Danh et al. (2009)
As, Cd, Pb	Oshunsanya et al. (2012)
As, Cu, Zn	Chiu et al. (2005)
B, Pb	Angin et al. (2008)
Cd	Chen et al. (2000), Aibibu et al. (2010)
Cd, Cu, Mn, Pb, Zn	Roongtanakiat and Chairoj (2001)
Cd, Cu, Pb	Jayashree et al. (2011)
Cd, Cu, Pb, Zn	Yang et al. (2003), Chena et al. (2004)
Cd, Pb	Minh and Khoa (2009)
Cd, Pb	Xia (2004)
Cd, Pb, Zn	Lai and Chen (2004), Zhuang et al. (2005), Roongtanakiat and Sanoh (2011)
Cr, Cu, Ni, Pb, Zn	Roongtanakiat et al. (2003)
Cr, Cu, Pb, Zn	Antiochia et al. (2007)
Cs, Sr	Singh et al. (2008)
Cu	Liu et al. (2009)
Cu, Fe, Mn, Pb, Zn	Roongtanakiat et al. (2007)
Cu, Fe, Mn, Zn	Roongtanakiat et al. (2008, 2009)
Cu, Pb, Zn	Chiu et al. (2006), Danh et al. (2011), Chen et al. (2012)
Pb	Chantachon et al. (2004), Wilde et al. (2005), Rotkittikhun et al. (2007), Rotkittikhun et al. (2007), Gupta et al. (2008), Andra et al. (2009, 2010, 2011), Punamiya et al. (2010)
Pb, Zn	Pang et al. (2003)

Table 7.12 Dye yielding plants capable of growing on contaminated soils

Plant name	Family	Parts used	Colour obtained
Abrus precatorius	Fabaceae	Seeds	Black
Acacia catechu	Mimosaceae	Bark	Brown/black
Acacia catechu var. *sundra*	Mimosaceae	Wood	Reddish brown
Acacia leucophloea	Mimosaceae	Bark, leaves	Red
Acacia nilotica	Mimosaceae	Bark and pods	Yellowish brown
Acacia nilotica	Mimosaceae	Seeds	Brown/black
Acanthophonax trifoliatum	Araliaceae	Fruit	Black
Achyranthes aspera	Amaranthaceae	Whole plant	Black-brown
Achyranthes bidentata	Amaranthaceae	Whole plant	Black-brown
Actaea acuminate	Ranunculaceae	Seeds	Reddish-green
Adenanthera pavonina	Mimosaceae	Wood	*Red*
Adhatoda vasica Nees	Acanthaceae	Leaves	Yellow
Adhatoda vasica Nees.	Acanthaceae	Leaf	Yellow
Adhatoda zeylanica.	Acanthaceae	Leaves	Yellow-green
Aegle marmelos	Rutaceae	Fruits	Yellow
Aegle marmelos	Rutaceae	Rind of the fruit	Reddish
Aesculus indica	Hippocastanaceae	Bark	Brown
Agrimonia pilosa.	Rosaceae	Roots	Yellow
Alnus glutinosa	Betulaceae	Bark	*Black*
Alnus nepalensis	Betulaceae	Bark	Red
Aloe barbadensis	Liliaceae	Whole plant	Red
Alpinia galanga	Zingiberaceae	Root, stalk	Yellow-brown
Althaea rosea	Malvaceae	Flowers	Red dye
Amaranthus hypocondriacus	Amaranthaceae	Arial parts	Red pigment
Ampelocissus latifolia	Vitaceae	Leaves	Black
Anacardium occidentale	Anacardiaceae	Pericarp	Light red
Annona reticulata.	Annonaceae	Fruit, shoots	Bluish black
Arnebia benthamii	Boraginaceae	Roots	Red
Artemisia japonica	Asteraceae	Leaves	Brown
Artemisia nilagirica	Asteraceae	Leaves	Brown
Artocarpus heterophyllus	Moraceae	Wood/fruits	Yellow
Artocarpus lakoocha	Moraceae	Wood/fruits	Yellow
Averrhoa carambola.	Oxalidaceae	Fruits	Yellow/brown
Azadirachta indica	Meliaceae	Bark	Brown
Bauhinia purpurea	Caesalpiniaceae	Bark	Purple colour
B. racemosa	Caesalpiniaceae	Bark	Light green
Bauhinia variegate.	Caesalpiniaceae	Flowers	Purple
Benthamidia capitata	Cornaceae	Fruits	Red
Berberis aristata	Berberidaceae	Bark roots	Yellow
Berberis asiatata	Berberidaceae	Bark/roots	Yellow
Berberis chitria	Berberidaceae	Bark/roots	Yellow
Beta vulgaris	Amaranthaceae	Roots	Red

(continued)

Table 7.12 (continued)

Plant name	Family	Parts used	Colour obtained
Bischofia javanica	Euphorbiaceae	Bark/seeds	Black
Bixa orellana	Bixaceae	Seeds	Red/pink
Bougainvillea glabra	Nyctaginaceae	Flower with ivory	Yellow brown
Brugmansia suaveolens	Solanaceae	Leaves	Green
Butea monosperma	Fabaceae	Dried flowers	Brilliant yellow dye
Butea superba	Fabaceae	Root	Red
Butea superba	Fabaceae	Root	Yellow
Caesalpinia sappan	Caesalpiniaceae	Wood and pod	Red
Capsicum annuum	Solanaceae	Fruits	Red
Careya arborea	Juglandaceae	Bark	Yellow
Carthamus tinctorious	Asteraceae	Flower	Yellow, red
Carthamus tinctorius.	Asteraceae	Flowers	Red and yellow
Cassia auriculata	Caesalpiniaceae	Flower, seeds	Yellow
Cassia fistula	Caesalpiniaceae	Bark/fruits	Brown
Cassia fistula	Caesalpiniaceae	Bark and sapwood	Red
Cassia tora	Caesalpiniaceae	Seeds	Blue
Casuarina equisetifolia	Casuarinaceae	Bark	Light reddish
Celtis australis	Ulmaceae	Bark	Yellow
Ceriops tagal	Rhizophoraceae	Bark	Black, brown or purple
Chrozophora tinctorial	Euphorbiaceae	Herb	Light green
Cinnamonum tamala	Lauraceae	Leaves	Brown
Cladonia verticullata	Cladoniaceae	Whole plant	Yellow-red
Commelina benghalensis	Commelinaceae	Juice of the flower	blue
Convallaria majalis	Liliaceae	Leaves and stalk	Green
Corylus jacquemontii	Betulaceae	Fruits rind	Camel
Crocus sativus	Iridaceae	Flower	Yellow, orange
Cupressus torulosa	Cupressaceae	Leaves	Green
Curcuma angustifolia	Zingiberaceae	Tubers	Yellow
C. aromatica	Zingiberaceae	Rhizome	Yellow
C. domestica	Zingiberaceae	Rhizome	Yellow
C.longa	Zingiberaceae	Rhizome	Yellow
Curcuma zedoaria	*Zingiberaceae*	Rhizome	Yellow
Cyperus scariosus	Cyperaceae	Roots	Brown
Daphne papyracea.	Thymelaeaceae	Bark/fruits	Red
Datisca cannabina Linn.	Datiscaceae	Roots	Yellowish-red
Dipterocarpus spp.	Diptero-carpaceae	Bark	Light brown
Elaeodendron glaucum	Celastraceae	Bark	Red
Emblica officinalis	Euphorbiaceae	Fruits	Brown
Engelhardtia spicata	Juglandaceae	Bark	Dark brown
Enonymus tingens.	Celastraceae	Bark	Yellow
Erythrina suberosa.	Fabaceae	Flower/bark	Dark brown

(continued)

Table 7.12 (continued)

Plant name	Family	Parts used	Colour obtained
Eugenia jambolana.	Myrtaceae	Bark, leaf	Red
Everniastrum cirrhatum	Parmeliaceae	Whole plant	Red brown
Galium aparine	Rubiaceae	Root	Purple
Garcinia mangostana	Clusiaceae	Fruit	Black
Geranium nepalense	Geraniaceae	Roots	Red
Geranium wallichianum	Geraniaceae	Roots	Red/brown
Grevillea robusta.	Proteaceae	Flowers	Yellow
Grewia optiva	Tiliaceae	Fruits	Yellow/orange
Grewia subinaequalis	Tiliaceae	Fruits	Yellow/orange
Haematoxylon campechianum	Caesalpiniaceae	Heart wood	Yellow
Hedychium spicatum	Zingiberacae	Rhizome	Yellow
Hippophae salicifolia	Elaegnaceae	Fruits	Yellow
Hypericum oblongifolium	Hypericaceae	Flowers	Yellow
Impatiens balsamina	Balsaminaceae	Flowers	Red
Impatiens balsamina	Balsaminaceae	Flower	Brown
Indigofera atropurpurea	Fabaceae	Flowers	Purple
Indigofera cassioides	Fabacae	Leaves/flowers	Blue
Indigofera heterantha	Fabacae	Flowers	Pink
Indigofera tinctoria	Fabacae	Leaves/flowers	Blue
Indigofera tinctoria	Fabaceae	Green crop	Blue
Isatis tinctoria	Brassicaceae	Leaves	Dark blue
Juglans regia	Juglandaceae	Bark/fruits	Camel-brown
Lannea coromandelica	Anacrdiaceae	Bark/resin	Yellow-brown
Largestomia parviflora	Lythraceae	Bark	Black
Lawsonia alba	Lythraceae	Leaves	Brown
Lawsonia inermis	Lythraceae	Leaves	Red/orange
Ligustrum vulgare.	Oleaceae	Mature berries	Blue
Madhuca indica	Sapotaceae	Bark	Reddish-yellow
Madhuca longifolia	Sapotaceae	Bark	Yellow-brown
Mahonia borealis	Berberidaceae	Bark/roots	Yellow
Mallotus philippensis	Euphorbiaceae	Fruits	Red
Mallotus phillippensis	Euphorbiaceae	Fruit capsules	Orange
Mangifera indica	Anacardiaceae	Bark/leaves	Yellow
Michelia champaca	Magnoliaceae	Wood	Yellow
Mimusops elengi	Sapotaceae	Seed	Yellow
Mirabilis jalapa	Nyctaginaceae	Flowers	Pink-red
Morinda citrifolia	Rubiaceae	Root bark	Dull red
Myrica esculenta	Lauraceae	Bark/fruits	Red yellow
Nardostachys grandiflora	Valerianaceae	Inflorescence	Red
Nyctanthes arbor-tristis	Oleaceae	Flower	Yellow
Nyctanthes arbourtritis	Oleaceae	Flowers	Yellow-orange
Nymphaea alba	Nymphaceae	Rhizome	Blue

(continued)

Table 7.12 (continued)

Plant name	Family	Parts used	Colour obtained
Onosma hispidum	Boraginaceae	Roots	Red
Oroxylum indicum	Bignoniaceae	Bark/fruits	Black
Osbeckia stellata	Melastomaceae	Fruits	Brown
Peristrophe piniculata	Acanthaceae	Whole plant	Greenish
Phlogacanthus thyrsiformis	Acanthaceae	Flowers	Orange-yellow
Pinus wallichiana	Pinaceae	Bark	Black
Pistacia khinjuk	Anacrdiaceae	Stem galls	Brown
Prinsepia utilis	Rosaceae	Fruits	Blue
Prunus cerasoides	Rosaceae	Fruits	Yellow
Prunus persica	Rosaceae	Leaves, root bark	Light yellow
Psidium guajava	Myrtaceae	Fruits	Black-brown
Pterocarpus marsupium	Cesalpiniaceae	Bark	Red
Pterocarpus santalinus	Cesalpiniaceae	Wood	Red
Punica granatum.	Punicaceae	Flowers/fruits	Yellow-red
Quercus infectoria	Fagaceae	Gall nuts	Light yellow
Rheum moorcroftianum	Polygonaceae	Roots	Yellow
Rheum webbianum	Polygonaceae	Roots	Yellow
Rhododendron arboretum	Ericaeace	Flower	Red
Rhododendron lepidotum	Ericaeace	Leaves/flowers	Pink-red
Rimelia reticulate	Parmeliaceae	Whole plants	Orange-yellow
Rubia cordifolia	Rubiaceae	Whole plant	Red-brown
Rubia cordifolia	Rubiaceae	Stem, root	Light Brown
Rubia tinctorum	Rubiaceae	Wood, root	red, pink
Rubus fructicosus.	Rosaceae	Berries	Brown
Rumex hastatus	Polygonaceae	Roots	Yellow-green
Rumex nepalensis	Polygonaceae	Roots	Yellow-green
Semecarpus anacardium	Anacardiaceae	Fruits	Black
Solanum lycopersicum	Solanaceae	Fruits	Red
Sophora mollis	Fabaceae	Roots/flowers	Brown/yellow
Symplocos paniculata	Symplocaceae	Bark/leaves	Yello
Symplocos ramosissima	Symplocaceae	Bark/Leaves	Yellow
Syzygium cuminii	Apiaceae	Bark and Leaves	Red
Tagetes erecta.	Asteraceae	Flowers	Yellow
Tagetes erecta.	Asteraceae	Flower	Yellow
Tamarindus indica.	Caesalpiniaceae	Leaves	Reddish-yellow
Taxus baccata Linn. *ssp. wallichiana*	Taxaceae	Bark	Red
Tectona grandis	Verbenaceae	Leaf/bark	Reddish
Terminalia alata	Combretaceae	Bark	Red/brown
Terminalia arjuna	Combretaceae	Bark	Red
Terminalia bellirica	Combretaceae	Fruits	Black
Terminalia chebula.	Combretaceae	Fruits	Yellow/black

(continued)

Table 7.12 (continued)

Plant name	Family	Parts used	Colour obtained
Toona hexandra var. gambleri	Meliaceae	Flowers/seeds	Yellow-brown
Toona serrata	Meliaceae	Flowers/seeds	Yellow
Usnia verticillata	Parmeliaceae	Whole plant	Red brown
Utrica dioica	Urticaceae	Roots	Brown-black
Ventilago denticulate	Rhamnaceae	Bark and roots	Violet
Woodfordia fruiticosa	Lythraceae	Flowers	Red Yellow
Wrightia arborea	Apocynaceae	Bark/leaves	Yellow-pal
Wrightia tinctoria	Apocynaceae	Seeds	Blue
Zanthoxylum armatnum	Rutaceae	Bark	Brown
Ziziphus mauritiana	Rhamnaceae	Leaves/bark	Pink/red

Table 7.13 Various groups of biomass resources and their possible scope for utilization as energy sources.

Biomass category	Products	Uses
Energy crops Industrial residues	Methyl esters, alcohol	Energy
	Fibrous waste from pulp, including black liquor	Energy, composite material
	Wet cellulosic industrial residues	Electricity and heat
Industrial products	Pellets, bio-oil (pyrolysis oil), ethanol, biodiesel	Electricity
High biomass Producing grasses	Biohydrogen	Electricity and heat, Proton exchange membrane fuel cell
Contaminated waste	Biodegradable waste Sewage sludge	Biogas

Table 7.14 Metal accumulation and adsorption capacity of *Eichornia crassipes* (Water hyacinth)

Eichornia crassipes	References
Ag, Al, As, Ba, Ca, Cd, Ce, Co, Cu, Dy, Er, Eu, Fe, Ga, Gd, Ge, Hg, Ho, K, La, Mg, Mn, Mo, Na, Nb, Nd, Ni, Pb, Pr, Rb, Sm, Sr, Tb, Th, TI, Tm, U, Y, Yb, Zn, Zr	Valitutto et al. (2006)
Ag, Cd, Cr, Cu, Hg, Ni, Pb, Zn	Odjegba and Fasidi (2006, 2007)
Al, As, Cd, Cr, Cu, Fe, Hg, Mn, Ni, Pb, Zn	Marchand et al. (2010)
Al, Cr, Cu,	Klumpp et al. (2002)
As	Alvarado et al. (2008), Rahman and Hasegawa (2011), Giri and Patel (2012)
As, Cd, Cr, Cu, Fe, Mn, Ni, Pb, V, Zn	Agunbiade et al. (2009)
Br, Ca, Cr, Cu, Fe, K, Mn, Ni, Pb, Rb, S, Sr, Ti, Zn	Tejeda et al. (2010)
Ca, Co, Cu, Fe, K, Mg, Mn	Cooley and Martin (1979)

(continued)

Table 7.14 (continued)

Eichornia crassipes	References
Cd	Wolverton and McDonald (1978), Cooley and Martin (1979), O'Keeffe et al. (1984), El-Enany and Mazen (1996), Das and Jana (1999), Maine et al. (2001), Das and jana (2004), di Toppi et al. (2007), de oliveira et al. (2009), Rana et al. (2011)
Cd, Co, Cr, Cu, Mn, Ni, Pb, Zn	Soltan and Rashed (2003)
Cd, Co, Cu, Hg, Pb, Zn	Buta et al. (2011)
Cd, Co, Pb	Oseni (2004)
Cd, Cr, Cu, Fe, Mn, Ni, Pb, Zn	Singh and Kalamdhad (2013)
Cd, Cr, Cu, Fe, Ni, Pb	Khan et al. (2009)
Cd, Cr, Cu, Fe, Ni, Zn	Mishra et al. (2008)
Cd, Cr, Cu, Fe, Zn	Mishra and Tripathi (2008)
Cd, Cr, Cu, Ni, Pb, Zn	Schorin et al. (1991)
Cd, Cr, Zn	Delgado et al. (1993)
Cd, Cu, Hg, Pb, Zn	Núñez et al. (2011)
Cd, Cu, Pb, Zn	Fayed and Abd-El-Shafy (1985), Yang (1997), Smolyakov (2012)
Cd, Cu, Zn	Smolyakov et al. (2010)
Cd, Mn, Cu, Hg, Pb	Mishra et al. (2008)
Cd, Pb, Zn	Mahamadi and Nharingo (2010)
Cd, Zn	Hardy and O'Keeffe (1985), Lu et al. (2004)
Ce	Chua (1998)
Co, Cs	Saleh (2012)
Co, Cr, Cu, Fe, Ni, Pb, Zn	Zaranyika et al. (1994)
Cr	Rulangaranga and Mugasha (2003), Faisal and Hasnain (2003), Mangabeira et al. (2004), Paivaa et al. (2009), Espinoza-Quiñones et al. (2009), Mangabeira et al. (2011)
Cr, Cu, Mn, Pb, Zn	Tiwari et al. (2007)
Cr, Hg	Jana (1988)
Cr, Ni, Zn	Maine et al. (2006), Maine et al. (2007), Maine et al. (2009)
Cu	Mokhtar et al. (2011), Abdelraheem et al. (2012)
Cu, Hg	Mishra et al. (2012)
Cu, Pb	Vesk and Allaway (1997)
Fe	Jayaweera et al. (2008)
Hg	Panda et al. (1988), Lenka et al. (1992), Ramlal et al. (2003), Skinner et al. (2007), Caldelas et al. (2009)
Ir, Os, Pd, Pt, Rh, Ru,	Farago and Parsons (1994)
Ni	Bres et al. (2012)
Pb	Hameed et al. (1997), Singh et al. (2012), Baruah et al. (2012)
Pb, Zn	Verma et al. (2005)
Se	Mane et al. (2011)

Table 7.15 Metal accumulation in *Alternanthera philoxeroides* (Alligator weed)

As	Li et al. (2011)
Ca, Cr, Cu, Fe, Mg, Mn, Na, Ni, Pb, Zn	Zuo et al. (2012)
Cd	Souza et al. (2009)
Cd, Cr, Cu, Mn, Pb, Zn	Zhi Zhong et al. (2011)
Cd, Cr, Cu, Fe, Ni, Pb, Zn	Lokeshwari and Chandrappa (2007)
Cd, Cr, Cu, Pb, Zn	Gu et al. (2004)
Cd, Cu, Pb, Zn,	Bi et al. (2007)
Cd, Pb, Zn	Liu et al. (2007a, b)
Cr	Mangabeira et al. (2010), Mangabeira et al. (2011)
Cr, Cu	Naqvi and Rizvi (2000)
Cs	Pinder et al. (2005, 2006)
Cu	Guo and Hu (2012)
Cu, Cr, Ni	Jian-Guo et al. (2010)
Cu, Pb, Zn	Li et al. (2010)
Cu, Zn	Tang et al. (2002)
Fe, Pb, Zn	Deng et al. (2009)
Hg	Hong-Wei et al. (2003)
Pb	Shabani et al. (2010)
Pb, Zn	Deng et al. (2006)
Zn	Yuan et al. (2009)

The present study highlights the potential of ornamentals, fiber and energy crops for phytoremediation, because they are non-edible and income generators. Further studies should focus on combining phytoremediation with enhancing soil fertility and maximizing pest management, together with investigation of additional species for phytoremediation which may also enhance economic benefits. The management of plant biomass applied in phytoremediation should be also studied. Use of phytoremediated phytomass for phytoproducts and boosting bioeconomy via cogeneration of value additions and value chain products has tremendous scope and should be a priority area of research.

Grison (2015) pioneered in turning metalliferous waste from plants into a resource through innovative technologies, processes and services. She and her team discovered an unprecedented concept in chemistry, namely 'ecocatalysis'. The development of this new concept created a paradigm shift in sustainable and green chemistry i.e. the metallic wastes are becoming new ecofriendly and efficient catalytic systems. Grison feels that combining phytoextraction and ecocatalysis has opened up a new vistas (E4) in greener chemistry i.e. Environmental, Ecological, Ethic and Economic (E4) opportunity.

Table 7.16 Metal accumulation in *Pistia stratiotes* (Water lettuce/Water cabbage)

Pistia stratiotes + Metals	References
Ag, Al, As, Ba, Ca, Cd, Ce, Co, Cu, Dy, Er, Eu, Fe, Ga, Gd, Ge, Hg, Ho, K, La, Mg, Mn, Mo, Na, Nb, Nd, Ni, Pb, Pr, Rb, Sm, Sr, Tb, Th, TI, Tm, U, Y, Yb, Zn, Zr	Valitutto et al. (2006)
Ag, Cd, Cr, Cu, Hg, Ni, Pb, Zn	Odjegba and Fasidi (2004)
Al, Ca, Cd, Co, Cr, Cu, Fe, K, Mg, Mn, Na, Ni, Pb, Zn	Lu et al. (2011)
As	Rahman and Hasegawa (2011)
Ca, Cd, Co, Cr, Cu, Fe, K, Mg, Mn, Na, Ni, Pb, Zn	Sridhar (1986)
Cd	Maine et al. (2001), di Toppi et al. (2007), Bhakta and Munekage (2008), Li et al. (2013)
Cd, Cr	Suñe et al. (2007)
Cd, Cr, Cu, Fe, Ni, Pb	Khan et al. (2009)
Cd, Cr, Cu, Fe, Ni, Zn	Upadhyay et al. (2007)
Cd, Cr, Cu, Fe, Zn	Mishra and Tripathi (2008)
Cd, Cu, Hg, Mn, Pb,	Mishra et al. (2008)
Cd, Cu, Hg, Pb, Zn	Núñez et al. (2011)
Cd, Cu, Ni, Pb, Zn	Miretzky et al. (2006)
Cr	Satyakala and Jamil (1992), Mainea et al. (2004), Sinha and Gupta (2005), Ganesh et al. (2008), Espinoza-Quiñones et al. (2008), Sinha et al. (2009), Sundaramoorthy et al. (2010)
Cr, Cu, Fe, Mn, Pb, Zn	Miretzky et al. (2004)
Cr, Cu, Mn, Ni, Pb, Zn	Tewari et al. (2008)
Cr, Ni, Zn	Mufarrege et al. (2010)
Cu	Upadhyay and Panda (2009)
Cu, Hg	Mishra et al. (2012)
Fe, Mn, Pb, Zn	Ndzomo et al. (1994)
Hg	Mhatre and Chaphekar (1985), De et al. (1985), Lenka et al. (1992), Mishra et al. (2009), Skinner et al. (2007)
Pb	Espinoza-Quiñones et al. (2009), Vesely et al. (2012)
Th, U	Cecal et al. (2003)

Table 7.17 Advantages and limitations of bioremediation (Prasad 2011)

Advantages	Limitations
In- situ	Limited to shallow soils, streams, and groundwater.
Passive	High concentrations of hazardous materials can be toxic to plants.
Solar driven/	Mass transfer limitations associated with other biotreatments.
Costs 10–20 % of mechanical treatments	Slower than mechanical treatments.
Transfer is faster than natural attenuation	Only effective for moderately hydrophobic contaminants.
High public acceptance	Toxicity and bioavailability of degradation products is not known.
Fewer air and water emissions	Contaminants may be mobilized into the groundwater.
Generate less secondary wastes	Potential contaminants enter food chain through animal consumption.
Soils remain in place and are usable following treatment	Unfamiliar to many regulators.
(i) Phytovolatilized contaminants could be transformed to less toxic forms (e.g. elemental mercury and dimethyl selenite gas) (ii) phytovolatilization accelerates degradation processes	The contaminant or a hazardous metabolite might accumulate in vegetation and be passed on in later products such as fruit or lumber. Low levels of metabolites have been found in plant.
(i) Phytostabilization it circumvents the removal of soil, (ii) It has a lower cost and is less disruptive than other more-vigorous soil remedial technologies, (iii) Revegetation enhances ecosystem restoration.	(i) The contaminants remain in place. (ii) The vegetation and soil may require long-term maintenance to prevent re-release of the contaminants and future leaching, (iii) require extensive fertilization or application of soil amendments (Zhu et al. 2004). (iv) Plant uptake of metals and translocation to the aboveground portion must be avoided, (v) The root zone must be monitored to prevent metal leaching.
In phytoextraction, the plant biomass containing the extracted contaminant can be a resource (phytoextraction). For example, biomass that contains selenium (Se), an essential nutrient, has been transported to areas that are deficient in Se and used for animal feed. In green house experiments, gold was harvested from plants.	(i) Metal hyper accumulators are generally s66 with a small biomass and shallow root systems, (ii) Plants harvested must be properly disposal. (iii) Phytoextraction studies conducted using hydroponically grown plants, with the contaminant added in solution, may not reflect actual conditions and results occurring in soil. (iv) Phytoextraction coefficients measured under field conditions are likely to be less than those determined in the laboratory.
(i) Rhizofiltration using terrestrial plants removes contaminants more efficiently than aquatic plants, (ii) This system can be either *in-situ* (floating rafts on ponds) or *ex situ* (an engineered tank system), (iii) An *ex situ* system can be placed anywhere because the treatment does not have to be at the original location of contamination.	(i) The pH of the influent solution may have to be continually adjusted to obtain optimum metals uptake, (ii) The chemical speciation and interaction of all species in the influent have to be understood and accounted for, (iii) A well-engineered system is required to control influent concentration and flow rate, (iv) The plants (especially terrestrial plants) may have to be grown in a greenhouse or nursery and then placed in the rhizofiltration system, (v) Periodic harvesting and plant disposal are required, (vi) Metal immobilization and uptake results from laboratory and greenhouse studies might not be achievable in the field.

Acknowledgements The author gratefully acknowledge the award of Pitamber Pant National Environment Fellowship for the year 2007 by the M/o Environment and Forests, GOI, New Delhi (MoEF Ref.No 17/3/2010-RE Dt 29-2-2012). The author is indebted to Dr. Eric Lichtfouse for invaluable suggestions on this Ms and inputs for improvement. Thanks are due to Dr K. Jayaram, Research Associate (until September 2013) for help in field work and literature collection. Thanks are also due to Dr G.V. Subrahmanyam, Advisor, Dr Tashi Wangdi, Addl. Director, MoEF, New Delhi for encouragement. Assistance received from several of my students and Dr Adarsh Kumar is thankfully acknowledged. Thanks are also due to TATA Steel, West Bokaro Coalfields for permission to conduct fieldwork related to mulberry cultivation on mine over dumps, Springer, Elsevier, John Wiley & Sons, CRC Press (Taylor & Francis), Vetiver network, Shrishti Eco-Research Institute, Pune Nyveli Lignite Corporation are gratefully acknowledged for valuable resources. Mr M.V. Bhaskar "Kadambari Consultants Pvt Ltd - NUALGI".

References

Abad M, Noguera P, Burés S (2001) National inventory of organic wastes for use as growing media for ornamental potted plant production: case study in Spain. Bioresour Technol 77:197–200. http://dx.doi.org/10.1016/S0960-8524(00)00152-8

Abbasi SA, Nipaney PC, Panholzer MB (1991) Biogas production from the aquatic weed pistia (*Pistia stratiotes*). Bioresour Technol 37:211–214. http://dx.doi.org/10.1016/0960-8524(91)90185-M

Abd El-Kader AA, Hussein MM, Alva AK (2012) Response of *Jatropha* on a clay soil to different concentrations of micronutrients. Am J Plant Sci 3:1376–1381. doi:10.4236/ajps.2012.310166

Abdelraheem WHM, Rabia MKM, Ismail NM (2012) Evaluation of copper speciation in the extract of *Eichhornia crassipes* using reverse and forward/CLE voltammetric titrations. Arab J Chem, http://dx.doi.org/10.1016/j.arabjc.2012.04.013, doi:10.1016/j.arabjc.2012.04.013#doilink

Abdul RM, Mutnuri L, Dattatreya PJ, Mohan DA (2012) Assessment of drinking water quality using ICP-MS and microbiological methods in the Bholakpur area, Hyderabad, India. Environ Monit Assess 184:1581–1592. doi:10.1007/s10661-011-2062-2

Abhilash PC, Yunus M (2011) Can we use biomass produced from phytoremediation? Biomass Bioenergy 35:1371–1372. http://dx.doi.org/10.1016/j.biombioe.2010.12.013

Abioye OP, Agamuthu P, Abdul Aziz A (2010) Phytoaccumulation of zinc and iron by *Jatropha curcas* grown in used lubricating oil-contaminated soil. Malays J Sci 29(3):207–213

Abioye OP, Agamuthu P, Abdul Aziz AR (2012) Phytotreatment of soil contaminated with used lubricating oil using *Hibiscus cannabinus*. Biodegradation 23:277–286

Abu-Darwish MS (2009) Essential oils yield and heavy metals content of some aromatic medicinal plants grown in Ash-Shoubak region, south of Jordan. Adv Environ Biol 3(3):296–301

Abu-Darwish MS, Al-Fraihat AH, Al-Dalain SYA, Afifi FU, Al-Tabbal JA (2011) Determination of essential oils and heavy metals accumulation in *Salvia officinalis* cultivated in three intra-raw spacing in Ash-Shoubak. Jordan Int J Agric Biol 13:981–985

Acharya S, Sharma DK, Joshi HC (2012) Phytotoxicity of zinc, chromium (VI) and cadmium in purging nut (*Jatropha curcas*) seedlings grown in hydroponics. Indian J Agric Sci 82(8):667–671

Adesodun JK, Atayese MO, Agbaje TA, Osadiaye BA, Mafe OF, Soretire AA (2010) Phytoremediation potentials of sunflowers (*Tithonia diversifolia* and *Helianthus annuus*) for metals in soils contaminated with zinc and lead nitrates. Water Air Soil Pollut 207:195–201

Adewole MB, Sridhar MKC, Adeoye GO (2010) Removal of heavy metals from soil polluted with effluents from a paint industry using *Helianthus annuus* L. and *Tithonia diversifolia* (Hemsl.) as influenced by fertilizer applications. Bioremediation J 14(4):169–179

Agamuthu P, Abioye OP, Abdul Aziz A (2010) Phytoremediation of soil contaminated with used lubricating oil using *Jatropha curcas*. J Hazard Mater 179:891–894. http://dx.doi.org/10.1016/j.jhazmat.2010.03.088

Agbogidi OM, Eruotor PG (2012) Morphological changes due to spent engine oil contamination and its heavy metal components of *Jatropha curcas* Linn seedlings. International conference on bioscience, biotechnology and healthcare sciences (ICBBHS '2012), 14–15 December, Singapore, pp 88–93

Agbogidi OM, Mariere AE, Ohwo OA (2013) Metal concentration in plant tissues of *Jatropha curcas* L grown in crude oil contaminated soil. J Sustain For 32(4):404–411

Aggarwal H, Goyal D (2007) Phytoremediation of some heavy metals by agronomic crops. Dev Environ Sci 5:79–98. http://dx.doi.org/10.1016/S1474-8177(07)05005-X

Agunbiade FO, Olu-Owolabi BI, Adebowale KO (2009) Phytoremediation potential of *Eichornia crassipes* in metal-contaminated coastal water. Bioresour Technol 100:4521–4526

Ahmad P, Nabi G, Ashraf M (2011) Cadmium-induced oxidative damage in mustard [*Brassica juncea* (L.) Czern.& Coss.] plants can be alleviated by salicylic acid. S Afr J Bot 77:36–44

Ahmadpour P, Nawi AM, Abdu A, Abdul-Hamid H, Singh DK, Hassan A, Majid NM, Jusop S (2010) Uptake of heavy metals by *Jatropha curcas* planted in soils containing sewage sludge. Am J Appl Sci 7(10):1291–1299

Ahmed A, Prasad Rao TVD, Reddy N, Satyanarayanan M (2011) Hydrogeochemistry of ground water around municipal solid waste dumping site in Hyderabad city, India. Oriental J Chem 27:1665–1670

Aibibu N, Liu Y, Zeng G, Wanga X, Chen B, Song H, Xua L (2010) Cadmium accumulation in *Vetiveria zizanioides* and its effects on growth, physiological and biochemical characters. Bioresour Technol 101:6297–6303

Al-Farraj AS, Al-Wabel MI (2007) Heavy metals accumulation of some plant species grown on mining area at Mahad AD'Dahas, Saudi Arabia. J Appl Sci 7(8):1170–1175

Ali B, Hayat S, Fariduddin Q, Ahmad A (2008) 24-Epibrassinolide protects against the stress generated by salinity and nickel in *Brassica juncea*. Chemosphere 72:1387–1392

Alia PKVSK, Saradhi PP (1995) Effect of Zinc on free radicals and proline in Brassica and Cajanus. Phytochemistry 39(1):45–47

Alvarado S, Guédez M, Lué-Merú M, Graterol N, Anzalone A, Arroyo CJ (2008) Arsenic removal from water by bioremediation with the aquatic plants water hyacinth (*Eichornia crassipes*) and lesser duckweed (*Lemna minor*). Biores Technol 99:8436–8440

Amirmoradi S, Moghaddam PR, Koocheki A, Danesh S, Fotovat A (2012) Effect of cadmium and lead on quantitative and essential oil traits of peppermint (*Mentha piperita* L.). Notulae Scientia Biologicae 4(4):101–109

Andra SS, Datta R, Sarkar D, Saminathan SKM, Mullens CP, Bach SBH (2009) Analysis of phytochelatin complexes in the lead tolerant vetiver grass [*Vetiveria zizanioides* (L.)] using liquid chromatography and mass spectrometry. Environ Pollut 157:2173–2183

Andra SS, Datta R, Sarkar D, Makris KC, Mullens CP, Sahi SV, Bach SBH (2010) Synthesis of phytochelatins in vetiver grass upon lead exposure in the presence of phosphorus. Plant Soil 326:171–185

Andra SS, Sarkar D, Saminathan SKM, Datta R (2011) Exchangeable lead from prediction models relates to vetiver lead uptake in different soil types. Environ Monit Assess 183:571–579

Angelova VR, Ivanova RV, Ivanov KI, Perifanova-Nemska MN, Uzunova GI (2012) Potential of sunflower (*Helianthus annuus* l.) for phytoremediation of soils contaminated with heavy metals. BALWOIS 2012 - Ohrid, Republic of Macedonia – 28 May, 2 June 2012

Angin I, Turan M, Ketterings QM, Cakici A (2008) Humic acid addition enhances B and Pb phytoextraction by Vetiver Grass (*Vetiveria zizanioides* (L.) Nash). Water Air Soil Pollut 188:335–343

Antiochia R, Campanella L, Ghezzi P, Movassaghi K (2007) The use of vetiver for remediation of heavy metal soil contamination. Anal Bioanal Chem 388:947–956

Aparna C, Saritha P, Himabindu V, Bhandari A, Anjaneyulu Y (2010) Evaluation of bioremediation effectiveness on sediments contaminated with industrial wastes. Int J Environ Sci 1:607–620

Araiza-Arvilla J, Jáuregui-Rincón J, Alférez-Chávez JM, Ponce-Andrade I (2006) Use of hydroponic sunflowers (*Helianthus annuus* L.) farming system for lead and chromium laying up. In "Protection 2006", July

Arias J, Peralta-Videa J, Ellzeyj VMN, Minghua R, Mokgalaka-Matlalan NS, Castillo-Michel H, Gardea-Torresdey J (2010) Plant growth and metal distribution In tissues of *Prosopis juliflora-velutina* grown on Chromium contaminated soil in the Presence of *glomus deserticola*. Environ Sci Technol 44:7272–7279

Ashfaq M, Ali S, Hanif MA (2009) Bioaccumulation of cobalt in silkworm (*Bombyx mori* L.) in relation to mulberry, soil and wastewater metal concentrations Process. Biochemistry 44:1179–1184

Atiq-Ur-Rehman S, Iqbal MZ (2008) level of heavy metals in the foliage of naturally growing plants collected from korangi and landhi industrial areas of Karachi city, Pakistan. Pak J Bot 40(2):785–789

Awotoye OO, Adewole MB, Salami AO, Ohiembor MO (2009) Arbuscular mycorrhiza contribution to the growth performance and heavy metal uptake of *Helianthus annuus* Linn. in pot culture. Afr J Environ Sci Technol 3(6):157–163

Awwokunmi EE, Asaolu SS, Ajayi OO, Adebayo OA (2012) The role of EDTA on heavy metals Phytoextraction by *Jatropha gossypifolia* grown on soil collected from dumpsites in Ekiti state Nigeria. Br J Environ Climate Change 2(2):153–162

Azad HN, Shiva AH, Malekpour R (2011) Toxic effects of lead on growth and some biochemical and ionic parameters of sunflower (*Helianthus annuus* L) seedlings. Current Res J Biol Sci 3(4):398–403

Azhar N, Ashraf MY, Hussain M, Hussain F (2006) phytoextraction of lead (Pb) by edta application through sunflower (*Helianthus annuus* L.) cultivation: seedling growth studies. Pak J Bot 38(5):1551–1560

Azhar N, Ashraf MY, Hussain M, Ashraf M, Ahmed R (2009) EDTA-induced improvement in growth and water relations of sunflower (*Helianthus annuus* L) plants grown in lead contaminated medium. Pak J Bot 41(6):3065–3074

Aziz EE, Gad N, Khaled SM (2011) Effect of cobalt on growth and chemical composition of peppermint plant grown in newly reclaimed soil. Aust J Basic Appl Sci 5(11):628–633

Babu AG, Shim J, Shea PJ, Oh B-T (2014) *Penicillium aculeatum* PDR-4 and Trichoderma sp. PDR-16 promote phytoremediation of mine tailing soil and bioenergy production with sorghum-sudangrass. Ecol Eng 69:186–191. http://dx.doi.org/10.1016/j.ecoleng.2014.03.055

Bada BS, Kalejaiye ST (2010) Response of Kenaf (*Hibiscus Cannabinus* L.) grown in different soil textures and lead concentrations. Res J Agric Biol Sci 6(5):659–664

Bali R, Siegele R, Harris AT (2010) Phytoextraction of Au: uptake, accumulation and cellular distribution in Medicago sativa and *Brassica juncea*. Chem Eng J 156:286–297

Bañuelos GS (2002) Irrigation of broccoli and canola with boron and selenium-laden effluent. J Environ Qual 31:1802–1808

Bañuelos GS (2006) Phytoproducts may be essential for sustainability and implementation of phyto-remediation. Environ Pollut 144:19–23. http://dx.doi.org/10.1016/j.envpol.2006.01.015

Bañuelos GS, Cardon G, Mackey B, Ben-Asher J, Wu L, Beuselinck P, Akohoueand S, Zambrzuski S (1993) Boron and selenium removal in boron-laden soils by four sprinkler irrigated plant species. J Environ Qual 22(4):786–792

Bañuelos GS, Ajwa HA, Mackey B, Wu L, Cook C, Akohoue S, Zambruzuski S (1997) Evaluation of different plant species used for phytoremediation of high soil selenium. J Environ Qual 26(3):639–646

Bareen F, Tahira SA (2011) Metal accumulation potential of wild plants in tannery effluent contaminated soil of Kasur, Pakistan: field trials for toxic metal cleanup using Suaeda fruticosa. J Hazard Mater 186:443–450

Baroni F, Boscagli A, Di Lella LA, Protano G, Riccobono F (2004) Arsenic in soil and vegetation of contaminated areas in southern Tuscany (Italy). J Geochem Explor 81:1–14

Baruah S, Hazarika KK, Sarma KP (2012) Uptake and localization of Lead in *Eichhornia crassipes* grown within a hydroponic system. Adv Appl Sci Res 3(1):51–59

Bauddh K, Singh RP (2012a) Growth, tolerance efficiency and phytoremediation potential of *Ricinus communis* (L.) and *Brassica juncea* (L.) in salinity and drought affected cadmium contaminated soil. Ecotoxicol Environ Safety 85:13–22

Bauddh K, Singh RP (2012b) Cadmium tolerance and its phytoremediation by two oil yielding plants *Ricinus communis* (L.) and *Brassica juncea* (L.) from the contaminated soil. Int J Phytoremediation 14(8):772–785

Baunthiyal M, Sharma V (2012) Phytoremediation of fluoride contaminated water and soil: a search for fluoride hyperaccumulators. J Agric Technol 8(6):1965–1978

Bhakta JN, Munekage Y (2008) Role of ecosystem components in Cd removal process of aquatic ecosystem. Ecol Eng 32:274–280

Bhat R, Kiran K, Arun AB, Karim AA (2010) Determination of mineral composition and heavy metal content of some nutraceutically valued plant products. Food Anal Methods 3:181–187

Bi X, Feng X, Yang Y, Li X, Sin GPY, Qiu G, Qian X, Li F, He T, Li P, Liu T, Fu Z (2007) Heavy metals in an impacted wetland system: a typical case from south western China. Sci Total Environ 387:257–268

Bi R, Schlaak M, Siefert E, Lord R, Connolly H (2011) Influence of electrical fields (AC and DC) on phytoremediation of metal polluted soils with rapeseed (*Brassica napus*). Chemosphere 83(3):318–326

Bitterli C, Bañuelos GS, Schulin R (2010) Use of transfer factors to characterize uptake of selenium by plants. J Geochem Explor 107:206–216

Bjelková M, Genčurová V, Griga M (2011) Accumulation of cadmium by flax and linseed cultivars in field-simulated conditions: a potential for phytoremediation of Cd-contaminated soils. Ind Crops Products 33:761–774. http://dx.doi.org/10.1016/j.indcrop.2011.01.020

Blagojević N, Damjanović-Vratnica B, Vukašinović-Pešić V, Đurović D (2009) Heavy metals content in leaves and extracts of wild-growing *Salvia officinalis* from Montenegro. Polish J Environ Stud 18(2):167–173

Bonanno G (2012) *Arundo donax* as a potential biomonitor of trace element contamination in water and sediment. Ecotoxicol Environ Saf 80:20–27. http://dx.doi.org/10.1016/j.ecoenv.2012.02.005

Bonanno G, Cirelli GL, Toscano A, Giudice RL, Pavone P (2013) Heavy metal content in ash of energy crops growing in sewage-contaminated natural wetlands: potential applications in agriculture and forestry? Science Total Environ 452–453(1):349–354. http://dx.doi.org/10.1016/j.scitotenv.2013.02.048

Bose S, Jain A, Rai V, Ramanathan AL (2008) Chemical fractionation and translocation of heavy metals in *Canna indica* L. grown on industrial waste amended soil. J Hazard Mater 160:187–193

Bosiacki M (2008) Accumulation of cadmium in selected species of ornamental plants. Acta Sci Pol Hortorum Cultus 7:21–31

Bosiacki M (2009a) Phytoextraction of cadmium and lead by selected cultivars of *Tagetes erecta* L. Part I. Contents of cd and pb in plants. Acta Sci Pol Hortorum Cultus 8:3–13

Bosiacki M (2009b) Phytoextraction of cadmium and lead by selected cultivars of *Tagetes erecta* L. Part II. Contents of cd and pb in plants. Acta Sci Pol Hortorum Cultus 8:15–26

Bres P, Crespo D, Rizzo P, La Rossa R (2012) Capacity of the macrophytes *Lemna minor* and *Eichhornia crassipes* to remove nickel. RIA/Research, April (in press)

Brosse N, Sannigrahi P, Ragauskas A (2009) Pretreatment of *Miscanthus x giganteus* using the ethanol organosolv process for ethanol production. Ind Eng Chem Res 48:8328–8334

Brown JE, Khodr H, Hider RC, Rice-Evans CA (1998) Structural dependence of flavonoid interactions with Cu^{2+} ions: implications for their antioxidant properties. Biochem J 330:1173–1178

Busuioc G, David I, Stirbu C, Iliescu N, Mocanu V (2009) Contributions to determination the affinity for heavy metals accumulation of some organs at *Helianthus annuus* L. Lucrări Ştiinţifice 52, seria Agronomie, pp 623–628

Buta E, Paulette L, Mihăiescu T, Buta M, Cantor M (2011) The influence of heavy metals on growth and development of *Eichhornia crassipes* species, cultivated in contaminated water. Not Bot Hortic Agrobo 39(2):135–141

Butler RA (2006) Why is oil palm replacing tropical rainforests? Why are biofuels fueling deforestation? See also: /http://news.mongabay.com/2006/0425-oil_palm.htmlS0425

Caldelas C, Iglesia-Turiño S, Araus JL, Bort J, Febrero A (2009) Physiological responses of Eichhornia crassipes (Mart.) Solms to the combined exposure to excess nutrients and Hg. Braz Soc Plant Physiol 21(1):01–12

Canakci M, Van Gerpen J (2001) Biodiesel production from oils and fats with high free fatty acids. Trans Am Soc Agric Eng [ASAE] 44:1429–1436

Cao A, Carucci A, Lai T, Colla PL, Tamburini E (2007) Effect of biodegradable chelating agents on trace metals phytoextraction with *Mirabilis jalapa* and its associated bacteria. Eur J Soil Biol 43:200–206

Cao L, Jiang M, Zeng Z, Dub A, Tan H, Liu Y (2008) Trichoderma atroviride F6 improves phytoextraction efficiency of mustard (*Brassica juncea* (L.) Coss. var. foliosa Bailey) in Cd, Ni contaminated soils. Chemosphere 71:1769–1773

Capuana M (2011) Heavy metals and woody plants – biotechnologies for phytoremediation. Biogeosci Forest 4:7–15

Carlson KD, Cunningham RL, Garcia WJ, Bagby MO, Kwolek WF (1982) Performance and trace metal content of Crambe and kenaf grown on sewage sludge-treated stripmine land. Environ Pollut Ser A, Ecol Biol 29:145–161

Carrier M, Loppinet-Serani A, Absalon C, Marias F, Aymonier C, Mench M (2011) Conversion of fern (*Pteris vittata* L.) biomass from a phyto-remediation trial in sub- and supercritical water conditions. Biomass Bioenergy 35:872–883

Carrier M, Loppinet-Serani A, Absalon C, Aymonier C, Mench M (2012) Degradation pathways of holocellulose, lignin and alpha-cellulose from *Pteris vittata* fronds in sub- and super critical conditions. Biomass Bioenergy 43:65–71

Cassina L, Tassi E, Pedron F, Petruzzelli G, Ambrosini P, Barbafieri M (2012) Using a plant hormone and a thioligand to improve phytoremediation of Hg-contaminated soil from a petrochemical plant. J Hazard Mater 231–232:36–42

Cecal A, Popa K, Caraus I, Cracium I (2003) Uranium and thorium uptake on hydrophilic plants. In: Merkel BJ, Planer-Friederich B, Wolkersdorfer C (eds) Uranium in the aquatic environment. Springer, Heidelberg, pp 479–488

Chaiyarat R, Suebsima R, Putwattana N, Kruatrachue M, Pokethitiyook P (2011) Effects of soil amendments on growth and metal uptake by *Ocimum gratissimum* grown in Cd/Zn-contaminated soil. Water Air Soil Pollut 214:383–392

Chakravarty B, Srivastava S (1992) Toxicity of some heavy metals in vivo and in vitro in Helianthus annuus. Mutat Res 283(4):287–294

Chami ZA, Amer N, Smets K, Yperman J, Carleer R, Dumontet S, Vangronsveld J (2014) Evaluation of flash and slow pyrolysis applied on heavy metal contaminated Sorghum bicolor shoots resulting from phytoremediation. Biomass Bioenergy 63:268–279. http://dx.doi.org/10.1016/j.biombioe.2014.02.027

Chand S, Pandey A, Patra DD (2012) Influence of nickel and lead applied in combination with vermiompost on growth and accumulation of heavy metals by Mentha arvensis Linn. cv. 'Kosi'. Indian J Nat Prod Resour 3(2):256–261

Chandra S, Misra V, Singh R, Lello Z (2011) EDTA-enhanced lead phytoremediation in sunflower (*Helianthus annuus* L.) hydroponic culture. Plant Soil 347(1/2):231

Chandran S, Niranjana V, Bennyjoseph (2012) Accumulation of heavy metals in wastewater irrigated crops in Madurai, India. J Environ Res Dev 6(3):432–438

Chantachon S, Kruatrachue M, Pokethitiyook P, Upatham S, Tantanasarit S, Soonthornsarathool V (2004) Phytoextraction and accumulation of lead from contaminated soil by vetiver grass: laboratory and simulated field study. Water Air Soil Pollut 154:37–55

Chaudhary DR, Ghosh A, Chikara J, Patolia JS (2009) Elemental content of three plant species growing on abandoned fly ash landfill. Indian Forester 135:393–402

Chaudhry TM, Hayes WJ, Khan AG, Khoo CS (1998) Phytoremediation-focusing on accumulator plants that remediate metal-contaminated soils. Aust J Ecotoxicol 4:37–51

Chaves LHG, de Mesquita EF, de Araujo DL, de França CP (2010) Acúmulo e Distribuição de Cobre e Zinco em Mamoneira Cultivar BRS Paraguaçu e Crescimento da Planta. Engenharia Ambiental: Pesquisa e Tecnologia 7(3):263–277

Chen H, Cutright T (2001) EDTA and HEDTA effects on Cd, Cr, and Ni uptake by *Helianthus annuus*. Chemosphere 45:21–28

Chen KF, Yeh TY, Lin CF (2012) Phytoextraction of Cu, Zn, and Pb enhanced by Chelators with Vetiver (*Vetiveria zizanioides*): hydroponic and pot experiments. ISRN Ecol 2012:1–12

Chen HM, Zheng CR, Tu C, Shen ZG (2000) Chemical methods and phytoremediation of soil contaminated with heavy metals. Chemosphere 41:229–234

Cheng SF, Choo YM, Ma AN, Chuah CH (2004) Kinetics study on transesterification of Palm oil. J Oil Palm Res 16(2):19–29

Chintakovid W, Visoottiviseth P, Khokiattiwong S, Lauengsuchonkul S (2008) Potential of the hybrid marigolds for arsenic phytoremediation and income generation of remediators in Ron Phibun District, Thailand. Chemosphere 70:1532–1537, ISSN 0045–6535, http://dx.doi.org/10.1016/j.chemosphere.2007.08.031

Chigbo C, Batty L, Bartlett R (2013) Interactions of copper and pyrene on phytoremediation potential *of Brassica juncea* in copper–pyrene co-contaminated soil. Chemosphere 90(10):2542–2548. doi:10.1016/j

Chisti Y (2008) Biodiesel from microalgae beats bioethanol. Trends Biotechnol 26:126–131

Chiu KK, Ye ZH, Wong MH (2005) Enhanced uptake of As, Zn, and Cu by *Vetiveria zizanioides* and *Zea mays* using chelating agents. Chemosphere 60:1365–1375

Chiu KK, Ye ZH, Wong MH (2006) Growth of *Vetiveria zizanioides* and *Phragmities australi*s on Pb/Zn and Cu mine tailings amended with manure compost and sewage sludge: a greenhouse study. Bioresour Technol 97:158–170

Chong CW, Chu LM (2007) Growth of vetivergrass for cutslope landscaping: effects of container size and watering rate. Urban Forestry Urban Greening 6:135–141

Chua H (1998) Bio-accumulation of environmental residues of rare earth elements in aquatic flora *Eichhornia crassipes* (Mart.) Solms in Guangdong Province of China. Sci Total Environ 214:79–85

Claudia, D'Innocenzo M, Jairo F (2012) Prospects of Miscanthus x giganteus for PAH phytoremediation: A microcosm study. Ind Crops Prod 36(1):276–281. ISSN 0926–6690, http://dx.doi.org/10.1016/j.indcrop.2011.10.030

Clemente R, Walker DJ, Bernal MP (2005) Uptake of heavy metals and As by Brassica juncea grown in a contaminated soil in Aznalcóllar (Spain): the effect of soil amendments. Environ Pollut 138:46–58

Clemente R, Almela C, Bernal MP (2006) A remediation strategy based on active phytoremediation followed by natural attenuation in a soil contaminated by pyrite waste. Environ Pollut 143:397–406. http://dx.doi.org/10.1016/j.envpol.2005.12.011

Conesa HM, Evangelou MW, Robinson BH, Schulin R (2012) A critical view of current state of phytotechnologies to remediate soils: still a promising tool? The Scientific World Journal Volume 2012, Article ID 173829, 10 pages doi:10.1100/2012/173829

Cooley TN, Martin DF (1979) A preliminary study of metal distribution in three water hyacinth biotypes. Water Res 13(4):343–348

Coscione AR, Berton RS (2009) Barium extraction potential by mustard, sunflower and castor bean. Sci Agric (Piracicaba, Braz.) 66(1):59–63

Cutright T, Gunda N, Kurt F (2010) Simultaneous hyperaccumulation of multiple heavy metals by *Helianthus annuus* grown in a contaminated sandy-loam soil. Int J Phytoremediation 12(6):562–573

Dalla Vecchia F, Rocca NL, Moro I, De Faveri S, Andreoli C, Rascio N (2005) Morphogenetic, ultrastructural and physiological damages suffered by submerged leaves of *Elodea canadensis* exposed to cadmium. Plant Sci 168:329–338

Dan TV, Raj KS, Saxena PK (2000) Metal tolerance of scented geranium (*Pelargonium* sp. 'Frensham'): effects of cadmium and nickel on chlorophyll fluorescence kinetics. Int J Phytoremediation 2:91–104

Danh LT, Truong P, Mammucari R, Tran T, Foster N (2009) Vetiver grass, *Vetiveria zizanioides*: a choice plant for phytoremediation of heavy metals and organic wastes. Int J Phytoremediation 11:664–691

Danh LT, Truong P, Mammucari R, Fostert N (2011) Economic incentive for applying vetiver grass to remediate lead, copper and zinc contaminated soils. Int J Phytoremediation 13(1):47–60

Darnoko D, Cheryman M (2000) Kinetics of palm oil transesterification in a batch reactor. J Am Oil Chem Soc 77(12):1263–1267

Das S, Jana BB (1999) Dose-dependent uptake and *Eichhornia*-induced elimination of cadmium in various organs of the freshwater mussel, *Lamellidens marginalis* (Linn.). Ecol Eng 12:207–229

Das S, Jana BB (2004) Distribution pattern of ambient cadmium in wetland ponds distributed along an industrial complex. Chemosphere 55:175–185

Dasaram B, Satyanarayanan M, Sudarshan V, Keshav Krishna A (2011) Assessment of soil contamination in Patancheru Industrial Area, Hyderabad, Andhra Pradesh, India. Res J Environ Earth Sci 3:214–220

Datta R, Quispe MA, Sarkar D (2011) Greenhouse study on the phytoremediation potential of vetiver grass, *Chrysopogon zizanioides* L., in arsenic-contaminated soils. Bull Environ Contam Toxicol 86:124–128

de Abreu CA, Coscione AR, Pires AM, Paz-Ferreiro J (2012) Phytoremediation of a soil contaminated by heavy metals and boron using castor oil plants and organic matter amendments. J Geochem Explor 123:3–7

De AK, Sen AK, Modak DP, Jana S (1985) Studies of toxic effects of Hg(II) on *Pistia stratiotes*. Water Air Soil Pollut 24:351–360

de Lorenzo V (2008) Systems biology approaches to bioremediation. Curr Opin Biotechnol 19:579–589

de Oliveira D, Di Luccio M, Faccio C, Dalla Rosa C, Bender JP, Lipke N, Amroginski C, Dariva C, De Oliveira JV (2005) Optimization of alkaline transesterification of soybean oil and castor oil for biodiesel production. Appl Biochem Biotech 122:553–560

de Oliveira JA, Cambraia J, de Sousa MV, Oliva MA (2009) Sulphate uptake and metabolism in water hyacinth and salvinia during cadmium stress. Aquat Bot 91:257–261

de Sá Salomão AL, Marques M, Severo RG, da Cruz Roque OC (2012) Engineered ecosystem for on-site wastewater treatment in tropical areas. Water Sci Technol 66:213–217

de Souza Costa ET, Guilherme LRG, de Melo EEC, Ribeiro BT, dos Santos B, Inácio E, da Costa SE, Faquin V, Hale BA (2012) Assessing the tolerance of castor bean to Cd and Pb for phytoremediation purposes. Biol Trace Elem Res 145:93–100

Delgado M, Bigeriego M, Guardiola E (1993) Uptake of Zn, Cr and Cd by water hyacinths. Water Res 27(2):269–272

Delplanque M, Collet S, Del Gratta F, Schnuriger B, Gaucher R, Robinson B (2013) Combustion of *Salix* used for phytoextraction: the fate of metals and viability of the processes. Biomass Bioenergy 49:160–170

Deng H, Ye ZH, Wong MH (2006) Lead and zinc accumulation and tolerance in populations of six wetland plants. Environ Pollut 141:69–80

Deng H, Ye ZH, Wong MH (2009) Lead, zinc and iron (Fe^{2+}) tolerances in wetland plants and relation to root anatomy and spatial pattern of ROL. Environ Exp Bot 65:353–362

Deo B, Nahak G, Sahu RK (2011) Studies on the uptake of heavy metals by selected plant species growing on coal mine spoils in sub-tropical regions of India. J Am Sci 7(1):26–34

di Toppi LS, Vurro E, Rossi L, Marabottini R, Musetti R, Careri M, Maffini M, Mucchino C, Corradini C, Badiani M (2007) Different compensatory mechanisms in two metal accumulating aquatic macrophytes exposed to acute cadmium stress in outdoor artificial lakes. Chemosphere 68:769–780

Ding ZH, Hu X (2012) Transfer of heavy metals (Cd, Pb, Cu and Zn) from roadside soil to ornamental plants in Nanjing, China. Adv Mater Res 356–360:3051–3054

Ding X, Jiang J, Wang Y, Wang W, Ru B (1994) Bioconcentration of cadmium in water hyacinth (*Eichhornia crassipes*) in relation to thiol group content. Environ Pollut 84:93–96

do Nascimento CWA, Amarasiriwardena D, Xing B (2006) Comparison of natural organic acids and synthetic chelates at enhancing phytoextraction of metals from a multi-metal contaminated soil. Environ Pollut 140:114–123. doi:10.1016/j.jhazmat.2010.03.088

Dowling DN, Doty SL (2009) Improving phyto-remediation through biotechnology. Curr Opin Biotechnol 20:204–206

Dwivedi S, Srivastava S, Mishra S, Dixit B, Kumar A, Tripathi RD (2008) Screening of native plants and algae growing on fly-ash affected areas near National Thermal Power Corporation, Tanda, Uttar Pradesh, India for accumulation of toxic heavy metals. J Hazard Mater 158(2–3):359–365. ISSN 0304–3894, http://dx.doi.org/10.1016/j.jhazmat.2008.01.081

El-Enany AE, Mazen AMA (1996) Isolation of Cd- Binding protein of water hyacinth (*Eichhornia crassipes*) grown in Nile river water. Water Air Soil Pollut 87:357–362

El Hage R, Chrusciel L, Desharnais L, Brosse N (2010) Effect of autohydrolysis of *Miscanthus* x *giganteus* on lignin structure and organosolv delignification. Bioresour Technol 101:9321–9329

El-Mashad HM, Zhang R, Avena-Bustillos RJ (2008) A two step process for biodiesel production from salmon oil. Biosyst Eng 99:220–227

Enache M, Dearden John C, Walker John D (2003) QSAR analysis of metal ion toxicity data in sunflower callus cultures (*Helianthus annuus* "Sunspot"). QSAR Comb Sci 22:234–240

Espinoza-Quiñones FR, da Silva EA, de Rizzutto MA, Palácio SM, Módenes AN, Szymanski N, Martin N, Kroumov AD (2008) Chromium ions phytoaccumulation by three floating aquatic macrophytes from a nutrient medium. World J Microbiol Biotechnol 24:3063–3070

Espinoza-Quiñones FR, Martin N, Stutz G, Tirao G, Palácio SM, Rizzutto MA, Módenes AN, Silva FG Jr, Szymanski N, Kroumov AD (2009) Root uptake and reduction of hexavalent chromium by aquatic macrophytes as assessed by high-resolution X-ray emission. Water Res 43:4159–4166

Faisal M, Hasnain S (2003) Synergistic removal of Cr (VI) by *Eichornia crassipes* in conjunction with bacterial strains. Pak J Biol Sci 6(3):264–268

Farago ME, Parsons PJ (1994) The effects of various platinum metal species on the water plant *Eichhornia crassipes* (Mart.). Chem Spec Bioavailability 6:1–12

Fariduddin Q, Yusuf M, Hayat S, Ahmad A (2009) Effect of 28-homobrassinolide on antioxidant capacity and photosynthesis in *Brassica juncea* plants exposed to different levels of copper. Environ Exp Bot 66:418–424

Fauziah SH, Wan Siti NurHajar WI, Agamuthu P (2011) Heavy metal accumulation in *Ipomoea* reptans and *Helianthus annuus*. Malays J Sci 30(1):36–44

Fayed SE, Abd-El-Shafy HI (1985) Accumulation of Cu, Zn, Cd, and Pb by aquatic macrophytes. Environ Int 11(1):77–87

Figueroa JAL, Wrobel K, Afton S, Caruso JA, Corona JFG, Wrobel K (2008) Effect of some heavy metals and soil humic substances on the phytochelatin production in wild plants from silver mine areas of Guanajuato, Mexico. Chemosphere 70:2084–2091

Flathman PE, Lanza GR (1998) Phytoremediation: current views on an emerging green technology. Soil Sediment Contam Int J 71:415–432

Fozia A, Muhammad AZ, Muhammad A, Zafar MK (2008) Effect of chromium on growth attributes in sunflower (*Helianthus annuus* L.). J Environ Sci (China) 20(12):1475–1480

Frérot H, Lefèbvre C, Gruber W, Collin C, Dos Santos A, Escarre J (2006) Specific interactions between local metallicolous plants improve the phytostabilization of mine soils. Plant and Soil 282:53–65

Gabriel IE, Patten T (1994) distribution of copper smelter emissions in Southeastern Arizona – using honey mesquite as a bioindicator. Water Air Soil Pollut 72:67–87

Galeş R, Ivănescu L, Zamfirache MM, Burducea M (2009) Effect of lead on seed germination and seedling development in some medicinal plants (1st note). Analele ştiinţifice ale Universităţii "Al. I. Cuza" Iaşi, Tomul LV, fasc. 2, s.II a. Biologie vegetală, 2009:27–31

Gallego S, Benavides M, Tomaro M (2002) Involvement of an antioxidant defence system in the adaptive response to heavy metal ions in *Helianthus annuus* L. cells. Plant Growth Regul 36:267–273

Gallego SM, Kogan MJ, Azpilicueta CE, Peña C, Tomaro ML (2005) Glutathione-mediated antioxidative mechanisms in sunflower (*Helianthus annuus* L.) cells in response to cadmium stress. Plant Growth Regul 46:267–276

Ganesh KS, Baskaran L, Rajasekaran S, Sumathi K, Chidambaram ALA, Sundaramoorthy P (2008) Chromium stress induced alterations in biochemical and enzyme metabolism in aquatic and terrestrial plants. Colloids Surf B Biointerfaces 63:159–163

Gao S, Yan R, Wu J, Zhang F, Wang S, Chen F (2009) Growth and antioxidant responses in *Jatropha curcas* cotyledons under lead stress. Z Naturforsch 64(c):859–863

Garcia JS, Grata PL, Azevedo RA, Arruda M (2006) Metal contamination effects on sunflower (*Helianthus annuus* L.) growth and protein expression in leaves during development. J Agric Food Chem 54:8623–8630

Garcia JS, Souza GHMF, Eberlin MN, Arruda MAZ (2009) Evaluation of metal-ion stress in sunflower (*Helianthus annuus* L.) leaves through proteomic changes. Metallomics 1:107–113

Gaur A, Adholeya A (2004) Prospects of arbuscular mycorrhizal fungi in phytoremediation of heavymetal contaminated soils. Curr Sci 86(4):528–534

Gaïda MM, Landoulsi NS, Rejeb MN, Smiti S (2013) Growth and photosynthesis responses of *Rosmarinus officinalis* L. to heavy metals at Bougrine mine. Afr J Biotechnol 12(2):150–161

Ghavri SV, Singh RP (2012) Growth, biomass production and remediation of copper contamination by Jatropha curcas plant in industrial wasteland soil. J Environ Biol 33:207–214

Ghnaya T, Zaier H, Baioui R, Sghaier S, Lucchini G, Sacchi GA, Lutts L, Abdelly C (2013) Implication of organic acids in the long-distance transport and the accumulation of lead in Sesuvium portulacastrum and *Brassica juncea*. Chemosphere 90:1449–1454

Ghosh M, Singh SP (2005) A review on phytoremediation of heavy metals and utilization of its byproducts. Appl Ecol Environ Res 3(1):1–18

Giansoldati V, Tassi E, Morelli E, Gabellieri E, Pedron F, Barbafieri M (2012) Nitrogen fertilizer improves boron phytoextraction by *Brassica juncea* grown in contaminated sediments and alleviates plant stress. Chemosphere 87:1119–1125

Giri A, Narasu L (2000) Transgenic hairy roots: recent trends and applications. Biotechnol Adv 18:1–22

Giri AK, Patel RK (2012) Phytoaccumulation potential and toxicity of arsenic ions by *Eichhornia Crassipes* in Hydroponic System. J Bioremed Biodegrad 3:2–6

Giordani C, Cecchi S, Zanchi C (2005) Phytoremediation of soil polluted by nickel using agricultural crops. Environ Manage 36:675–681

Giovanni A, Pecchioni N, Rabaglio M, Allavena A (1997) Characterization of ornamental *Datura* plants transformed by *Agrobacterium rhizogenes*. In Vitro Cell Dev Biol–Plant 33:101–106

Glanze W (1996) In: Mosby C (ed) Mosby medical encyclopedia, Revised Edition ed. St. Louis 1996.

Govil PK, Reddy GLN, Krishna AK (2001) Contamination of soil due to heavy metals in the Patancheru Industrial Development Area, Andhra Pradesh, India. Environ Geol 41:461–469

Govil PK, Sorlie JE, Murthy NN, Sujatha D, Reddy GLN, Rudolph-Lund K, Krishna AK, Rama Mohan K (2008) Soil contamination of heavy metals in the Katedan Industrial Development Area, Hyderabad, India. Environ Monit Assess 140:313–323

Govil PK, Sorlie JE, Sujatha D, Krishna AK, Murthy NN, Rama Mohan K (2012) Assessment of heavy metal pollution in lake sediments of Katedan Industrial Development Area, Hyderabad, India. Environ Earth Sci 66:121–128

Grant TD, Montes-Bayón M, LeDuc D, Fricke MW, Terry N, Caruso JA (2004) Identification and characterization of Se-methyl selenomethionine in *Brassica juncea* roots. J Chromatogr A 1026:159–166

Griga M, Bjelkova M (2013) Flax (Linum usitatissimum L.) and Hemp (Cannabis sativa L.) as fibre crops for phytoextraction of heavy metals: biological, agro-technological and economical point of view. In: Gupta DK (ed) Plant-based remediation processes, soil biology 35, pp 199–237. doi:10.1007/978-3-642-35564-6_11

Grison C (2015) Combining phytoextraction and ecocatalysis: a novel concept for greener chemistry, an opportunity for remediation. Environ Sci Pollut Res 22:5589–5591. doi:10.1007/s11356-014-3169-0

Gu Chao, Li Lei, He Chiquan (2004) Accumulation of heavy metals by *Alternanthera philoxeroides* and *Commelina communis*. J Shanghai Univ (Natural Science Edition) 10(6):626–629

Guo W, Hu ZH (2012) Effects of stolon severing on the expansion of *Alternanthera philoxeroides* from terrestrial to contaminated aquatic habitats. Plant Species Biol 27:46–52

Gupta AK, Sinha S (2006) Role of *Brassica juncea* (L.) Czern. (var. Vaibhav) in the phytoextraction of Ni from soil amended with fly ash: selection of extractant for metal bioavailability. J Hazard Mater B136:371–378

Gupta AK, Sinha S (2007) Assessment of single extraction methods for the prediction of bioavailability of metals to *Brassica juncea* L. Czern. (var. Vaibhav) grown on tannery waste contaminated soil. J Hazard Mater 149:144–150

Gupta DK, Srivastava A, Singh VP (2008) EDTA enhances lead uptake and facilitates phytoremediation by vetiver grass. J Environ Biol 29(6):903–906

Gurunadha Rao VVS, Yoshida M, Prakash BA, Chandrasekhar SVN, Mahesh Kumar K (2004) Environmental impact of human activities to urban lake sediments: Potentially Toxic Elements (PTEs) contamination in Hussainsagar Lake, Hyderabad. The 11th national symposium on hydrology National Institute of Hydrology, Roorkee (India), 22–23 November

Gurunadha Rao VVS, Jain CK, Prakash BA, Mahesh Kumar K (2008) Heavymetal speciation study of sediments in Hussainsagar lake, Greater Hyderabad, India. In: Sengupta M, Dalwani R (eds) 2008. Proceedings of Taal 2007: the 12th world lake conference: 2098–2104

Hadad HR, Maine MA, Mufarrege MM, Del Sastre MV, Di Luca GA (2011) Bioaccumulation kinetics and toxic effects of Cr, Ni and Zn on *Eichhornia crassipes*. J Hazard Mater 190:1016–1022

Hall JL (2002) Cellular mechanisms for heavy metal detoxification and tolerance. J Exp Bot 53:1–11

Hameed PS, Shaheed K, Somasundaram SSN (1997) Bioaccumulation of 210Pb in the Kaveri River Ecosystem, India. J Environ Radioactivity 37(1):17–27

Hammer D, Kayser A, Keller C (2003) Phytoextraction of Cd and Zn with *Salix viminalis* in field trials. Soil Use Manage 19:187–192

Haque N, Peralta-Videa JR, Duarte-Gardea M, Gardea-Torresdey JL (2009) Differential effect of metals/metalloids on the growth and element uptake of mesquite plants obtained from plants grown at a copper mine tailing and commercial seeds. Bioresour Technol 100:6177–6182. http://dx.doi.org/10.1016/j.biortech.2009.06.090

Hardy JK, O'Keeffe DH (1985) Cadmium uptake by the water hyacinth: effects of root mass, solution volume, complexers and other metal ions. Chemosphere 14(5):417–426

Hattori H, Kuniyasu K, Chiba K, Chino M (2006) Effect of chloride application and low soil pH on cadmium uptake from soil by plants. Soil Sci Plant Nutr 52:89–94

Hayat S, Ali B, Hasan SA, Ahmad A (2007) Brassinosteroid enhanced the level of antioxidants under cadmium stress in *Brassica juncea*. Environ Exp Bot 60:33–41

Hegde RS, Fletcher JS (1996) Influence of plant growth stage and season on the release of root phenolics by mulberry as related to development of phytoremediation technology. Chemosphere 32:2471–2479

Hernandez-Viezcas JA, Castillo-Michel H, Servin AD, Peralta-Videa JR, Gardea-Torresdey JL (2011) Spectroscopic verification of zinc absorption and distribution in the desert plant *Prosopis juliflora*-velutina (velvet mesquite) treated with ZnO nanoparticles. Chem Eng J 170:346–352

Herrera-Rodríguez MB, Pérez-Vicente R, Maldonado JM (2007) Expression of asparagine synthetase genes in sunflower (*Helianthus annuus*) under various environmental stresses. Plant Physiol Biochem 45(1):33–38

Hinchman RR, Negri MC, Gatliff EG (1995) Phytoremediation: using green plants to clean up contaminated soil, groundwater, and wastewater. Argonne National Laboratory Hinchman, Applied Natural Sciences, Inc, 1995, http://www.treemediation.com/Technical/Phytoremediation_1998.pdf

Hiroyuki K, Mio W, Takahiko H (2005) Phytoremediation with Kenaf (*Hibiscus cannabinus*) for cadmium – contaminated paddy field in Southwest area of Japan. Jpn J Soil Sci Plant Nutr 76(1):27–34

Hladun KR, Parker DR, Trumble JT (2011) Selenium accumulation in the floral tissues of two Brassicaceae species and its impact on floral traits and plant performance. Environ Exp Bot 74:90–97

Ho WM, Ang LH, Lee DK (2008) Assessment of Pb uptake, translocation and immobilization in kenaf (*Hibiscus cannabinus* L.) for phytoremediation of sand tailings. J Environ Sci 20:1341–1347

Hong-Wei Z, Guo-Xin S, Jing-Yao C, Lei Z (2003) Relieving to Hg ~ (2+) injury of *Alternanthera philoxeroides* antioxidant enzyme system by sprayed 6-BA. Acta Ecologica Sinica 23(2):387–392

Hong-Xia Y, Wei L, Bing L, Wei W, Hui-Juan Z, Deng-Yun C (2009) Speciation analysis of cadmium in Indian mustard (*Brassica juncea*) by size exclusion chromatography-high performance liquid chromatography-inductively coupled plasma mass spectrometry. Chin J Anal Chem 37(10):1511–1514

Hristozkova M, Geneva M, Istancheva I, Boychinova M, Djonova E (2015) Aspects of mycorrhizal colonization in adaptation of sweet marjoram (*Origanum majorana* L.) grown on industrially polluted soil. Turkish J Biol 29:1–8

Hsiao KH, Kao PH, Hseu ZY (2007) Effects of chelators on chromium and nickel uptake by *Brassica juncea* on serpentine-mine tailings for Phytoextraction. J Hazard Mater 148:366–376

Huang H, Ning Yu, Lijun Wang, Gupta DK, Zhenli He, Kai Wang, Zhiqiang Zhu, Xingchu Yan, Tingqiang Li, Xiao-e Yang (2011) The phytoremediation potential of bioenergy crop Ricinus communis for DDTs and cadmium co-contaminated soil. Bioresour Technol 102:11034–11038. http://dx.doi.org/10.1016/j.biortech.2011.09.067

Huiping Xiao, Shuiping Cheng, Zhenbin Wu (2010) Microbial community variation in phytoremediation of triazophos by *Canna indica* Linn. in a hydroponic system. J Environ Sci 22:1225–1231. http://dx.doi.org/10.1016/S1001-0742(09)60242-4

Husain Q, Husain M, Kulshrestha Y (2009) Remediation and treatment of organopollutants mediated by peroxidases: a review. Crit Rev Biotechnol 29(2):94–119

Hussein FH, Khalifa RKM, El-Mergawi RA, Youssef AA (2006) Utilization of treated municipal wastewater for growing some aromatic plants to produce volatile oils and study its nutritional status in arid region. In: The 2nd international conference on water resources & arid environment, pp 1–16

Imada S, Yamanaka N, Tamai S (2009) Effects of salinity on the growth, Na partitioning, and Na dynamics of a salt-tolerant tree, *Populus alba* L. J Arid Environ 73:245–251. http://dx.doi.org/10.1016/j.jaridenv.2008.10.006

Imelouane B, Tahri M, Elbastrioui M, Aouinti F, Elbachiri A (2011) Mineral contents of some medicinal and aromatic plants growing in eastern Morocco. J Mater Environ Sci 2(2):104–111

Iqbal MZ, Sherwani AK, Shafiq M (1999) vegetation characteristics and trace metals (Cu, Zn and Pb) in soils along the super highways near Karachi, Pakistan. Studia Bot Hung 29:79–86, issue3/index.html

Jadhav AN, Deo S, Inam F, Ali Z (2012) Biological and chemical sciences elemental analysis (Mineral and Heavy metal) composition of *Phyllanthus amarus, Jatropha gossypifolia* and *Ruta graveolens*. Res J Pharm 3(3):43–48

Jadia CD, Fulekar MH (2008) Phytoremediation: the application OF vermicompost to remove zinc, cadmium, copper, nickel and lead by sunflower plant. Environ Eng Manage J 7(5):547–558

Jain CK, Gurunadha Rao VVS, Prakash BA, Mahesh Kumar K, Yoshida M (2010) Metal fractionation study on bed sediments of Hussainsagar Lake, Hyderabad, India. Environ Monit Assess 166:57–67

Jaiswal SK, Prakash R, Acharya R, Reddy AVR, Prakash NT (2012) Selenium content in seed, oil and oil cake of Se hyperaccumulated Brassica juncea (Indian mustard) cultivated in a seleniferous region of India. Food Chem 134:401–404

Jamal SN, Iqbal MZ, Athar M (2006) Effect of aluminum and chromium on the growth and germination of mesquite (*Prosopis juliflora* swartz.) DC. Int J Enviorn Sci Technol 3(2):173–176

Jamil S, Abhilash PC, Singh N, Sharma PN (2009) *Jatropha curcas*: a potential crop for phytoremediation of coal fly ash. J Hazard Mater 172:269–275. http://dx.doi.org/10.1016/j.jhazmat.2009.07.004

Jana S (1988) Accumulation of Hg and Cr by three aquatic species and subsequent changes in several physiological and biochemical plant parameters. Water Air Soil Pollut 38:105–109

Jansson C, Wullschleger SD, Kalluri UC, Tuskan GA (2010) Phytosequestration: carbon sequestration by plants and the prospects of genetic engineering. Bioscience 60:685

January MC, Cutright TJ, Keulen HV, Wei R (2008) Hydroponic phytoremediation of Cd, Cr, Ni, As, and Fe: can *Helianthus annuus* hyperaccumulate multiple heavy metals? Chemosphere 70:531–537

Jayashree S, Rathinamala J, Lakshmanaperumalsamy P (2011) Determination of heavy metal removal efficiency of *Chrysopogon zizanioides* (Vetiver) using textile wastewater contaminated soil. J Environ Sci Technol 4(5):543–551

Jayaweera MW, Kasturiarachchi JC, Kularatne RKA, Wijeyekoon SLJ (2008) Contribution of water hyacinth (*Eichhornia crassipes* (Mart.) Solms) grown under different nutrient conditions to Fe-removal mechanisms in constructed wetlands. J Environ Manage 87(2008):450–460

Jeonng GT, Park DH (2008) Lipase catalyzed transestirification of rapeseed oil for biodiesel production with tert-butanol. Appl Biochem Biotechnol 148:131–139

Jian-Guo L, Guang-Hui L, Wan-Chen S, Jia-Kuan X, De-Ke W (2010) Variations in uptake and translocation of Copper, Chromium and Nickel among nineteen wetland plant species. Pedosphere 20(1):96–103

Junkang Guo J, Renwei Feng, Yongzhen Ding, Ruigang Wang (2014) Applying carbon dioxide, plant growth-promoting rhizobacterium and EDTA can enhance the phytoremediation efficiency of ryegrass in a soil polluted with zinc, arsenic, cadmium and lead. J Environ Manage 141:1–8. http://dx.doi.org/10.1016/j.jenvman.2013.12.039

Juwarkar AA, Yadav SK, Kumar P, Singh SK (2008) Effect of biosludge and biofertilizer amendment on growth of *Jatropha curcas* in heavy metal contaminated soils. Environ Monit Assess 145:7–15

Jyothi NVV, Mouli PC, Reddy S, Reddy J (2003) Determination of zinc, copper, lead and cadmium in some medicinally important leaves by differential pulse anodic stripping analysis. J Trace Elem Med Biol 17(2):79–83

Kahakachchi C, Boakye HT, Uden PC, Tyson JF (2004) Chromatographic speciation of anionic and neutral selenium compounds in Se-accumulating *Brassica juncea* (Indian mustard) and in selenized yeast. J Chromatogr A 1054:303–312

Kaiser J, Galiová M, Novotný K, Červenka R, Reale L, Novotný J, Liška M, Samek O, Kanický V, Hrdlička A, Stejskal K, Adam V, Kizek R (2009) Mapping of lead, magnesium and copper accumulation in plant tissues by laser-induced breakdown spectroscopy and laser-ablation inductively coupled plasma mass spectrometry. Spectrochim Acta B: Atomic Spectrosc 64:67–73. http://dx.doi.org/10.1016/j.sab.2008.10.040

Kanchi S, Saraswathi K, Naidu NV (2012) Voltammetric method for manganese analysis in Indian traditional leafy vegetables and medicinal plants collected around Tirupati Town, a famous pilgrim center in India: The Catalytic Hydrogen Wave (CHW) Technique. Food Anal Methods 5:69–81

Kanwar MK, Bhardwaj R, Arora P, Chowdhary SP, Sharma P, Kumar S (2012) Plant steroid hormones produced under Ni stress are involved in the regulation of metal uptake and oxidative stress in *Brassica juncea* L. Chemosphere 86:41–49

Katayama H, Banba N, Sugimura Y, Tatsumi M, Kusakari S, Oyama H, Nakahira A (2013) Subcellular compart-mentation of strontium and zinc in mulberry idioblasts in relation to phytoremedia-tion potential. Environ Exp Bot 85:30–35

Keller C, Ludwig C, Davoli F, Wocheke J (2005) Thermal treatment of metal-enriched biomass produced from heavy metal phytoextration. Environ Sci Technol 39:3359–3367

Khalifa RKM, Abouziena HF, El-Mergawi RA, Youssef AA (2011) Nutritional status of some aromatic plants grown to produce volatile oils under treated municipal wastewater irrigation. Aust J Basic Appl Sci 5(12):2999–3007

Khan D (2007) Effects of cadmium on germination and seedling growth of *Prosopis juliflora* (Swartz) DC. – a potential metallophyte. Int J Biol Biotechnol 4(2–3):133–147

Khan S, Ahmad I, Shah MT, Rehman S, Khaliq A (2009) Use of constructed wetland for the removal of heavy metals from industrial wastewater. J Environ Manage 90:3451–3457

Khanam A, Singh BS (2012) Determination of heavy metal ions in selected medicinal plants of Agra. In: Chemistry of phytopotentials: health, energy and environmental perspectives, pp 289–291

Khouja HR, Abba S, Lacercat-Didier L, Daghino S, Doillon D, Richaud P et al (2013) *OmZnT1* and *OmFET*, two metal transporters from the metal-tolerant strain Zn of the ericoid mycorrhizal fungus Oidiodendron maius, confer zinc tolerance in yeast. Fungal Genet Biol 52:53–64

Kiliç CC, Kukul YS, Anaç D (2008) Performance of purslane (*Portulaca oleracea* L.) as a salt-removing crop. Agric Water Manag 95(7):854–858. ISSN 0378–3774. http://dx.doi.org/10.1016/j.agwat.2008.01.019

Klumpp A, Bauer K, Franz-Gerstein C, de Menezes M (2002) Variation of nutrient and metal concentrations in aquatic macrophytes along the Rio Cachoeira in Bahia (Brazil). Environ Int 28:165–171

Koc H, Sari H (2009) Trace metal contents of some medicinal, aromatic plants and soil samples in the Mediterranean region, Turkey. J Appl Chem Res 8:52–57

Kötschau A, Büchel G, Einax JW, Mirgorodsky D, Meißner R, von Tümpling W, Merten D (2014) Element contents in shoots of sunflower (*Helianthus annuus*): prediction versus measuring. Chemie der Erde - Geochemistry. Available online 22 March 2014, http://dx.doi.org/10.1016/j.chemer.2014.02.006

Kumar GP, Yadav SK, Thawale PR, Singh SK, Juwarkar AA (2008) Growth of Jatropha curcas on heavy metal contaminated soil amended with industrial wastes and Azotobacter – a greenhouse study. Bioresour Technol 99:2078–2082

Kumar N, Bauddh K, Kumar S, Dwivedi N, Singh DP, Barman SC (2013) Accumulation of metals in weed species grown on the soil contaminated with industrial waste and their phytoremediation potential. Ecol Eng 61:491–495. http://dx.doi.org/10.1016/j.ecoleng.2013.10.004

Kumari SP, Joanny K, Jesudas L, Kumar R (2012) Comparative study of metal uptake by medicinal plants growing in natural and polluted area of Eastern Ghats. Drug Invent Today 4(5):375–380

Kurosawa K, Egashira K, Tani M, Jahiruddin M, Zofar A, Moslehuddin AZM, Rahman ZM (2008) Groundwater–soil–crop relationship with respect to arsenic contamination in farming villages of Bangladesh – a preliminary study. Environ Pollut 156:563–565

La Rocca N, Andreoli C, Giacometti GM, Rascio N, Moro I (2009) Responses of the Antarctic microalga *Koliella antartica* (Trebouxiophyceae, Chlorophyta) to cadmium contamination. Photosynthetica 47:471–479

Lai HY, Chen ZS (2004) Effects of EDTA on solubility of cadmium, zinc, and lead and their uptake by rainbow pink and vetiver grass. Chemosphere 55:421–430

Lee S, Leustek T (1999) The affect of cadmium on sulfate assimilation enzymes in *Brassica juncea*. Plant Sci 141:201–207

Lee JH, Hossner LR, Attrep M Jr, Kung KS (2002) Comparative uptake of plutonium from soils by *Brassica juncea* and *Helianthus annuus*. Environ Pollut 120:173–182

Lenka M, Panda KK, Panda BB (1992) Monitoring and assessment of mercury pollution in the vicinity of a chloralkali plant. iv. Bioconcentration of mercury in *in situ* aquatic and terrestrial plants at Ganjam, India. Arch Environ Contam Toxicol 22:195–202

Lesage E, Meers E, Vervaeke P, Lamsal S, Hopgood M, Tack FM, Verloo MG (2005) Enhanced phytoextraction: II. Effect of EDTA and citric acid on heavy metal uptake by *Helianthus annuus* from a calcareous soil. Int J Phytoremediation 7(2):143–152

Lewandowski I, Schmidt U, Londo M, Faaij A (2006) The economic value of the phytoremediation function – assessed by the example of cadmium remediation by willow (Salix ssp). Agric Syst 89:68–89. http://dx.doi.org/10.1016/j.agsy.2005.08.004

Li M, Yue-Jin Wu, Zeng-Liang Yu, Guo-Ping Sheng, Han-Qing Yu (2009) Enhanced nitrogen and phosphorus removal from eutrophic lake water *by Ipomoea aquatica* with low-energy ion implantation. Water Res 43:1247–1256. http://dx.doi.org/10.1016/j.watres.2008.12.013

Li YH, Hu HY, Liu JC, Wu GL (2010) Distribution and mobility of copper, zinc and lead in plant-sediment systems of Quanzhou Bay estuary, China. Int J Phytoremediation 12(3):291–305

Li H, Ye ZH, Wei ZJ, Wong MH (2011) Root porosity and radial oxygen loss related to arsenic tolerance and uptake in wetland plants. Environ Pollut 159:30–37

Li Y, Zhang S, Jiang W, Liu D (2013) Cadmium accumulation, activities of antioxidant enzymes, and malondialdehyde (MDA) content in *Pistia stratiotes* L. Environ Sci Pollut Res 20:1117–1123

Li Z, Huiping Xiao, Shuiping Cheng, Liping Zhang, Xiaolong Xie, Zhenbin Wu (2014) A comparison on the phytoremediation ability of triazophos by different macrophytes. J Environ Sci 26:315–322. http://dx.doi.org/10.1016/S1001-0742(13)60417-9

Liang J, Yang Z, Tang L, Xu Y, Wang S, Chen F (2012) Growth performance and tolerance responses of jatropha (*Jatropha curcas*) seedling subjected to isolated or combined cadmium and lead stresses. Int J Agric Biol 14:861–869

Lim JM, Salido AL, Butcher DJ (2004) Phytoremediation of lead using Indian mustard (*Brassica juncea*) with EDTA and electrodics. Microchem J 76:3–9

Lim CY, Chia-Lung Chen, Jing-Yuan Wang (2013) A strategy for urban outdoor production of high-concentration algal biomass for green biorefining. Bioresour Technol 135:175–181. ISSN 0960–8524, http://dx.doi.org/10.1016/j.biortech.2012.10.028

Linger P, Müssig J, Fischer H, Kobert J (2002) Industrial hemp (*Cannabis sativa* L.) growing on heavy metal contaminated soil: fibre quality and phytoremediation potential. Ind Crops Prod 16:33–42. http://dx.doi.org/10.1016/S0926-6690(02)00005-5

Liphadzi MS, Kirkham MB (2006) Availability and plant uptake of heavy metals in EDTA-assisted phytoremediation of soil and composted biosolids. S Afr J Bot 72:391–397, http://dx.doi.org/10.1016/j.sajb.2005.10.010

Liphadzi MS, Kirkham MB, Mankin KR, Paulsen GM (2003) EDTA-assisted heavy-metal uptake by poplar and sunflower grown at a long-term sewage-sludge farm. Plant and Soil 257:171–182

Liu J, Dong Y, Xu H, Wang D, Xu J (2007a) Accumulation of Cd, Pb and Zn by 19 wetland plant species in constructed wetland. J Hazard Mater 147:947–953

Liu Y-G, Fei Ye, Guang-ming Zeng, Ting Fan, Lei Meng, Hua-shan Yuan (2007b) Effects of added Cd on Cd uptake by oilseed rape and pai-tsai co-cropping. Trans Nonferrous Metals Soc China, 17:846–852. http://dx.doi.org/10.1016/S1003-6326(07)60186-1

Liu X, Shen Y, Lou L, Ding C, Cai Q (2009) Copper tolerance of the biomass crops Elephant grass (*Pennisetum purpureum* Schumach), Vetiver grass (*Vetiveria zizanioides*) and the upland reed (*Phragmites australis*) in soil culture. Biotechnol Adv 27:633–640

Liu J, Changqun Duan, Xuehong Zhang, Yinian Zhu, Xiaoyan Lu (2011) Potential of *Leersia hexandra* Swartz for phytoextraction of Cr from soil. J Hazard Mater 188:85–91. http://dx.doi.org/10.1016/j.jhazmat.2011.01.066

Liu L Liangliang Hu, Jianjun Tang, Yuefang Li, Qian Zhang, Xin Chen (2012) Food safety assessment of planting patterns of four vegetable-type crops grown in soil contaminated by electronic waste activities. J Environ Manage 93:22–30. http://dx.doi.org/10.1016/j.jenvman.2011.08.021

Liua D, Jianga W, Liua C, Xina C, Hou W (2000) Uptake and accumulation of lead by roots, hypocotyls and shoots of Indian mustard [*Brassica juncea* (L.)]. Bioresour Technol 71:273–277

Lokeshwari H, Chandrappa GT (2007) Effects of heavy metal contamination from anthropogenic sources on Dasarahalli tank, India. Lakes Reservoirs: Res Manage 12:121–128

Lombi E, Gerzabek MH, Hborak O (1998) Mobility of heavy metals in soil and their uptake by sunflowers grown at different contamination levels. Agronomie 18:361–371

Lotfy SM, Abdel Bary SAA, El-Naka EA, Abdel Sabour MF (2009) Influence of some chelators on the phytoextraction ability of sunflower, *Helianthus annuus* for cd and pb metals in polluted soil. Zagazig J Agric Res 36(3):541–558

Lu X, Kruatrachue M, Pokethitiyook P, Homyok K (2004) Removal of cadmium and zinc by water hyacinth, *Eichhornia crassipes*. ScienceAsia 30:93–103

Lu Q, He ZL, Graetz DA, Stoffella PJ, Yang X (2011) Uptake and distribution of metals by water lettuce (*Pistia stratiotes* L.). Environ Sci Pollut Res 18:978–986

Luhach J, Chaudhry S (2012) Phytoremediation potential of *Jatropha curcas* for removal of heavy metals from refinery sludge. Int J Sci Eng Res 3(10):1–5

Lydakis-simantiris N, Skoula M, Fabian M, Naxakis G (2012) Cultivation of medicinal and aromatic plants in heavy metal contaminated soils -exploitation with caution. In: CRETE, 3rd international conference on industrial and hazardous waste management. pp 1–8

Machender G, Dhakate R, Prasanna L, Govil PK (2011) Assessment of heavy metal contamination in soils around Balanagar industrial area, Hyderabad, India. Environ Earth Sci 63:945–953

Machender G, Dhakate R, Mallikharjuna Rao ST, Mangaraja Rao B, Prasanna L (2012) Heavy metal contamination in sediments of Balanagar industrial area, Hyderabad, Andra Pradesh, India. Arabian J Geosci 7:513–525. doi:10.1007/s12517-012-0759-3

Madejón P, Murillo JM, Marañón T, Cabrera F, Soriano MA (2003) Trace element and nutrient accumulation in sunflower plants two years after the Aznalcóllar mine spill. Sci Total Environ 307:239–257. http://dx.doi.org/10.1016/S0048-9697(02)00609-5

Madrid F, Liphadzi MS, Kirkham MB (2008) EDTA-assisted phytostabilization by barley roots contaminated with heavy metals. In: Hartemink AE, McBratney AB, Naidu R (eds) Developments in soil science, vol 32. Elsevier, pp 697–718. http://dx.doi.org/10.1016/S0166-2481(07)32029-1

Mahamadi C, Nharingo T (2010) Competitive adsorption of Pb^{2+}, Cd^{2+} and Zn^{2+} ions onto *Eichhornia crassipes* in binary and ternary systems. Bioresour Technol 101:859–864

Maine MA, Duarte MV, Suñ NL (2001) Cadmium uptake by floating macrophytes. Water Res 35(11):2629–2634

Maine MA, Suñe N, Hadad H, Sánchez G, Bonetto C (2006) Nutrient and metal removal in a constructed wetland for wastewater treatment from a metallurgic industry. Ecol Eng 2(6):341–347

Maine MA, Suñe N, Hadad H, Sánchez G, Bonetto C (2007) Removal efficiency of a constructed wetland for wastewater treatment according to vegetation dominance. Chemosphere 68:1105–1113

Maine MA, Suñe N, Hadad H, Sánchez G, Bonetto C (2009) Influence of vegetation on the removal of heavy metals and nutrients in a constructed wetland. J Environ Manage 90:355–363

Mainea MA, Suñe NL, Lagger SC (2004) Chromium bioaccumulation: comparison of the capacity of two floating aquatic macrophytes. Water Res 38:1494–1501

Majid NM, Islam MM, Riasmi Y (2012) Heavy metal uptake and translocation by *Jatropha curcas* L. in sawdust sludge contaminated soils. Aust J Crop Sci 6(5):891–898

Makris KC, Shakya KM, Datta R, Sarkar D, Pachanoor D (2007) High uptake of 2,4,6-trinitrotoluene by vetiver grass – potential for phytoremediation? Environ Pollut 146:1–4

Malawska M, Wiłkomirski B (2001) An analysis of soil and plant *(Taraxacum officinale)* contamination with heavy metals and polycyclic aromatic hydro-carbons (PAHs) in the area of the railway junction Iława Główna, Poland. Water Air Soil Pollut 127:339–349

Malenčić D, Kevrešan Ž, Popović M (2005) Heavy metals content in some medicinal and aromatic wild growing plants from the Frushka Gora Mountain. Annals of The Faculty of Engineering Hunedoara – 2005 Tome III. Fascicole 1. pp 37–40

Malik A (2007) Environmental challenge *vis a vis* opportunity: the case of water hyacinth. Environ Int 33:122–138

Manciulea A, Ramsey MH (2006) Effect of scale of Cd heterogeneity and timing of exposure on the Cd uptake and shoot biomass, of plants with a contrasting root morphology. Sci Total Environ 367:958–967

Mane PC, Bhosle AB, Kulkarni PA (2011) Biosorption and biochemical study on water hyacinth (*Eichhornia crassipes*) with reference to selenium. Arch Appl Sci Res 3(1):222–229

Mangabeira PAO, Lamperti A, de Almeida AAF, Oliveira AH, Escaig F, Severo MIG, Silva D d C, Saloes M, Mielke MS, Lucena ER, Martins MC, Santana KB, Gavrilov KL, Galle P, Levi-Setti R (2004) Accumulation of chromium in root tissues of *Eichhornia Crassipes* (Mart.) Solms. in Cachoeira river – Brazil. Appl Surf Sci 231–232(2004):497–501

Mangabeira PA, de Almeida AAF, Silva DC, Oliveira AH, Fernandes VF, Ferreira AS, dos Santos Júnior AJ, Galle P (2010) Bioaccumulation of chromium in *Alternanthera philoxeroides* using ion microscopy (SIMS). Microsc Microanal 16(Suppl 2):398–399

Mangabeira PA, Ferreira AS, de Almeida AAF, Fernandes VF, Lucena E, Souza VL, dos Santos Júnior AJ, Oliveira AH, Grenier-Loustalot MF, Barbier F, Silva DC (2011) Compartmentalization and ultrastructural alterations induced by chromium in aquatic macrophytes. Biometals 24:1017–1026

Mangkoedihardjo S, Surahmaida (2008) *Jatropha curcas* L. for phytoremediation of lead and cadmium polluted soil. World Appl Sci J 4(4):519–522

Mangkoedihardjo S, Ratnawati R, Alfianti N (2008) Phytoremediation of hexavalent chromium polluted soil using *Pterocarpus indicus* and *Jatropha curcas* L. World Appl Sci J 4(3):338–342

Marchand L, Mench M, Jacob DL, Otte ML (2010) Metal and metalloid removal in constructed wetlands, with emphasis on the importance of plants and standardized measurements: a review. Environ Pollut 158:3447–3461

Marchiol L, Fellet G, Perosa D, Zerbi G (2007) Removal of trace metals by *Sorghum bicolor* and *Helianthus annuus* in a site polluted by industrial wastes: a field experience. Plant Physiol Biochem 45:379–387

Marillia E-F, Francis T, Falk KC, Smith M, Taylor DC (2014) Palliser's promise: *Brassica carinata*, an emerging western Canadian crop for delivery of new bio-industrial oil feedstocks. Biocatalysis Agric Biotechnol 3:65–74. ISSN 1878–8181, http://dx.doi.org/10.1016/j.bcab.2013.09.012

Martin RS, Mather TA, Pyle DM, Day JA, Witt MLI, Collins SJ, Hilton RG (2010) Major and trace element distributions around active volcanic vents determined by analyses of grasses: implications for element cycling and bio-monitoring. Bull Volcanol 72:1009–1020

Mayora ON, Njeri KP, Rang'ondi OE, Sarima CJ (2012) The potential of Zea mays, Commelina bengelensis, *Helianthus annuus* and *Amaranthus* hybridus for phytoremediation of waste water. Ambiente & Água – Interdiscip J Appl Sci 7(3):51

McCutcheon SC, Schnoor JL (eds) (2003) Phytoremediation – transformation and control of contaminants. Wiley Interscience, Hoboken, pp 985

Meera M, Agamuthu P (2011) phytoextraction of As and Fe using *Hibiscus cannabinus* L. from soil polluted with landfill leachate. Int J Phytoremediation 14:186–199

Meers E, Ruttens A, Hopgood M, Lesage E, Tack MFG (2005) Potential of *Brassic rapa, Cannabis sativa, Helianthus annuus* and *Zea mays* for phytoextraction of heavy metals from calcareous dredged sediment derived soils. Chemosphere 61:561–572

Meers E, Van Slycken S, Adriaensen K, Ruttens A, Vangronsveld J, Du Laing G, Witters N, Thewys T, Tack FMG (2010) The use of bio-energy crops (*Zea mays*) for 'phytoattenuation' of heavy metals on moderately contaminated soils: a field experiment. Chemosphere 78:35–41. http://dx.doi.org/10.1016/j.chemosphere.2009.08.015

Meka PK, Tripathi V, Singh RP (2007) Synthesis of biodiesel fuel from safflower oil using various reaction parameters. J Oleo Sci 56(1):9–12

Melo VF, Gabardo J, Souza LCP, Reissmann CB, de Andrade MG (2009) Heavy metals in soils of a lead mining and metallurgy area: I – Phytoextraction. Rev Bras Ciênc Solo 33(6):1879–1888

Meng DK, Chen J, Yang ZM (2011) Enhancement of tolerance of Indian mustard (*Brassica juncea*) to mercury by carbon monoxide. J Hazard Mater 186:1823–1829

Meyers DER, Auchterlonie GJ, Webb RI, Wood B (2008) Uptake and localisation of lead in the root system of *Brassica juncea*. Environ Pollut 153:323–332

Mhatre GN, Chaphekar SB (1985) The effect of mercury on some aquatic plants. Environ Pollut A, Ecol Biol 39(3):207–216

Minglin L, Yuxiu Z, Tuanyao C (2005) Identification of genes up-regulated in response to Cd exposure in Brassica juncea L. Gene 19(363):151–158

Minh VV, Khoa LV (2009) Phytoremediation of cadmium and lead contaminated soil types by vetiver grass. VNU J Sci, Earth Sci 25:98–103

Miretzky P, Saralegui A, Fernández Cirelli A (2004) Aquatic macrophytes potential for the simultaneous removal of heavy metals (Buenos Aires, Argentina). Chemosphere 57:997–1005

Miretzky P, Saralegui A, Cirelli AF (2006) Simultaneous heavy metal removal mechanism by dead macrophytes. Chemosphere 62:247–254

Mishra VK, Tripathi BD (2008) Concurrent removal and accumulation of heavy metals by the three aquatic macrophytes. Bioresour Technol 99:7091–7097

Mishra VK, Tripathi BD, Kima KH (2009) Removal and accumulation of mercury by aquatic macrophytes from an open cast coal mine effluent. J Hazard Mater 172:749–754

Mishra VK, Upadhyay AR, Pandey SK, Tripathi BD (2008) Concentrations of heavy metals and aquatic macrophytes of Govind Ballabh Pant Sagar an anthropogenic lake affected by coal mining effluent. Environ Monit Assess 141:49–58

Mishra S, Mohanty M, Pradhan C, Patra HK, Das R, Sahoo S (2012) Physico-chemical assessment of paper mill effluent and its heavy metal remediation using aquatic macrophytes—a case study at JK Paper mill, Rayagada. India Environ Monit Assess 185:4347–4359. doi:10.1007/s10661-012-2873-9

Mobin M, Khan NA (2007) Photosynthetic activity, pigment composition and antioxidative response of two mustard (Brassica juncea) cultivars differing in photosynthetic capacity subjected to cadmium stress. J Plant Physiol 164:601–610

Mokgalaka-Matlala NS, Flores-Tavińon E, Castillo-Michel H, Peralta-Videa JR, Gardea-Torresdey JL (2008) Toxicity of arsenic (III) and (V) on plant growth, element uptake, and total amylolytic activity of mesquite (*Prosopis juliflora* x *P. velutina*). Int J Phytoremediation 10:47–60

Mokgalaka-Matlala NS, Flores-Tavizón E, Castillo-Michel H, Peralta-Videa JR, Gardea-Torresdey JL (2009) Arsenic tolerance in mesquite (*Prosopis* sp.): low molecular weight thiols synthesis and glutathione activity in response to arsenic. Plant Physiol Biochem 47:822–826

Mokhtar H, Morad N, Fizri FFA (2011) Hyperaccumulation of copper by two species of aquatic plants. In: 2011 international conference on environment science and engineering IPCBEE, vol 8. IACSIT Press, Singapore

Moodley KG, Baijnath H, Southway-Ajulu FA, Maharaj S, Chetty SR (2007) Determination of Cr, Pb and Ni in water, sludge and plants from settling ponds of sewage treatment works. Water SA 33:723–728

Mufarrege MM, Hadad HR, Maine MA (2010) Response of Pistia stratiotes to heavy metals (Cr, Ni, and Zn) and phosphorous. Arch Environ Contam Toxicol 58:53–61

Mulligan CN, Youn RN, Gibbs B (2001) Heavy metal removal from Sediments by biosurfactants. J Hazard Mater 85:111–125

Nagaraju A, Prasad KSS (1998) Growth of *Prosopis juliflora* on Pegmatite tailings from Nellore Mica Belt, Andhra Pradesh, India. Environ Geol 36(3–4):320

Najafi F, Khavari-Nejad RA, Hasanjanzadeh F (2011) The physiological responses of sunflower (Helianthus annuus L.) to NiSO4. Afr J Plant Sci 5(3):201–206

Nakwanit S, Visoottiviseth P, Khokiattiwong S, Sangchoom W (2011) Management of arsenic-accumulated waste from constructed wetland treatment of mountain tap-water. J Hazard Mater 185:1081–1085

Naqvi SM, Rizvi SA (2000) Accumulation of chromium and copper in three different soils and bioaccumulation in an aquatic plant, *Alternanthera philoxeroides*. Bull Environ Contami Toxicol 65:55–61

Nathan O, Njeri KP, Rang'ondi OE, Sarima CJ (2012) The potential of Zea mays, Commelina bengelensis, *Helianthus annuus* and *Amaranthus hybridus* for phytoremediation of waste water. Revista Ambiente & Água – Interdiscip J Appl Sci 7(3):51–60. http://dx.doi.org/10.4136/ambi-agua.684

Navari-Izzo F, Quartacci MF, Pinzino C, Dalla VF, Sgherri C (1998) Thylakoid-bound and stromal antioxidative enzymes in wheat treated with excess of copper. Physiol Plant 104:630–638

Navari-Izzo F, Pinzino C, Quartacci MF, Sgherri C (1999) Superoxide and hydroxyl radical generation, and superoxide dismutase in PSII membrane fragments from wheat. Free Radic Res 31:S3–S9

Naveed NH, Batool AI, Rehman FU, Hameed U (2010) Leaves of roadside plants as bioindicator of traffic related lead pollution during different seasons in Sargodha, Pakistan. Afr J Environ Sci Technol 4(11):770–774

Naveed NH, Batool AI, Hameed U, Ali A, Rehman MFU, Sher M, Ali S, Faiz S (2012) Biomonitoring of the traffic related heavy metal pollution using roadside plants as possible bioindicators during different seasons. Asian J Chem 24(10):4661–4664

Nazir A, Malik RN, Ajaib M, Khan N, Siddiqui MF (2011) hyperaccumulators of heavy metals of industrial areas of Islamabad and Rawalpindi. Pak J Bot 43(4):1925–1933

Ndzomo GT, Ndoumou DO, Awah MT (1994) Effect of Fe^{2+}, Mn^{2+}, Zn^{2+} and Pb^{2+} on H^+/K^+ fluxes in excised *Pistia stratiotes* roots. Biol Plant 36(4):591–597

Nehnevajova E, Herzig R, Federer G, Erismann KH, Schwitzguébel JP (2005) Screening of sunflower cultivars for metal phytoextraction in a contaminated field prior to mutagenesis. Int J Phytoremediation 7(4):337–349

Nehnevajova E, Herzig R, Federer G, Erismann KH, Schwitzguébel JP (2007) Chemical mutagenesis–a promising technique to increase metal concentration and extraction in sunflowers. Int J Phytoremediation 9(2):149–165

Niu Z, Sun L, Sun T (2009) Response of root and aerial biomass to phytoextraction of Cd and Pb by sunflower, castor bean, alfalfa and mustard. Adv Environ Biol 3(3):255–262

Nouairi I, Ammar WB, Youssef NB, Daoud DBM, Ghorbal MH, Zarrouk M (2006) Comparative study of cadmium effects on membrane lipid composition of Brassica juncea and Brassica napus leaves. Plant Sci 170:511–519

NRC (1985) Meat and poultry inspection. The scientific basis of the nation's program. National Academy Press, Washington, DC

Núñez SER, Negrete JLM, Rios JEA, Hadad HR, Maine MA (2011) Hg, Cu, Pb, Cd, and Zn accumulation in macrophytes growing in tropical wetlands. Water Air Soil Pollut 216:361–373

Núñez-López RA, Meas Y, Gama SC, Borges RO, Olguín EJ (2008) Leaching of lead by ammonium salts and EDTA from *Salvinia minima* biomass produced during aquatic phytoremediation. J Hazard Mater 154(I):623–632. http://dx.doi.org/10.1016/j.jhazmat.2007.10.101

O'Keeffe DH, Hardy JK, Rao RA (1984) Cadmium uptake by the water hyacinth: effects of solution factors. Environ Pollut Ser A, Ecol Biol 34(2):133–147

Odjegba VJ, Fasidi IO (2004) Accumulation of trace elements by Pistia stratiotes: implications for phytoremediation. Ecotoxicology 13:637–646

Odjegba VJ, Fasidi IO (2006) Effects of heavy metals on some proximate composition of *Eichhornia crassipes*. J Appl Sci Environ Manage 10(1):83–87

Odjegba VJ, Fasidi IO (2007) Phytoremediation of heavy metals by Eichhornia crassipes. Environmentalist 27:349–355

Olivares AR, Carrillo-González R, González-Chávez MDCA, Soto Hernández RM (2013) Potential of castor bean (Ricinus communis L.) for phytoremediation of mine tailings and oil production. J Environ Manage 114:316–323

Oseni AO (2004) Phytoremediation of selected heavy metals from aqueous media by water hyacinth (*Eichhornia crassipes* (Mart.) Solms. M.Tech degree thesis submitted to The Federal university of Technology Akuro, Ondo State, Nigeria

Oshunsanya SO, Oluwasemire KO, Ogunwumi KS (2012) The use of vetiver grass slips in removing heavy metal contamination of dumpsite in Ibadan Metropolis. Scholarly J Agric Sci 2(6):115–118

Ozcan MM, Akbulut M (2007) Estimation of minerals, nitrate and nitrite contents of medicinal and aromatic plants used as spices, condiments and herbal tea. Food Chem 106:852–858

Page K, Harbottle MJ, Cleall PJ, Hutchings TR (2014) Heavy metal leaching and environmental risk from the use of compost-like output as an energy crop growth substrate. Sci Total Environ 487:260–271. http://dx.doi.org/10.1016/j.scitotenv.2014.04.021

Pagliano C et al (2006) Evidence for PSII-donor-side damage and photoinhibition induced by cadmium treatment on rice (*Oryza sativa* L.). J Photochem Photobiol B: Biol 84:70–78

Paivaa LB, de Oliveira JG, Azevedo RA, Ribeiro DR, da Silva MG, Vitória AP (2009) Ecophysiological responses of water hyacinth exposed to Cr^{3+} and Cr^{6+}. Environ Exp Bot 65:403–409

Pan W, Chuan Wu, Shengguo Xue, William Hartley (2014) Arsenic dynamics in the rhizosphere and its sequestration on rice roots as affected by root oxidation. J Environ Sci 26:892–899. http://dx.doi.org/10.1016/S1001-0742(13)60483-0

Pandey VC (2012a) Invasive species based efficient green technology for phytoremediation of fly ash deposits. J Geochem Explor 123:13–18. http://dx.doi.org/10.1016/j.gexplo.2012.05.008

Pandey VC (2012b) Phytoremediation of heavy metals from fly ash pond by *Azolla caroliniana*. Ecotoxicol Environ Saf 82:8–12. http://dx.doi.org/10.1016/j.ecoenv.2012.05.002

Pandey VC (2013) Suitability of *Ricinus communis* L. cultivation for phytoremediation of fly ash disposal sites. Ecol Eng 57:336–341. http://dx.doi.org/10.1016/j.ecoleng.2013.04.054

Pandey V, Dixit V, Shyam R (2005) Antioxidative responses in relation to growth of mustard (*Brassica juncea* cv. Pusa Jaikisan) plants exposed to hexavalent chromium. Chemosphere 61:40–47

Pang J, Chan GSY, Zhang J, Liang J, Wong MH (2003) Physiological aspects of vetiver grass for rehabilitation in abandoned metalliferous mine wastes. Chemosphere 52:1559–1570

Park JBK, Craggs RJ, Shilton AN (2011) Wastewater treatment high rate algal ponds for biofuel production. Bioresour Technol 102:35–42

Parth V, Murthy NN, Saxena PR (2009) Hydro-geochemistry around waste disposal site at Pattulaguda, Hyderabad, India. Goldschmidt Conf Abstr 2009:A998

Parth V, Murthy NN, Saxena PR (2011) Assessment of heavy metal contamination in soil around hazardous waste disposal sites in Hyderabad city (India): natural and anthropogenic implications. J Environ Res Manage 2:027–034

Patel A, Patra DD (2014) Influence of heavy metal rich tannery sludge on soil enzymes vis-à-vis growth of *Tagetes minuta*, an essential oil bearing crop. Chemosphere 112:23–332. http://dx.doi.org/10.1016/j.chemosphere.2014.04.063

Patrick Audet P (2014) Arbuscular mycorrhizal fungi and metal phytoremediation: ecophysiological complementarity in relation to environmental stress. In: Ahmad P, Rasool S (eds) Emerging technologies and management of crop stress tolerance. Academic Press, San Diego, pp 133–160. ISBN 9780128008751, http://dx.doi.org/10.1016/B978-0-12-800875-1.00006-5

Paun A, Neagoe A, Baciu I (2012) The effects of arbuscular mychorrizal fungi on the transfer of heavy metals and oxidative stress related parameters in sunflower exposed to multi-element pollution. Rev Chim (Bucharest) 63(2):146–152

Pavel P-B, Puschenreiter M, Wenzel WW, Diacu E, Barbu CH (2014) Aided phytostabilization using *Miscanthus sinensis* giganteus on heavy metal-contaminated soils. Sci Total Environ 479–480:125–131. http://dx.doi.org/10.1016/j.scitotenv.2014.01.097

Pedron F, Petruzzelli G, Barbafieri M, Tassi E (2013) Remediation of a mercury-contaminated industrial soil using bioavailable contaminant stripping. Pedosphere 23(1):104–110

Pellegrineschi A, Damon JP, Valtorta N, Paillard N, Tepfer D (1994) Improvement of ornamental characters and fragrance production through genetic transformation by Agrobacterium rhizigenes. Biotechnology 12:64–68

Peña A, Mingorance MD, Guzmán I, Sánchez L, Fernández-Espinosa AJ, Valdés B, Rossini-Oliva S (2014) Protecting effect of recycled urban wastes (sewage sludge and wastewater) on ryegrass against the toxicity of pesticides at high concentrations. J Environ Manag 142:23–29. http://dx.doi.org/10.1016/j.jenvman.2014.04.002

Pérez-López R, Márquez-García B, Abreu MM, Nieto JM, Córdoba F (2014) *Erica andevalensis* and *Erica australis* growing in the same extreme environments: phytostabilization potential of mining areas. Geoderma 230–231:194–203. http://dx.doi.org/10.1016/j.geoderma.2014.04.004

Pinder JE III, Hintona TG, Whicker FW (2005) The influence of a whole-lake addition of stable cesium on the remobilization of aged ^{137}Cs in a contaminated reservoir. J Environ Radioact 80:225–243

Pinder JE III, Hintona TG, Whicker FW (2006) Foliar uptake of cesium from the water column by aquatic macrophytes. J Environ Radioact 85:23–47

Płażek A, Dubert F, Kościelniak J, Tatrzańska M, Maciejewski M, Gondek K, Żurek G (2014) Tolerance of *Miscanthus giganteus* to salinity depends on initial weight of rhizomes as well as high accumulation of potassium and proline in leaves. Ind Crops Products 52:278–285. http://dx.doi.org/10.1016/j.indcrop.2013.10.041

Poonam, Bhardwaj R, Sharma R, Handa N, Kaur H, Kaur R, Sirhindi G, Thukral AK (2014) Prospects of field crops for phytoremediation of contaminants. In: Ahmad P, Rasool S (eds) Emerging technologies and management of crop stress tolerance. Academic Press, San Diego, pp 449–470. http://dx.doi.org/10.1016/B978-0-12-800875-1.00019-3

Prabavathi R, Mathivanan V, Selvisabanayagam (2011) Analysis of concentration and accumulation of heavy metal cadmium in four selected terrestrial plants. Int J Dev Res 1(4):027–030

Prasad KVSK, Saradhi PP, Sharmila P (1999) Concerted action of antioxidant enzymes and curtailed growth under zinc toxicity in *Brassica juncea*. Environ Exp Bot 42(1):1–10

Prasad MNV (2004a) Phytoremediation of metals and radionuclides in the environment: the case for natural hyperaccumulators, metal transporters, soil amending chelators and transgenic plants. In: Prasad MNV (ed) Trace metal stress in plants: from biomolecules to ecosystems, 2nd edn. Springer/Narosa, Heidelberg/New Delhi, pp 345–392

Prasad MNV (2004b) Phytoremediation of metals in the environment for sustainable development. Proc Indian Natl Sci Acad 70:71–98

Prasad MNV (2007a) Phytoremediation in India. In: Willey N (ed) Phytoremediation methods and reviews. Humana Press, Totowa

Prasad MNV (2007b) Sunflower (*Helinathus annuus* L.): a potential crop for environmental industry. Helia 30(46):167–174

Prasad MNV (2007c) Aquatic plants for phytotechnology. In: Singh SN, Tripathi RD (eds) Environmental bioremediation technologies. Springer, New York, pp 259–274

Prasad MNV (2008) Trace elements in traditional healing plants – remedies or risks. In: Prasad MNV (ed) Trace elements as contaminants and nutrients: consequences in ecosystems and human health. Wiley, New York, pp 137–160

Prasad MNV (2011) A state-of-the-art report on bioremediation, its applications to contaminated sites in India. M/o of Environment & Forests, GOI, New Delhi. 90

Prasad MNV (2012) Exploitation of weeds and ornamentals for bioremediation of metalliferous substrates in the era of climate change. In: Ahmad P, Prasad MNV (eds) Environmental adaptations and stress tolerance of plants in the era of climate change. doi:10.1007/978-1-4614-0815-4_23

Prasad MNV (2013) Metallophytes – properties, functions and applications. The Botanica 62, 63:62–63, 17–26

Prasad MNV (2014) Engineered phyto-covers as natural caps for containment of hazardous mine and municipal solid waste dump sites – possible energy sources. In: Öztürk M et al (eds) Phytoremediation for green energy. doi:10.1007/978-94-007-7887-0_3

Prasad MNV (ed) (2015) Bioremediation and bioeconomy. Elsevier, USA (in press)

Prasad MNV, Freitas H (2003) Metal hyperaccumulation in plants – Biodiversity prospecting for phyto- remediation technology. Electron J Biotechnol 6:75–321, Online electronic journal http://www.ejbiotechnology.info/content/vol6/

Prasad MNV, Rajendra Prasad (2012) Nature's cure for cleanup of contaminated environment – a review of bioremediation strategies. Rev Environ Health 28:181–189. doi:10.1515/reveh-2012-0028

Prasad MNV, Nakbanpote W (2015) Integrated management of mine waste using biogeo-technologies focusing Thai mines. In: Thangavel P, Sridevi G (eds) Environmental sustainability. Role of green technologies. doi:10.1007/978-81-322-2056-5_14

Prasad MNV, Greger M, Smith BN (2001) Aquatic macrophytes. In: Prasad MNV (ed) Metals in the environment – analysis by biodiversity. Marcel Dekker Inc, New York, pp 259–288

Prasad MNV, Freitas H, Fraenzle S, Wuenschmann S, Markert B (2010) Knowledge explosion in phytotechnologies for environmental solutions. Environ Pollut 158(1):18–23

Premarathna HMPL, Hettiarachchi GM, Indraratne SP (2011) Trace metal concentration in crops and soils collected from intensively cultivated areas of Sri Lanka. Pedologist 2011:230–240

Princewill-Ogbonna IL, Ogbonna PC (2011) Heavy metal content in soil and medicinal plants in high traffic urban area. Pak J Nutr 10(7):618–624

Puhui Ji P, Tieheng Sun, Yufang Song, M. Leigh Ackland, Yang Liu (2011) Strategies for enhancing the phytoremediation of cadmium-contaminated agricultural soils by Solanum nigrum L. Environ Pollut 159(3):762–768. ISSN 0269–7491, http://dx.doi.org/10.1016/j.envpol.2010.11.029

Qadir S, Qureshi MI, Javed S, Abdin MZ (2004) Genotypic variation in phytoremediation potential of Brassica juncea cultivars exposed to Cd stress. Plant Sci 167:1171–1181, ISSN 0168–9452, http://dx.doi.org/10.1016/j.plantsci.2004.06.018

Quartacci MF, Cosi E, Navari-Izzo F (2001) Lipids and NADPH-dependent superoxide production in plasma membrane vesicles from roots of wheat grown under copper deficiency and excess. J Exp Bot 152:67–75

Quartacci MF, Baker AJM, Navari-Izzo F (2005) Nitrilotriacetate- and citric acid-assisted Phytoextraction of cadmium by Indian mustard (*Brassica juncea* (L.) Czernj, Brassicaceae). Chemosphere 59:1249–1255

Quartacci MF, Argilla A, Baker AJM, Navari-Izzo F (2006) Phytoextraction of metals from a multiply contaminated soil by Indian mustard. Chemosphere 63:918–925

Qureshi MI, D'Amici GM, Fagioni M, Rinalducci S, Zolla L (2010) Iron stabilizes thylakoid protein–pigment complexes in Indian mustard during Cd-phytoremediation as revealed by BN-SDS-PAGE and ESI-MS/MS. J Plant Physiol 167:761–770

Rahi TS, Singh K, Singh B (2013) Screening of sodicity tolerance in Aloe vera: an industrial crop for utilization of sodic lands. Ind Crops Products 44:528–533. ISSN 0926–6690, http://dx.doi.org/10.1016/j.indcrop.2012.10.001

Rahman MA, Hasegawa H (2011) Aquatic arsenic: phytoremediation using floating macrophytes. Chemosphere 83:633–646

Rai PK, Kumar G (2010) The genotoxic potential of two heavy metals in inbred lines of maize (Zea mays L.). Turk J Bot 34:39–46

Rai UN, Sinha S, Tripathi RD, Chandra P (1995) Wastewater treatability potential of some aquatic macrophytes: removal of heavy metals. Ecol Eng 5:5–12

Rai UN, Pandey K, Sinha S, Singh A, Saxena R, Gupta DK (2004) Revegetating fly ash landfills with *Prosopis juliflora* L.: impact of different amendments and Rhizobium inoculation. Environ Int 30:293–300

Rajkumar M, Freitas H (2008a) Effects of inoculation of plant-growth promoting bacteria on Ni uptake by Indian mustard. Bioresour Technol 99:3491–3498

Rajkumar M, Freitas H (2008b) Influence of metal resistant-plant growth-promoting bacteria on the growth of *Ricinus communis* in soil contaminated with heavy metals. Chemosphere 71(5):834–842, ISSN 0045–6535

Rajkumar M, Nagendran R, Lee KJ, Lee WH, Kim SZ (2006) Influence of plant growth promoting bacteria and Cr^{6+} on the growth of Indian mustard. Chemosphere 62:741–748

Rajkumar M, Ma Y, Freitas H (2008) Characterization of metal-resistant plant-growth promoting Bacillus weihenstephanensis isolated from serpentine soil in Portugal. J Basic Microbiol 48:500–508

Rajkumar M, Prasad MNV, Freitas H, Ae N (2009) Biotechnological applications of serpentine soil bacteria for phytoremediation of trace metals. Crit Rev Biotechnol 29(2):120–130

Rajkumar M, Ae N, Prasad MNV, Freitas H (2010) Potential of siderophore producing bacteria for improving heavy-metal phytoextraction. Trends Biotechnol 28(3):142–149

Rajkumar M, Sandhya S, Prasad MNV, Freitas H (2012) Perspectives of plant-associated microbes in heavy metal phytoremediation. Biotechnol Adv 30:1562–1574

Rakhshaee R, Giahi M, Pourahmad A (2009) Studying effect of cell wall's carboxyl-carboxylate ratio change of *Lemna minor* to remove heavy metals from aqueous solution. J Hazard Mater 163(1):165–173

Ramadhs AS, Jayaraj S, Muraleedharan C (2004) Use of vegetable oils as I.C engine fuels: a review. Renew Energy 29:727–742

Ramlal PS, Bugeny FWB, Kling GW, Nriagu JO, Rudd JWM, Campbell LM (2003) Mercury concentrations in water, sediment, and biota from lake Victoria, East Africa. J Great Lakes Res 29(2):283–291

Rana S, Jana J, Bag SK, Mukherjee (Roy) S, Biswas JK, Ganguly (Lahiri) S, Sarkar (Paria) D, Jana BB (2011) Performance of constructed wetlands in the reduction of cadmium in a sewage treatment cum fish farm at Kalyani, West Bengal, India. Ecol Eng 37:2096–2100

Rascio N, Vecchia FD, La Rocca N, Barbato R, Pagliano C, Raviolo M, Gonnelli C, Gabbrielli R (2008) Metal accumulation and damage in rice (cv. Vialone nano) seedlings exposed to cadmium. Environ Exp Bot 62:267–278

Reichenauer TG, Germida JJ (2008) Phytoremediation of organic contaminants in soil and groundwater. ChemSusChem 1:708–717

Rocio M, Elvira E, Pilar Z, María-José S (2013) Could an abandoned mercury mine area be cropped? Environ Res 125:150–159. ISSN 0013–9351, http://dx.doi.org/10.1016/j.envres.2012.12.012

Romeiro S, Lagôa AMMA, Furlani PR, de Abreu CA, de Abreu MF, Erismann NM (2006) Lead uptake and tolerance of *Ricinus communis* L. Braz J Plant Physiol 18(4):483–489

Rosselli W, Keller C, Boschi K (2003) Phytoextraction capacity of tree growing on a metal contaminated soil. Plant Soil 256:265–272

Roodi MM, Said MABM, Honari H (2012) Phytoremediation using the influence of aromatic crop on heavy-metal polluted soil, a review. Adv Environ Biol 6(10):2663–2668

Roongtanakiat N, Chairoj P (2001) Uptake potential of some heavy metals by vetiver grass. Kasetsart J (Nat Sci) 35:46–50

Roongtanakiat N, Sanoh S (2011) Phytoextraction of zinc, cadmium and lead from contaminated soil by vetiver grass. Kasetsart J (Nat Sci) 45:603–612

Roongtanakiat N, Tangruangkiat S, Meesat R (2007) Utilization of Vetiver grass (*Vetiveria zizanioides*) for removal of heavy metals from industrial wastewaters. Sci Asia 33:397–403

Roongtanakiat N, Osotsapar Y, Yindiram C (2008) Effects of soil amendment on growth and heavy metals content in Vetiver grown on iron ore tailings. Kasetsart J (Nat Sci) 42:397–406

Roongtanakiat N, Osotsapar Y, Yindiram C (2009) Influence of heavy metals and soil amendments on Vetiver (*Chrysopogon zizanioides*) grown in Zinc mine soil. Kasetsart J (Nat Sci) 43:37–49

Roongtanakiat N, Nirunrach T, Chanyotha S, Hengchaovanich D (2003) Uptake of heavy metals in landfill leachate by vetiver grass. Kasetsart J (Nat Sci) 37:168–175

Rotkittikhun P, Chaiyarat R, Kruatrachue M, Pokethitiyook P, Baker AJM (2007) Growth and lead accumulation by the grasses *Vetiveria zizanioides* and *Thysanolaena maxima* in lead-contaminated soil amended with pig manure and fertilizer: a glasshouse study. Chemosphere 66:45–53

Ruiz JM, Rivero RM, Romero L (2007) Comparative effect of Al, Se, and Mo toxicity on NO3– assimilation in sunflower (*Helianthus annuus* L.) plants. J Environ Manage 83:207–212. http://dx.doi.org/10.1016/j.jenvman.2006.03.001

Rulangaranga ZK, Mugasha AL (2003) Effects of Chromium uptake on the growth characteristics of *Eichhornia crassipes* (Mart.) Solms. Tanz J Sci 29:109–125

Sabudak T, Seren G, Kaykioglu G, Dincer AR (2007) determination of trace elements in soil and Sunflower (*Helianthus annuus* L.) plant parts. Fresenius Environ Bull 16(10):1274–1278

Sahi SV, Bryant NL, Sharma NC, Singh SR (2002) Characterization of a lead hyperaccumulator shrub, Sesbania drummondii. Environ Sci Technol 36:4676–4680

Saini P, Khan S, Baunthiyal M, Sharma V (2012) Organ-wise accumulation of fluoride in *Prosopis juliflora* and its potential for phytoremediation of fluoride contaminated soil. Chemosphere 89:633–635

Saleh HM (2012) Water hyacinth for phytoremediation of radioactive waste simulate contaminated with cesium and cobalt radionuclides. Nucl Eng Des 242:425–432

Samardjieva KA, Pissarra J, Castro PM, Tavares F (2011) Insights into phytoremediation solutions for environmental recovery. Recent Pat Biotechnol 5(1):25–39

Santos-Jallath J, Castro-Rodríguez A, Huezo-Casillas J, Torres-Bustillos L (2012) Arsenic and heavy metals in native plants at tailings impoundments in Queretaro, Mexico. Phys Chem Earth 37–39:10–17

Sarin R, Sharma M (2007) Jatropha Palm biodiesel blends: an optimum mix for Asia. Fuel 86:1365–1371

Sas-Nowosielska A, Kucharski R, Małkowski E, Pogrzeba M, Kuperberg JM, Kryński K (2004) Phytoextraction crop disposal—an unsolved problem. Environ Pollut 128:373–379. http://dx.doi.org/10.1016/j.envpol.2003.09.012

Sas-Nowosielska A, Kucharski R, Pogrzeba M, Malkowski E (2008) Soil remediation scenarios for heavy metal contaminated soil. In: Simeonov L, Sargsyan V (eds) Soil chemical pollution, risk assessment, remediation and security. Springer Science+Business Media B.V, pp 301–307

Satyakala G, Jamil K (1992) Chromium-induced biochemical changes in *Eichhornia crassipes* (Mart) Solms and *Pistia stratiotes L.* Bull Environ Contam Toxicol 48:921–928

Saxena PK, Raj SK, Dan T, Perras MR, Vettakkorumaka-nakav NN (1999) Phytoremediation of heavy metal contaminated and polluted soils. In: Prasad MNV, Hagemayer J (eds) Heavy metal stress in plants – from molecules to ecosystems. Springer, Berlin, pp 305–329

Saxena M, Bisht R, Roy SD, Sopory SK, Bhalla-Sarin N (2005) Cloning and characterization of a mitochondrial glyoxalase II from *Brassica juncea* that is upregulated by NaCl, Zn, and ABA. Biochem Biophys Res Commun 336:813–819

Schafer HJ, Greiner S, Rausch T, Haag-Kerwer A (1997) In seedlings of the heavy metal accumulator *Brassica juncea* Cu^{2+} differentially affects transcript amounts for y-glutamylcysteine synthetase (y-ECS) and metallothionein (MT2). FEBS Lett 404:216–220

Schiavon M, Pittarello M, Pilon-Smits EAH, Wirtz M, Hell R, Malagoli M (2012) Selenate and molybdate alter sulfate transport and assimilation in *Brassica juncea* L. Czern.: implications for phytoremediation. Environ Exp Bot 75:41–51, http://dx.doi.org/10.1016/j.envexpbot.2011.08.016

Schmidke I, Stephan UW (1995) Transport of metal micronutrients in the phloem of castor bean (*Ricinus communis*) seedlings. Physiol Plant 95(1):147–153

Schmidt U (2003) Enhancing phytoextraction: the effect of chemical plant accumulation and soil manipulation on mobility, plant accumulation and leaching of heavy metals. J Environ Qual 32:1939–1954

Schneider T, Haag-Kerwer A, Maetz M, Niecke M, Povh B, Rausch T, Schuûler A (1999) Micro-PIXE studies of elemental distribution in Cd-accumulating *Brassica juncea* L. Nucl Instrum Methods Phys Res B 158:329–334

Schorin H, de Benzo ZA, Bastidas C, Velosa M, Marcano E (1991) The use of water hyacinths to determine trace metal concentrations in the tropical Morichal Largo River, Venezuela. Appl Geochem 6(2):195–200

Scora RW, Chang AC (1997) Essential oil quality and heavy metal concentrations of peppermint grown on municipal sludge amended soil. J Environ Qual 26:975–979

Scow KM, Hicks KA (2005) Natural attenuation and enhanced bioremediation of organic contaminants in groundwater. Curr Opin Biotechnol 16:246–253

Sebastiani L, Scebba F, Tognetti R (2004) Heavy metal accumulation and growth responses in poplar clones Eridano (*Populus deltoides* × *maximowiczii*) and I-214 (*P.* × *euramericana*) exposed to industrial waste. Environ Exp Bot 52:79–88. ISSN 0098–8472, http://dx.doi.org/10.1016/j.envexpbot.2004.01.003

Sekhar KC, Chary NS, Kamala CT, Venkateswara Rao J, Balaram V, Anjaneyulu Y (2003) Risk assessment and pathway study of arsenic in industrially contaminated sites of Hyderabad: a case study. Environ Int 29:601–611

Sekhar KC, Kamala CT, Chary NS, Balaram V, Garcia G (2005) Potential of *Hemidesmus indicus* for phytoextraction of lead from industrially contaminated soils. Chemosphere 58:507–514

Sekhar KC, Chary NS, Kamala CT, Vairamani M, Anjaneyulu Y, Balaram V, Sorlie JE (2006) environmental risk assessment studies of heavy metal contamination in the industrial area of Kattedan, India—a case study. Human Ecol Risk Assess 12:408–422

Sekhar KC, Kamala CT, Chary NS, Balaram V, Garcia G (2005) Potential of Hemidesmus indicus for phytoextraction of lead from industrially contaminated soils. Chemosphere 58:507–514, http://dx.doi.org/10.1016/j.chemosphere.2004.09.022

Seleiman MF, Santanen A, Stoddard FL, Mäkelä P (2012) Feedstock quality and growth of bioenergy crops fertilized with sewage sludge. Chemosphere 89:1211–1217. ISSN 0045–6535, http://dx.doi.org/10.1016/j.chemosphere.2012.07.031

Senthilkumar P, Prince WS, Sivakumar S, Subbhuraam CV (2005) Prosopis juliflora–a green solution to decontaminate heavy metal (Cu and Cd) contaminated soils. Chemosphere 60(10):1493–1496

Seth CS, Chaturvedi PK, Misra V (2008) The role of phytochelatins and antioxidants in tolerance to Cd accumulation in *Brassica juncea* L. Ecotoxicol Environ Saf 71:76–85

Seth CS, Misra V, Singh RR, Zolla L (2012) EDTA-enhanced lead phytoremediation in sunflower (*Helianthus annuus* L.) hydroponic culture. Plant Soil 347:231–242

Sgherri C, Cosi E, Navari-Izzo F (2003) Phenols and antioxidative status of *Raphanus sativus* grown in copper excess. Physiol Plant 118:21–28

Shabani N, Mahajan DM, Gunale VR, Sayadi MH (2010) comparative assessment of *Alternanthera philoxeroides* and *Hygrophila shoolis* in lead phytoextraction from soil. Pollut Res 29(3):483–488

Sharmin SA, Iftekhar Alam, Kyung-Hee Kim, Yong-Goo Kim, Pil Joo Kim, Jeong Dong Bahk, Byung-Hyun Lee (2012) Chromium-induced physiological and proteomic alterations in roots of *Miscanthus sinensis*. Plant Sci 187:113–126. http://dx.doi.org/10.1016/j.plantsci.2012.02.002

Shelef O, Gross A, Rachmilevitch S (2012) The use of Bassia indica for salt phytoremediation in constructed wetlands. Water Res 46:3967–3976. http://dx.doi.org/10.1016/j.watres.2012.05.020

Shelmerdine PA, Black CR, McGrath SP, Young SD (2009) Modelling phytoremediation by the hyperaccumulating fern, *Pteris vittata*, of soils historically contaminated with arsenic. Environ Pollut 157:1589–1596, http://dx.doi.org/10.1016/j.envpol.2008.12.029

Sheng X, Leni Sun, Zhi Huang, Linyan He, Wenhui Zhang, Zhaojin Chen (2012) Promotion of growth and Cu accumulation of bio-energy crop (Zea mays) by bacteria: implications for energy plant biomass production and phytoremediation. J Environ Manage 103:58–64. http://dx.doi.org/10.1016/j.jenvman.2012.02.030

Shekhar C (2012) Nature cure: bioremediation as a sustainable solution for polluted sites. Chem Biol 19:307–308

Shi G, Cai Q (2009) Cadmium tolerance and accumulation in eight potential energy crops. Biotechnol Adv 27:555–561

Shiyab S, Chen J, Han FX, Monts DL, Matta FB, Gu M, Sua Y (2009) Phytotoxicity of mercury in Indian mustard (Brassica juncea L.). Ecotoxicol Environ Saf 72:619–625

Shtangeeva I, Laiho JV-P, Kahelin H, Gobran GR (2004) Phytoremediation of metal-contaminated soils. Symposia papers presented before the Division of Environmental Chemistry. American Chemical Society, Anaheim, CA, USA. http://ersdprojects.science.doe.gov/workshoppdfs/california 2004/p050.pdf

Shukla OP, Juwarkar AA, Singh SK, Khan S, Rai UN (2011) Growth responses and metal accumulation capabilities of woody plants during the phytoremediation of tannery sludge. Waste Manage 31:115–123

Siegel BZ, Lasconia M, Yaeger E, Siegel SM (1984) The phytotoxicity of mercury vapor. Water Air Soil Pollut 23:15–24

Sinegani AAS, Khalilikhah F (2008) Phytoextraction of lead by Helianthus annuus: effect of mobilising agent application time. Plant Soil Environ 54(10):434–440

Singh G (1995) An agroforestry practice for the development of salt lands using *Prosopis juliflora* and *Leptochloa fusca*. Agrofor Syst 29:61–75

Singh AK, Fernando S (2006) Catalyzed fast-transesterification of Soybean oil using ultrasonication. American Society of Agricultural Engineers, ASAE Annual Meeting, Paper # 066220

Singh J, Kalamdhad AS (2013) Assessment of bioavailability and leachability of heavy metals during rotary drum composting of green waste (Water hyacinth). Ecol Eng 52:59–69

Singh D, Tiwari A, Gupta R (2012) Phytoremediation of lead from wastewater using aquatic plants. J Agric Technol 8(1):1–11

Singh S, Melo JS, Eapen S, D'Souza SF (2008) Potential of vetiver (*Vetiveria zizanioides* L. Nash) for phytoremediation of phenol. Ecotoxicol Environ Saf 71:671–676

Singh S, Saxena R, Pandey K, Bhatt K, Sinha S (2004) Response of antioxidants in sunflower (*Helianthus annuus* L.) grown on different amendments of tannery sludge: its metal accumulation potential. Chemosphere 57:1663–1673

Singh R, Singh DP, Kumar N, Bhargava SK, Barman SC (2010) Accumulation and translocation of heavy metals in soil and plants from fly ash contaminated area. J Environ Biol 31:421–430

Singh K, Singh B, Tuli R (2013) Sodic soil reclamation potential of *Jatropha curcas*: a long-term study. Ecol Eng 58:434–440. http://dx.doi.org/10.1016/j.ecoleng.2013.07.006

Singh K, Singh B, Verma SK, Patra DD (2014) *Jatropha curcas*: a ten year story from hope to despair. Renewable Sustain Energy Rev 35:356–360, http://dx.doi.org/10.1016/j.rser.2014.04.033

Sinha S, Gupta AK (2005) Translocation of metals from fly ash amended soil in the plant of Sesbania cannabina L. Ritz: effect on antioxidants. Chemosphere 61:1204–1214. http://dx.doi.org/10.1016/j.chemosphere.2005.02.063

Sinha S, Saxena R, Singh S (2002) Comparative studies on accumulation of Cr from metal solution and tannery effluent under repeated metal exposure by aquatic plants: its toxic effects. Environ Monit Assess 80(1):17–31

Sinha S, Basant A, Malik A, Singh KP (2009) Multivariate modeling of chromium-induced oxidative stress and biochemical changes in plants of *Pistia stratiotes* L. Ecotoxicology 18:555–566

Sinha VK, Yadav V, Pandey A (2011) Phytoextraction of lead through *Helianthus annuus* and its effects on the growth of *Cajanus cajan*. Curr Bot 2(2):01–07

Skinner K, Wright N, Porter-Goff E (2007) Mercury uptake and accumulation by four species of aquatic plants. Environ Pollut 145:234–237

Smith SL, Thelen KD, MacDonald SJ (2013) Yield and quality analyses of bioenergy crops grown on a regulatory brownfield. Biomass Bioenergy 49:123–130. ISSN 0961–9534, http://dx.doi.org/10.1016/j.biombioe.2012.12.017

Smolyakov BS (2012) Uptake of Zn, Cu, Pb, and Cd by water hyacinth in the initial stage of water system remediation. Appl Geochem 27:1214–1219

Smykalova I, Vrbova M, Tejklova E, Vetrovcova M, Griga M (2010) Large scale screening of heavy metal tolerance in flax/linseed (*Linum usitatissimum* L.) tested in vitro. Ind Crops Products 32:527–533. http://dx.doi.org/10.1016/j.indcrop.2010.06.027

Solhi M, Hajabbasi MA, Shareatmadari H (2005) Heavy metals extraction potential of sunflower (*Helianthus annuus*) and Canola (*Brassica napus*). Caspian J Environ Sci 3(1):35–42

Solís-Domínguez FA, Valentín-Vargas A, Chorover J, Maier RM (2011) Effect of arbuscular mycorrhizal fungi on plant biomass and the rhizosphere microbial community structure of mesquite grown in acidic lead/zinc mine tailings. Sci Total Environ 409:1009–1016

Soltan ME, Rashed MN (2003) Laboratory study on the survival of water hyacinth under several conditions of heavy metal concentrations. Adv Environ Res 7:321–334

Soudek P, Tykva R, Vanek T (2004) Laboratory analyses of ^{137}Cs uptake by sunflower, reed and poplar. Chemosphere 55:1081–1087

Soudek P, Petrová Š, Benešová D, Vaněk T (2010) Phytoextraction of toxic metals by sunflower and corn plants. J Food Agric Environ 8(3&4):383–390

Souza VL, da Costa SD, Santana KB, Mielke MS, de Almeida AAF, Mangabeira PAO, Rocha EA (2009) Efeitos do cádmio na anatomia e na fotossíntese de duas macrófitas aquáticas. Acta Bot Bras 23(2):343–354

Sridhar MKC (1986) Trace element composition of Pistia stratiotes L. in a polluted lake in Nigeria. Hydrobiologia 131:273–276

Sridhar BBM, Diehl SV, Han FX, Monts DL, Su Y (2005) Anatomical changes due to uptake and accumulation of Zn and Cd in Indian mustard (*Brassica juncea*). Environ Exp Bot 54:131–141

Srivastava NK, Srivastava AK (2010) Influence of some heavy metals on growth, alkaloid content and composition in *Catharanthus roseus* L. Indian J Pharm Sci 72:775–778

Srivastava S, Verma PC, Chaudhry V, Singh N, Abhilash PC, Kumar KV, Sharma N, Singh N (2013) Influence of inoculation of arsenic-resistant *Staphylococcus arlettae* on growth and arsenic uptake in Brassica juncea (L.) Czern. Var. R-46. J Hazard Mater 262:1039–1047. doi:10.1016/j.jhazmat.2012.08.019

Stanbrough R, Chuaboonmee S, Palombo EA, Malherbe F, Bhave M (2013) Heavy metal phytoremediation potential of a heavy metal resistant soil bacterial isolate, Achromobacter sp. Strain AO22. APCBEE Procedia 5:502–507. http://dx.doi.org/10.1016/j.apcbee.2013.05.085

Stancheva I, Geneva M, Hristozkova M, Boychinova M, Markovska Y (2009) Essential oil variation of *Salvia officinalis* (l.), grown on heavy metals polluted soil. Biotechnol Biotechnol. EQ. 23/2009/SE, Special Edition/on-Line. XI anniversary scientific conference, 120 Years of Academic Education In Biology, 45 Years Faculty of Biology, pp 373–376

Stancheva I, Geneva M, Yonova P, Markovska Y (2011) Accumulation of Cd, Pb and Zn in Tribulus terrestris L grown on industrially polluted soil and plant antioxidant response. Adv Environ Biol 5(2):300–306

Stephan UW, Schmidke I, Pich A (1994) Phloem translocation of Fe, Cu, Mn, and Zn in *Ricinus* seedlings in relation to the concentrations of nicotianamine, an endogenous chelator of divalent metal ions, in different seedling parts. Plant Soil 165:181–188

Stephan UW, Schmidke I, Pich A (1995) Phloem translocation of Fe, Cu, Mn, and Zn in *Ricinus* seedlings in relation to the concentrations of nicotianamine, an endogenous chelator of divalent metal ions, in different seedling parts. Iron Nutrition in Soils and Plants, Dev Plant Soil Sci 59:43–50

Subramanian R, Gayathri S, Rathnavel C, Raj V (2012) Analysis of mineral and heavy metals in some medicinal plants collected from local market. Asian Pac J Trop Biomed 2:S74–S78

Sundaramoorthy P, Chidambaram A, Ganesh KS, Unnikannan P, Baskaran L (2010) Chromium stress in paddy: (i) Nutrient status of paddy under chromium stress(ii) Phytoremediation of chromium by aquatic and terrestrial weeds. C R Biologies 333:597–607. doi:10.1016/j.crvi.2010.03.002

Suñe N, Sánchez G, Caffaratti S, Maine MA (2007) Cadmium and chromium removal kinetics from solution by two aquatic macrophytes. Environ Pollut 145:467–473

Suneela M, Radha Krishna G, Vamsi Krishna K, Manga Sai V, Bhargav V, Syama Sundara Reddy K, Srinivas DSS, Srinivas JS (2008) Water and Sediment analysis of Hussain Sagar Lake, Hyderabad. In: Sengupta M, Dalwani R (eds) Proceedings of Taal 2007: the 12th world lake conference, pp 304–306

Szőllősi R, Varga IS, Erdei L, Mihalik E (2009) Cadmium induced oxidative stress andantioxidative mechanismsmin germinating Indian mustard (*Brassica juncea* L.) seeds. Ecotoxicol Environ Saf 72:1337–1342

Szymaska M, Matraszek R (2005) Reaction of the sunflower (*Helianthus annuus* L.) to nickel conditioned by the way of metal penetration. Acta Sci Pol Hortorum Cultus 4(1):139–152

Tandy S, Schulin R, Nowack B (2006) The influence of EDDS on the uptake of heavy metals in hydroponically grown sunflowers. Chemosphere 62:1454–1463

Tang Q, Moussa S, Cehui M, Jean-Louis M (2002) Simultaneous sludge stabilization and metal removal by metal hyper-accumulator plants. 17th WCSS, 14–21 August 2002, Thailand

Tejeda S, Zarazúa G, Ávila-Pérez P, Carapia-Morales L, Martínez T (2010) Total reflection X-ray fluorescence spectrometric determination of elements in water hyacinth from the Lerma River. Spectrochimica Acta Part B 65:483–488

Tewari A, Singh R, Singh NK, Rai UN (2008) Amelioration of municipal sludge by Pistia stratiotes L.: role of antioxidant enzymes in detoxification of metals. Bioresour Technol 99:8715–8721

Thangavel P, Sridevi G (eds) (2015) Environmental sustainability: role of green technologies. doi:10.1007/978-81-322-2056-5_14

Tiwari S, Dixit S, Verma N (2007) An effective means of biofiltration of heavy metal contaminated water bodies using aquatic weed Eichhornia crassipes. Environ Monit Assess 129:253–256

Tomé FV, Rodríguez PB, Lozano JC (2008) Elimination of natural uranium and ^{226}Ra from contaminated waters by rhizofiltration using *Helianthus annuus* L. Sci Total Environ 393:351–357

Tomé FV, Blanco Rodríguez P, Lozano JC (2009) The ability of *Helianthus annuus* L. and *Brassica juncea* to uptake and translocate natural uranium and ^{226}Ra under different milieu conditions. Chemosphere 74:293–300. http://dx.doi.org/10.1016/j.chemosphere.2008.09.002

Topalov VD (1962) Peppermint. In: Topalov VD (ed) Essential oil and medicinal crops. Hr. G. Danov Press, Plovdiv, pp 88–104

Truong P (1999) Vetiver grass technology for mine rehabilitation. Pacific Rim Vetiver Network Tech Bull 2:1–19

Truong P (2000) Vetiver grass technology for environmental protection. In: Vetiver and the environment, the second international vetiver conference. China, America, Thailand

Truong P (2003) Vetiver system for water quality improvement. Proceeding of 3rd international vetiver conference, 6–9 October, Guangzhou, China, pp 61–74

Türker OC, Harun Böcük, Anıl Yakar (2013) The phytoremediation ability of a polyculture constructed wetland to treat boron from mine effluent. J Hazard Mater 252–253(15):132–141. ISSN 0304–3894, http://dx.doi.org/10.1016/j.jhazmat.2013.02.032

Ullah R, Bakht J, Shafi M, Iqbal M, Khan A, Saeed M (2011) Phyto-accumulation of heavy metals by sunflower (*Helianthus annuus* L.) grown on contaminated soil. Afr J Biotechnol 10(75):17192–17198

Upadhyay RK, Panda SK (2009) Copper-induced growth inhibition, oxidative stress and ultrastructural alterations in freshly grown water lettuce (*Pistia stratiotes* L.). C R Biol 332:623–632

Upadhyay AR, Mishra VK, Pandey SK, Tripathi BD (2007) Biofiltration of secondary treated municipal wastewater in a tropical city. Ecol Eng 30:9–15

Usha B, Venkataraman G, Parida A (2009) Heavy metal and abiotic stress inducible metallothionein isoforms from *Prosopis juliflora* (SW) D.C. show diVerences in binding to heavy metals in vitro. Mol Genet Genomics 281:99–108

Usha R, Vasavi A, Thishya K, Jhansi Rani S, Supraja P (2011) phytoextraction of lead from industrial effluents by Sunflower (*Helianthus annuus*.L). Rasayan J Chem 4(1):8–12

Valitutto RS, Sella SM, Silva-Filho EV, Pereira RG, Miekeley N (2006) Accumulation of metals in macrophytes from water reservoirs of a power supply plant, Rio de Janeiro State. Brazil Water Air Soil Pollut 178:89–102

Vamerali T, Bandiera M, Mosca G (2011) In situ phytoremediation of arsenic- and metal-polluted pyrite waste with field crops: effects of soil management. Chemosphere 83:1241–1248. http://dx.doi.org/10.1016/j.chemosphere.2011.03.013

Vamerali T, Bandiera M, Lucchini P, Dickinson NM, Mosca G (2014) Long-term phytomanagement of metal-contaminated land with field crops: integrated remediation and biofortification. Eur J Agron 53:56–66. http://dx.doi.org/10.1016/j.eja.2013.11.008

van der Ent A, Baker AJ, Reeves RD, Chaney RL, Anderson CW, Meech JA, Erskine PD, Simonnot MO, Vaughan J, Morel JL, Echevarria G, Fogliani B, Rongliang Q, Mulligan DR (2015) Agromining: farming for metals in the future? Environ Sci Technol 2015(49):4773–4780. doi:10.1021/es506031u

Varun M, D'Souza R, Pratas J, Paul MS (2012) Phytoextraction potential of *Prosopis juliflora* (Sw.) DC. with specific reference to lead and cadmium. Bull Environ Contam Toxicol 87:45–49

Verma VK, Gupta RK, Rai JPN (2005) Biosorption of Pb and Zn from pulp and paper industry effluent by water hyacinth (*Eichhornis crassipes*). J Sci Ind Res 64:778–781

Vesely T, Neuberg M, Trakal L, Szakova J, Tlustoa P (2012) Water lettuce Pistia stratiotes L. response to lead toxicity. Water Air Soil Pollut 223:1847–1859

Vesk PA, Allaway WG (1997) Spatial variation of copper and lead concentrations of water hyacinth plants in a wetland receiving urban run-off. Aquatic Bot 59:33–44

Vicente G, Martinez M, Aracil J (2004) Integrated biodiesel production: a comparison of different homogenous catalysts systems. Bioresour Technol 92:297–305

Viezcas JAS (2009) Toxicity of zinc oxide and cerium oxide nanoparticles to mesquite (*Prosopis juliflora-velutina*). ETD collection for University of Texas, El Paso. Paper AAI1473870. http://digital commons.utep.edu/dissertations/AAI1473870

Vikram Reddy M, Sagar Babu K, Balaram V, Satyanarayanan M (2012) Assessment of the effects of municipal sewage, immersed idols and boating on the heavy metal and other elemental pollution of surface water of the eutrophic Hussainsagar Lake (Hyderabad, India). Environ Monit Assess 184:1991–2000

Vwioko DE, Anoliefo GO, Fashemi SD (2006) Metal concentration in plant tissues of *Ricinus communis* L. (castor oil) grown in soil contaminated with spent lubricating oil. J Appl Sci Environ Manage 10(3):127–134

Walker DJ, Clemente R, Roig A, Bernal MP (2003) The effects of soil amendments on heavy metal bioavailability in two contaminated Mediterranean soils. Environ Pollut 122:303–312

Watanabe T, Osaki M (2002) Mechanism of adaptation to high aluminium condition in native plant species growing in acid soils: a review. Commun Soil Sci Plant Anal 33:1247–1260

Wilde EW, Brigmon RL, Dunn DL, Heitkamp MA, Dagnan DC (2005) Phytoextraction of lead from firing range soil by Vetiver grass. Chemosphere 61:1451–1457

Witters N, Mendelsohn R, Van Passel S, Van Slycken S, Weyens N, Schreurs E, Meers E, Tack F, Vanheusden B, Vangronsveld J (2012) Phytoremediation, a sustainable remediation technology? II: Economic assessment of CO2 abatement through the use of phytoremediation crops for renewable energy production. Biomass Bioenergy 39:470–477. http://dx.doi.org/10.1016/j.biombioe.2011.11.017

Wolverton BC, McDonald RC (1978) Bioaccumulation and detection of trace levels of cadmium in aquatic systems by *Eichhornia crassipes*. Environ Health Perspect 27:161–164

Wood TK (2008) Molecular approaches in bioremediation. Curr Opin Biotechnol 19:572–578

Wu LH, Luo YM, Xing XR, Christie P (2004) EDTA-enhanced phytoremediation of heavy metal contaminated soil with Indian mustard and associated potential leaching risk. Agr Ecosyst Environ 102:307–318

Wu F, Wanqin Yang, Jian Zhang, Liqiang Zhou (2010) Cadmium accumulation and growth responses of a poplar (Populus deltoids& Populus nigra) in cadmium contaminated purple soil and alluvial soil. J Hazard Mater 177:268–273. http://dx.doi.org/10.1016/j.jhazmat.2009.12.028

Wu Q, Wang S, Thangavel P, Li Q, Zheng H, Bai J, Qiu R (2011) Phytostabilization potential of *Jatropha Curcas* L. in polymetallic acid mine tailings. Int J Phytoremediation 13(8):788–804

Xia HP (2004) Ecological rehabilitation and phytoremediation with four grasses in oil shale mined land. Chemosphere 54:345–353

Xiaohai L, Yuntao G, Khan S, Gang D, Aikui C, Li L, Lei Z, Zhonghan L, Xuecan X (2008) Accumulation of Pb, Cu, and Zn in native plants growing on contaminated sites and their potential accumulation capacity in Heqing, Yunnan. J Environ Sci 20:1469–1474

Xiu-Zhen H, Dong-Mei Z, Dan-Dan L, Ping J (2012) Growth, cadmium and zinc accumulation of ornamental sunflower (*Helianthus annuus* L.) in contaminated soil with different amendments. Pedosphere 22(5):631–639

Xu SY, Chen YX, Wu WX, Wang KX, Lin Q, Liang XQ (2006) Enhanced dissipation of phenanthrene and pyrene in spiked soils by combined plants cultivation. Sci Total Environ 63:206–215. ISSN 0048–9697, http://dx.doi.org/10.1016/j.scitotenv.2005.05.030

Yadav SK, Juwarkar AA, Phani Kumar G, Thawale PR, Singh SK, Chakrabarti T (2009) Bioaccumulation and phyto-translocation of arsenic, chromium and zinc by *Jatropha curcas* L.: impact of dairy sludge and biofertilizer. Bioresour Technol 100(20):4616–4622. ISSN 0960–8524, http://dx.doi.org/10.1016/j.biortech.2009.04.062

Yadav AK, Kumar N, Sreekrishnan TR, Satya S, Bishnoi NR (2010) Removal of chromium and nickel from aqueous solution in constructed wetland: mass balance, adsorption–desorption and FTIR study. Chem Eng J 160:122–128

Yadav AK, Abbassi R, Kumar N, Satya S, Sreekrishnan TR, Mishra BK (2012) The removal of heavy metals in wetland microcosms: effects of bed depth, plant species, and metal mobility. Chem Eng J 211–212:501–507

Yang Z (1997) The potential of using Eichhornia crassipes and Ipomoea aquatica as biomonitors in assessing heavy metal pollution in the Mae Kha canal the Mae Ping river in Chiang Mai, Northern Thailand. J Sci Soc Thailand 23:371–380

Yang B, Shu WS, Ye ZH, Lan CY, Wong MH (2003) Growth and metal accumulation in vetiver and two Sesbania species on lead/zinc mine tailings. Chemosphere 52:1593–1600

Yathavakilla SVK, Shah M, Mounicou S, Caruso JA (2005) Speciation of cationic selenium compounds in *Brassica juncea* leaves by strong cation-exchange chromatography with inductively coupled plasma mass spectrometry. J Chromatogr A 1100:153–159

Ye WL, Asaduzzaman Khan M, McGrath SP, Zhao F-J (2011) Phytoremediation of arsenic contaminated paddy soils with *Pteris vittata* markedly reduces arsenic uptake by rice. Environ Pollut 159:3739–3743, http://dx.doi.org/10.1016/j.envpol.2011.07.024

Yuan QH, Shi GX, Zhao J, Zhang H, Xu QS (2009) Physiological and proteomic analyses of *Alternanthera philoxeroides* under Zinc stress. Russ J Plant Physiol 56(4):495–502

Yurekli F, Kucukbay Z (2003) Synthesis of phytochelatins in *Helianthus annuus* is enhanced by cadmium nitrate. Acta Bot Croat 62(1):21–25

Zaier H, Ghnaya T, Lakhdar A, Baioui R, Ghabriche R, Mnasri M, Sghair S, Lutts S, Abdelly C (2010) Comparative study of Pb-phytoextraction potential in Sesuvium portulacastrum and *Brassica juncea*: tolerance and accumulation. J Hazard Mater 183:609–615

Zalesny Jr. RS, Wiese AH, Bauer EO, Riemenschneider DE (2009) Ex situ growth and biomass of Populus bioenergy crops irrigated and fertilized with landfill leachate. Biomass Bioenergy 33(1):62–69. ISSN 0961–9534, http://dx.doi.org/10.1016/j.biombioe.2008.04.012

Zaranyika MF, Mutoko F, Murahwa H (1994) Uptake of Zn, Co, Fe and Cr by water hyacinth (*Eichhornia crassipes*) in Lake Chivero, Zimbabwe. Sci Total Environ 153(1–2):117–121

Zengin FK, Kirbag S (2007) Effects of copper on chlorophyll, proline, protein and abscisic acid level of sunflower (*Helianthus annuus* L.) seedlings. J Environ Biol 28(3):561–566

Zhang Q, Varenyam Achal, Yatong Xu, Wei-Ning Xiang (2014a) Aquaculture wastewater quality improvement by water spinach (*Ipomoea aquatica* Forsskal) floating bed and ecological benefit assessment in ecological agriculture district. Aquacultural Eng 60:48–55. http://dx.doi.org/10.1016/j.aquaeng.2014.04.002

Zhang X, Xuehong Zhang, Bo Gao, Zhian Li, Hanping Xia, Haifang Li, Jian Li (2014b) Effect of cadmium on growth, photosynthesis, mineral nutrition and metal accumulation of an energy crop, king grass (*Pennisetum americanum* × *P. purpureum*). Biomass Bioenergy 67:179–187. http://dx.doi.org/10.1016/j.biombioe.2014.04.030

Zheljazkov VD, Nelson NE (1996) Effect of heavy metals on peppermint and cornmint. Plant Soil 178:59–66

Zheljazkov VD, Warman PR (2004) Phytoavailability and fractionation of copper, manganese, and zinc in soil following application of two composts to four crops. Environ Pollut 131(2):187–195

Zheljazkov VD, Jeliazkova EA, Kovacheva N, Dzhurmanski A (2008a) Metal uptake by medicinal plant species grown in soils contaminated by a smelter. Environ Exp Bot 64:207–216. http://dx.doi.org/10.1016/j.envexpbot.2008.07.003

Zheljazkov VD, Craker LE, Xing B, Nielsen NE, Wilcox A (2008b) Aromatic plant production on metal contaminated soils. Sci Total Environ 395:51–62. http://dx.doi.org/10.1016/j.scitotenv.2008.01.041

Zheng ZC, Li TX, Zeng FF, Zhang XZ, Yu HY, Wang YD, Liu T (2013) Accumulation characteristics of and removal of nitrogen and phosphorus from livestock wastewater by *Polygonum hydropiper*. Agric Water Manage 117:19–25. http://dx.doi.org/10.1016/j.agwat.2012.10.017

Zhi Zhong S, Li W, Man J, YanHua S (2011) *Alternanthera philoxeroides* might be used for bioremediating heavy metal contaminated soil. Genom Appl Biol 30(5):614–619

Zhi-xin N, Li-na S, Tie-heng S, Yu-shuang L, Hong W (2007) Evaluation of phytoextracting cadmium and lead by sunflower, ricinus, alfalfa and mustard in hydroponic culture. J Environ Sci 19:961–967

Zhu Y-G, Chen S-B, Yang J-C (2004) Effects of soil amendments on lead uptake by two vegetable crops from a lead-contaminated soil from Anhui, China. Environ Int 30:351–356. http://dx.doi.org/10.1016/j.envint.2003.07.001

Zhuang P, Ye ZH, Lan CY, Xie ZW, Shu WS (2005) Chemically assisted phytoextraction of heavy metal contaminated soils using three plant species. Plant and Soil 276:153–162

Zuo S, Mei H, Ye L, Wang J, Ma S (2012) Effects of water quality characteristics on the algicidal property of *Alternanthera philoxeroides* (Mart.) Griseb. in an aquatic ecosystem. Biochem Syst Ecol 43:93–100

Chapter 8
Growth and Defense Metabolism of Plants Exposed to Ultraviolet-B Radiation

Rima Kumari, M.N.V. Prasad, and S.B. Agrawal

Abstract Increases in ultraviolet (UV) radiation at the Earth's surface due to the depletion of the stratospheric ozone layer are major concern for researchers, posing interest in the mechanisms of various effects on organisms. Although UV radiation comprises only a small fraction of the total solar radiation that is incident at the earth's surface, it has the greatest energy per unit wavelength and, thus, the greatest potential to damage the biosphere. Intensified current and projected UV-B radiation is responsible for multiple detrimental effects on plant growth and metabolism via interfering with biological processes. In plants, these effects include lipid oxidation, DNA damage, protein destruction which often causes heritable mutations. High UV-B levels introduce DNA damage, number of different lesions, predominantly cyclobutane pyrimidine dimer and pyrimidine (6–4) pyrimidinone products in the genome. It affects physiological processes, including the photosynthetic apparatus; causes loss of crop yield and productivity. Plants developed a number of repair or tolerance mechanisms to counteract the damage caused by UV or any other stressors.

Here we reviewed the structural and biochemical defense strategy of plants in response to Ultraviolet-B radiation. The major points are as follows (i) in brief focus on global scenario of ozone depletion and consequent trend of increase in UV-B irradiation level; (ii) Effect of UV-B on shoot morphology, canopy structure (anatomical changes) could alter the stand photosynthesis, attributed to loss of Biomass and yield (iii) UV-B might play role as a signal molecule to up-regulate the expression of gene involve in multiple developmental response i.e. photomorphogenic response and antioxidants defense (iv) Some kinds of UV-B screening pigments such as carotenoids, flavonoids protect plant cells against intensive penetration of UV-B radiation. (v) Activation of non enzymatic antioxidants as well as enzymes by the Ultraviolet-B factors is discussed.

R. Kumari (✉) • M.N.V. Prasad
Department of Plant Sciences, University of Hyderabad, Hyderabad 500047, India
e-mail: rimabotany@gmail.com

S.B. Agrawal
Laboratory of Air Pollution & Global Climate Change, Department of Botany, Banaras Hindu University, Varanasi 221005, India

E. Lichtfouse (ed.), *Sustainable Agriculture Reviews*,
Sustainable Agriculture Reviews 17, DOI 10.1007/978-3-319-16742-8_8

Keywords UV-B radiation • Ozone depletion • Plant growth • Morphology • Photosynthetic pigments • UV-B absorbing compounds • Oxidative stress • Antioxidants

Abbreviations

UV-B	Ultraviolet-B
O_3	Ozone
ROS	Reactive Oxygen Species
DNA	Deoxyribonucleic acid
PSII	Photosystem II
$O_2^{\cdot-}$	Superoxide ion
$^1O_2$	Singlet Oxygen
H_2O_2	Hydrogen peroxide
·OH	Hydroxyl radicals
CPDs	Cyclobutane-type pyrimidine dimers
Rubisco	Ribulose-1, 5-bisphosphate carboxylase/oxygenase
PAL	Phenylalanine ammonia-lyase
SOD	Superoxide dismutase
CAT	Catalase
POD	Peroxidases
APX	Ascorbate peroxidase
MDHAR	Monodehydro-ascorbate reductase
DHAR	Dehydro-ascorbate reductase
GR	Glutathione reductase
GSH	Glutathione

8.1 Introduction

Increases in ultraviolet (UV)-B radiation reaching the Earth's surface have become a major climate change global concern in past few decades (Caldwell et al. 2007; Madronich et al. 1995; UNEP 2005). This increase is due to man made uses of nitrogen, oxygen and hydrocarbon compound, fluorin, chlorin, compounds i.e. Chlorofluorocarbons, which appeared in the 1930s having the main characteristic of high affinity of reaction with ozone and acts as natural filter of UV-B radiation. This initiated the partial destruction of the stratospheric ozone layer detected in the 1970s (WMO 2007), and its function of filtrating Ultraviolet-B light is weakened. As a consequence, the level of Ultraviolet-B irradiation within wavelength ranges 280–315 nm; penetrating the biosphere has been increased. In recent era, the ongoing depletion of the stratospheric ozone (O_3) layer and corresponding increase in ultraviolet-B (UV-B) radiation on the Earth's surface are a major global issue because it poses a serious threat to life due to alteration of diverse range of biological and ecological processes (Caldwell et al. 2007; Zepp et al. 2011). Thus, governments and scientists are highly worried with global changes concerned with increased level of UV-B radiation on Earth. Although the effects might be the most dramatic in the Antarctic, there are also significant signs of increasing levels of UV-B radiation in the temperate and tropical regions (McKenzie et al. 2011). The highest daily erythemal doses of UV occur in the tropical zone far exceed that incident at higher latitudes as solar elevation is a strong determinant of UV and direct transmission of UV radiation occur in tropics. Since ambient levels of UV-B radiation in the tropics are already high, any further enhancement in UV-B due to further decrease in O_3 level is of great concern for tropical regions including India lies in the low O_3 belt and is expected to receive high flux of UV-B radiation.

To better understand the cumulative effects of UV-B at ecosystem level, here we compiled the in general UV-B response on higher plants, found to cause photomorphogenic, biochemical as well as genetic changes in plants.

8.2 Stratospheric Ozone Level and UV-B Irradiance

In spite of the current efforts to restrict the production of O_3 depleting substances, thinning of the stratospheric O_3 layer and increased penetration of UV-B radiation to the Earth's surface will continue for decades (Weatherhead et al. 2005). The study of the biological effects of ultraviolet-B radiation having wavelength ranged in 280–315 nm, has attracted considerable attention during the last few decades as atmospheric O_3 remains depleted and the annual average O_3 loss is approximately 3 % globally (Executive Summary 2003). The average area of the O_3 hole observed over Antarctica in September 2006 was 27.5 million square kilometers, the largest ever observed. The lowest concentrations of O_3 ever observed over Antarctica, revealing that the O_3 hole was the deepest in 2006 since ever been seen (Keil et al. 2006). The global trend of stratospheric O_3 depletion and resultant increase in UV-B radiation has been confirmed in mid-latitude Japan (Sasaki et al. 2002), in New Zealand in the 1990s (McKenzie et al. 1999) and at high and mid-latitudes in both hemispheres. Although complicated by substantial daily, seasonal and annual fluctuations that may be up to 20 %, a gradual downward trend in the amount of stratospheric O_3 has been measured in temperate and polar climate zones over the last two decades leading to 3–6 % decline per decade at mid-latitudes. In the last two decades, there has been an average of 7 % increase in biologically active UV-B radiation in northern mid-latitudes due to depletion of the stratospheric Ozone layer (Yang et al. 2007; Gilbert et al. 2009). A significant increase of 4–7 % per decades in surface erythemal UV radiation has occurred in the mid-latitude areas of both hemispheres of the Earth (Madronich et al. 1995). Although solar UV-B irradiance data collection for the higher latitudes has been carried out intensively at many locations from earlier studies, not much report are available for the equatorial/tropical regions.

Goddard Institute for Space Studies (GISS) it has been estimated that relative to trends at 1979–1992 in both Northern and Southern Hemispheric zone, the maximum increase in the annual UV dose would be 14 % in 2010–2020 in northern, whereas a 40 % of this enhancement is expected between in the Southern Hemisphere (Kumari and Agrawal 2010).

8.3 Ozone Trend in India

The total Ozone trend studies based on data obtained from Total Ozone Mapping Spectrometer (TOMS) on board Nimbus-7 (1979–1993) and Earth Probe (EP) (1997–2003) missions showed a decreasing trend of Total Ozone in the northern Indian region compared with other parts of India where the trend is almost stable for the period 1978–1993. While, total ozone content was found to be increased during 1996–2000 (Singh et al. 2002). Sahoo et al. (2005) have also observed the trends of ozone decline in northern regions of India and analyze that total ozone data during the period of 1997–2003 showed more decline in total ozone at several locations over the western, eastern and central parts of the Indo-Gangetic basin such as Delhi, Kanpur, Varanasi, Patna and Kolkata. The causes of this declining trend of total ozone content over the Indo-Gangetic basin are attributed due to increasing pollution e.g. dust and sulphate aerosols loading, that have been shown to react with ozone content in the Indo-Gangetic basin (Bonasoni et al. 2001). The rate of higher decline of ozone in recent years over the northern parts of India covering Indo-Gangetic basin is a serious threat in these region.

8.4 UV-B Response in Plants

UV-B radiation causes multiple responses on plants; depending on perception of UV-B, its signaling response and its ultimate effects on expression of UV-B specific gene and these responses are dependent on its fluence rate. UV-B acts as environmental stress factor causing cellular damage, it also acts as an informational signal that regulates photo morphogenesis and UV-protective responses in plants (Jansen et al. 1998; Frohnmeyer and Staiger 2003; Agrawal et al. 2009). UV-B induces oxidative stress on plants; induces molecular damage to plants These responses can vary between different species, communities and ecosystems and are highly dependent on the level of UV irradiance (Bornman 1989).

8.5 UV-B Perception, Signal Transduction and Gene Expression in Plants

UV-B radiation causes a diverse and specific set of responses at the level of gene regulation at different UV-B fluence rates. Exposure to high amounts of UV-B causes tissue necrosis and induces the expression of stress associated genes. Exposure to UV-B radiation leads to reduction in expression and synthesis of key photosynthetic proteins, including the chlorophyll *a/b*-binding proteins (*Lhcb*), the D1 polypeptide of photosystem II (*psbA*), Rubisco etc. (Jordan 1996). Low fluence rates of UV-B promote the expression of a range of genes involved in UV-B protection (Frohnmeyer and Staiger 2003). It have been shown to trigger the expression of genes that participate in several protective pathways, including DNA repair, detoxification of reactive oxygen species and the production of protective pigments of phenylpropanoid origin, such as flavonoids (Brosche and Strid 2003). Genes that encode for enzymes of the phenylpropanoid pathway have been shown to be up-regulated at the transcription level, leading to the biosynthesis of flavonoids and other UV-B protecting compounds. Many other defense related genes are also 'switched on' by UV-B radiation (Jordan 2002).

Therefore, it is important to understand the mechanisms of UV-B perception and signal transduction in plants and to understand how these processes lead to the regulation of gene expression. However, remarkably little knowledge is available about the mechanisms by which plants perceive UV-B radiation or the signal-transduction mechanisms by which UV-B regulates gene expression. Plants use different receptors for different wavelength of solar light as specific triggers for multiple developmental responses in plants. The photoreceptor phytochrome is thought to be an important photoreceptor that phosphorylates the initial substrate of signaling transduction pathway of gene regulation (Agrawal et al. 2009). G protein, calmodulin and cGMP may be also important participant in signal transduction pathway. However, photomorphogenetic responses on plants are mediated by various known photoreceptors; phytochrome, cryptochrome and phototropin (Frohnmeyer and Staiger 2003; Agrawal et al. 2009). These are specific receptors i.e. phytochromes mediate responses to red and far-red light, cryptochromes and phototropins mediate responses to UV-A/blue light. Photoreception systems mediate responses to UV-B may be occurred by a novel class of photoreceptor but specifically these are yet unidentified.

Highly energetic photons in the UV-B range are absorbed by the chromophoric groups of many biologically important molecules such as chlorophylls and quinones. It can act as photosensitizers for the production of reactive oxygen species. UV-B specific induction of several UVB-responsive genes is regulated at the level of transcription, the product of which in turn can affect the expression of target gene expression (Jordan 1996; Agrawal et al. 2009).

Despite revealing the structure revelation and underlying mechanism of UV-B perception and signalling, important questions remain about the UVR signaling pathway remain to be clarified.

8.6 Plant Growth and Morphogenic Response to UV-B

Commonly observed UV-B effects on plants include reduced growth and yield, altered photosynthesis (Reddy et al. 2004; Singh et al. 2009a); change in plant morphology, leaf area, leaf waxes, and epidermal layers (Liakoura et al. 2003; Laposi et al. 2005) damage to deoxyribonucleic acid (DNA) and altered physiological functions mainly photosystem II (PSII) efficiency of plants (Yang et al. 2007). Many studies have shown deleterious UV effects on plants including physiological damage to photosynthetic apparatus, alteration in protein content and enzyme activity, effects on membrane lipids and change in leaf chemistry (Rathore et al. 2003; Xu et al. 2008; Laposi et al. 2009). Enhanced UV-B may affect plant development, particularly biomass distribution and reproduction stage (Gao et al. 2003; Kakani et al. 2003a, b; Urban et al. 2006; Yao et al. 2007). However, the primary effect of enhanced UV-B radiation on plants seems to be changes in carbon partitioning from growth pools to secondary metabolic pathways, such as biosynthesis of phenolic secondary metabolites, which may have ameliorating effects on plants to UV-B stress (Ryan and Hunt 2005).

Photomorphogenetic response on plants includes inhibition of hypocotyl elongation, increased axillary branching and change in biomass allocation, leaf weight ratio and root shoot ratio (Barnes et al. 1996). At high UV-B dose, stress response of UV-B on plants includes damage and necrosis to plants (Gerhardt et al. 2005). Some of the sub-processes that are regulated by light signals include changes in structure and form, such as; seed germination, leaf expansion, stem elongation, flower initiation and pigment synthesis. Most of these responses appear to be adaptive. Some of these morphogenic changes may diminish the exposure of cells or leaves to UV-B radiation. For example, inhibition of hypocotyl elongation has been hypothesized to minimize UV-B exposure of the emerging seedling until it has accumulated UV screening pigments (Ballaré et al. 1996). Leaf thickening is thought to diminish the number of cell layers exposed to ambient UV-B, since these wavelengths penetrate only the upper cell layers of a leaf (Cen and Bornman 1993). Similarly, the reduction of apical dominance such as decreased shoot length and increased axillary branching, reduction of leaf expansion, leaf curling will, within a canopy, diminish exposure of plant to direct sunlight.

Photomorphogenic UV-B signaling is mediated by the UV-B-specific component UV Resistance Locus8 (UVR8). Both UVR8 and Constitutive Photomorphogenesis1 are required for UV-B induced expression of the Elongated Hypocotyl5 transcription factor, which plays a central role in the regulation of genes involved in photomorphogenic UV-B responses (Jenkins 2009). Oravecz et al. (2006) identified as a crucial promoter of the photomorphogenic UV-B response in *Arabidopsis*.

In this section, detrimental response of UV-B in higher plants exposed to enhanced UV-B radiation were summarized and discussed. The greater reduction in morphological parameters in also illustrated a higher sensitivity of plants to UV-B compared to others.

8.7 Visible Injury Symptoms

Visible symptoms of possible UV-B injury i.e. necrotic spots and curled leaf edges, bronzing and chlorosis are damage response, observed in most of the cases at the higher levels of UV-B exposure (Kakani et al. 2003b). On continued exposure to UV-B, leaves become involuted or cup shaped i.e. cupping, being desiccated and dry up and ultimately resultant into early senescence (Santos et al. 1993). Landry et al. (1995) have assessed the enhanced UV-B induced injury in different mutants of *Arabidopsis thaliana* and observed inhibition in growth and causes tissue discoloration for *fahl* mutant at both *0.6* and 0.9 kJ UV-B m^{-1} h^{-2}, even at lowest fluence of 0.4 kJ UV-B m^{-1} h^{-1}, upward leaf curling was observed. Nara and Takeuchi (2002) observed initial visible symptoms of leaf injury i.e. bronzing and necrosis of parts of leaves under the condition of higher solar UV radiation at the middle latitude in the northern hemisphere just after 4 days of UV-B irradiation in tobacco leaves. Singh et al. (2009a) in a case study with *Amaranthus tricolor* reported the plants exposed to elevated level UV-B dose i.e. +7.2 kJ m^{-2} d^{-1}, showed interveinal chlorosis on leaves which later changed into necrotic and desiccated areas.

8.8 Growth

Inhibitory response of enhanced UV-B radiation on growth and development have been studied on wheat cultivars (Li et al. 2000; Agrawal et al. 2004b), on maize (Correia et al. 2000), cotton (Gao et al. 2003), on tertiary buckwheat (Yao et al. 2007), pea (Agrawal and Mishra 2009), okra plants (Kumari et al. 2009a) and others. Growth reduction was due to destruction of growth promoting hormone indole acetic acid *via* its conversion into various photo oxidative products (Tevini and Teramura 1989; Hopkins et al. 2002). The inhibitory effects on plant growth and development are mainly due to enhanced oxidative stress caused by UV-B (Jansen et al. 1998). UV-B irradiation may also affect the light activated protein kinase which is the key regulator of plant development including growth, leaf expansion and pigment synthesis (Stratmann et al. 2000). Various studies on plant-UV-B response state that growth inhibition at the whole-plant level often correlates with reduced leaf expansion, which appears to be more sensitive to UV-B radiation than photosynthesis per unit leaf area. Caldwell and Flint (1994) concluded that morphological changes and reduced growth were more common response to elevated UV-B levels than those of reduced photosynthesis. In plants exposed to enhanced UV-B radiation throughout their development, a reduction in photosynthetic productivity is usually associated with a reduced ability to intercept light i.e. smaller leaf area, and not due to an inhibition in photosynthetic competence (Allen et al. 1998). So, the growth rate of a plant mainly depends on the surface area of leaves that perceived solar radiation and the rate of dry matter increase per unit of leaf area i.e. net assimilates production.

UV-B studies on growth response of plants have revealed inter as well as intra-species variations in magnitude and type of response to enhanced UV-B radiation that may also depend on experimental conditions as well as level of UV-B intensity. A detail data on several studies concerned with UV-B induced changes on growth as well as anatomical modification have been given in Table 8.1. Several studies and publications on past few decades indicated that only about one-third of plant species appears to shows deleterious effects of enhanced UV-B levels on growth characteristics, whereas many species exhibited no such effects (Bassman et al. 2002; Turtola et al. 2006). These studies suggest that some species are already well adapted to enhanced level of UV-B radiation.

Table 8.1 Effects of UV-B radiation on growth, morphological and leaf anatomical traits in various plant species under different environmental conditions

Plants	UV-B $_{BE}$ dose (kJ m^{-2} d^{-1})	Photosynthetic active radiation (μmol photons m^{-2} s^{-1})	Experimental condition	Growth parameters affected	Reference	Country
Pisum sativum	Ambient + UV-B (+7.1)	1,100–1,200	Field condition	All growth parameter negatively affected.	Agrawal and Mishra (2009)	India
				Decrease in shoot length by 19.4 %.		
Abelmoschus esculentus	Ambient + UV-B (+1.8)	1,200–1,500	Field condition	Reductions of plant height by 33.6 %, leaf area 60.8 % and leaf area ratio up to 48.8 %	Kumari et al. (2009a)	India
				Non-significant variation in relative growth rate		
				Increase in net primary productivity by 7.7 %		
Crotalaria juncea L.	Ambient + UV-B (12.2)	Ambient	Field condition	Curling of leaves and waxy coating on leaf surface, reduction in shoot length up to 50 % and leaf area by 5.6 %	Balakrishnan et al. (2005)	India
Acacia & *Eucalyptus* sp.	–UV-B Ambient	5.1 % reduction from solar PAR	Growth chamber	Reduction in plant height and specific leaf area, increase in leaf thickness (3–10 %) in both species	Liu et al. (2005)	Australia
Solanum tuberosum L.	–UV-B	450	Growth chamber	Increase of leaf thickness	Santos et al. (2004)	Portugal
	UV-B (7.83)					

Betula pendula	Control (130–170 kJ m^{-2})	318–338	Field condition	No change in height or specific leaf area	Kostina et al. (2001)	Finland
				Increase in leaf expansion by 25 %		
	UV-B (167–212 kJ m^{-2})			No significance change in relative growth rate		
Pinus sylvestris & *Picea abies*	Ambient + UV-B (30 % above ambient)	–	Field condition	No significant effect on growth of seedlings.	Turtola et al. (2006)	Finland
Gossypium hirsutum L.	–UV-B Ambient (8)	–	Growth chamber	Reduction in plant height by 47 %, decrease in internode length, leaf area and branch length	Kakani et al. (2003a)	USA
	+UV-B (16)			Increase in stomatal number by 36 and 65 % at ambient and + UV-B on adaxial phase		
				Decrease in leaf thickness		
				Formation of dense and interwoven wax		
Saussurea superba & *Gentiana straminea* L.	Ambient + UV-B (15.8)	1,800	Field condition	Significant increase in leaf thickness and leaf width in both species and decrease in leaf length	Shi et al. (2004)	China
Pseudotsuga menziesii	Ambient + sUV-B (2x and 3x of ambient)	ranged from 600–1,800	Growth chamber & Field condition	Increase in plant height: significantly in Ist year at both dose, in IInd year at 2x. No significant difference at 3x sUV-B dose	Bassman et al. (2002)	USA

(continued)

Table 8.1 (continued)

Plants	UV-B $_{BE}$ dose (kJ m^{-2} d^{-1})	Photosynthetic active radiation (μmol photons m^{-2} s^{-1})	Experimental condition	Growth parameters affected	Reference	Country
Spirodela polyrhiza	Control (0.4 mW/cm^2)	2,500 lux	Growth chamber	Dose dependent visible injury: severe injury in form of chlorosis and necrosis	Farooq et al. (2005)	India
	UV-B doses (0.72 and 1.44 J)			Decrease in frond growth by 30.76 and 46.15 % and total frond area by 29.33 and 51.55 % at respective test doses		
Ttiticum aestivum	Ambient + UV-B (+7.1)	1,100–1,200	Field condition	Reduction in shoot and root length by 7 and 19.7 % and decline in absolute growth rate	Agrawal et al. (2004b)	India
Amaranthus tricolor	Ambient + UV-B (+7.1)	Approx. 1,500	Field condition	Interveinal chlorosis, severe injury of necrosis and desiccated areas.	Singh et al. (2009a)	India
				Decrease in plant height (7.4 %) and leaf area (up to 40.6 %) and increase in number of leaves by 61.9 %		

Fagopyrum tataricum	–UV-B (Ambient)	Ranged 87–93 & 82–90 of Ambient	Field condition	Dose dependent reduction in growth parameters. Decrease in plant height (34.9 %), stem diameter (34.4 %) and leaf area (40 %).	Yao et al. (2006)	China
	+UV-B (+5.3, +8.5)			Increase in specific leaf weight (16.7 %) and primary branching (26.6 %).		
Jaborosa magellanica	Control + UV-B (14.21)	1,200	Green house	Reduction in leaf area (92 %) and leaf length (72 %). Alteration of plant architecture.	Cuadra et al. (2004)	Chile
				Reduction in relative growth rate		
Chenopodium quinoa (var. Cristalina and Chucapaca)	–UV-B	Ambient	Field condition	Increase in cotyledon area and seedling height, reductions of leaf number in Cristalina variety	Gonzalez et al. (2009)	Argentina
	+UV–B			Non significant affect on leaf thickness		
Acorus calamus L.	Ambient + UV-B (+1.8, +3.6)	Ambient	Field condition	Decrease in plant height (up to 20.4 %) and leaf area (approx. 40 %).	Kumari et al. (2009b)	India
				Increase in branching and leaf number		

(continued)

Table 8.1 (continued)

Plants	UV-B $_{BE}$ dose (kJ m^{-2} d^{-1})	Photosynthetic active radiation (μmol photons m^{-2} s^{-1})	Experimental condition	Growth parameters affected	Reference	Country
Vigna unguiculata & Crotalaria juncea	Control UV-B (1.0, 1.4, 4.7 and 6.0)	–	Field condition	Increase in leaf thickness by 70 % in *V. unguiculata*. No effect on *C. juncea*.	Selvakumar (2008)	India
				Leaf injury (chlorosis): Dose dependent response in *V. unguiculata*.		
				In *C. juncea*: injury noticed at higher dose.		
Cucumis sativus	Control (0.05 W m^{-2}), UV-B (0.05, 0.15 & 0.60 W m^{-2})	60–80	–	No effect on growth and morphology at low dose, Decreased growth and delayed germination, bronzing and glazing at higher UV-B dose	Fukuda et al. (2008)	Japan
				Medium dose UV-B induced wax production		
Trifolium repens	Control + UV-B (13.3 kJ^{-1} m^{-2})	–	Field condition	Reducing leaf size, leaf number and stolon elongation	Hofmann and Campbell (2011)	New Zealand
Ocimum sanctum	Control (–UV-B)	–	Growth chamber	Increased leaf area growth	Sakalauskaite et al. (2013)	Lithuania
	UV-B (2 & 4 kJ m^{-2} d^{-1})					

Abbreviations and symbols used:
+UV-B (supplemental UV-B above ambient); –UV-B (UV-B excluded) – informations not available

8.9 Leaf Ultrastructure and Anatomical Changes

At the structural level, increase in epidermal cell layer thickening is the most common adaptive response (Liu et al. 2005), increased length of inner leaf cells or increases in cell number, both palisade and spongy mesophyll, influence the penetration and spectral distribution of UV radiation across a leaf. Increased leaf thickness assessed through increase in specific leaf weight, usually reported in many studies, alleviating the transmittance of UV-B penetration to inner target tissues (Tevini and Teramura 1989). Thicker leaves enable plants to attenuate more UV-B radiation and hence to protect palisade layers from deleterious effects of irradiation. The thick leaves were associated with increases in the thickness of the upper epidermis, spongy parenchyma and also spongy intercellular space (Kakani et al. 2003b). Nevertheless, UV-induced ultrastructural changes in birch indicate that increase in leaf thickness did not completely provide protection for palisade mesophyll against UV-B radiation in birch seedling (Kostina et al. 2001). Santos et al. (2004) reported no change in structural integrity of leaf cells in potato cultivar by UV-B exposure. They observed changes only at ultrastructural level that includes appearance of paracrystalline inclusions in peroxisome, constricted plastid and increased thylakoids.

Leaf expansion and anatomical development are altered by UV-B in case of many agronomic and tree species. Diversified anatomical responses of UV-B have been reported in several plant studies (Table 8.1). For example, UV-B radiation caused increased leaf thickness in *Brassica napus* L. (Cen and Bornman 1993) whereas, Staxen and Bornman (1994) reported that *Petunia* hybrid plants grown without UV-B irradiance had thicker leaves than UV-B-irradiated plants. Supplemental UV-B applied to sweetgum (*Liquidambar styraciflua*) seedlings had no effect on final leaf size but rates of leaf elongation and accumulation of leaf area was slower in leaves exposed to daily UV-B supplementation rates of 3 kJ UV-B (Dillenburg et al. 1995). Kostina et al. (2001) reported changes at the ultrastructural level in response to enhanced UV-B exposure in birch seedling *(Betula pendula)* and observed the rise to numerous cytoplasmic lipid bodies and abnormal membrane whorls. The lipid accumulations may indicate accelerated cell senescence at higher UV-B irradiance (Wulff et al. 1996). Strong damaging effect of UV-B might be disruption of thylakoid membrane structure (increase thylakoid swelling, dilation of thylakoid membranes) which may breaks the structure of stroma, disrupt the integrity of membranes (He et al. 1994; Kostina et al. 2001; Kakani et al. 2003b).

This part descript the alteration and modification in cell structural integrity, suggesting that this may encompassing defence response enables plants to cope with UV-B aggression.

8.10 Biomass and Yield Changes in Response to UV-B

Change in biomass in response to UV-B represents the long term integration of cumulative changes in biochemical and physiological functions of plant, which ultimately bring about significant effect on biomass productivity. Since supplemental UV-B irradiance detrimentally affected plant height, leaf area and carbon assimilation rate, the biomass and yield are negatively affected. Several field supplementation studies have indicated that UV-B radiation deleteriously affected the biomass production and overall growth in a large number of plant species (Agrawal et al. 2004a, b, 2006; Yao et al. 2007, 2008). Several other studies shown reduction in biomass accumulation under UV-B exposure as in *Glycine max* (Feng et al. 2003), *Triticum aestivum* (Zheng et al. 2003; Rathore et al. 2003), *Gossypium hirsutum* (Gao et al. 2003), *Pisum sativum* (Agrawal and Mishra 2009), *Amaranthus tricolor* (Singh et al. 2009a), *Abelmoschus esculentus* (Kumari et al. 2009a) under field conditions or under green house condition (Tosserams et al. 1997; Cuadra et al. 2004), may be the amount of carbon gained by plants not directly utilized in biomass accumulation while in turn cost into maintenance, repair, construction and increased respiration due to UV-B stress (Monsi 1968). Several studies have identified the UV-B sensitivity of many cultivars of plants on the basis of yield response (Smith et al. 2000).

However effect of UV-B on biomass reduction depends on inter and intra-specific variability their sensitivity to UV-B radiation, experimental conditions, duration of UV-B exposure (Searles et al. 2001) A review by Kakani et al. (2003b) on effects of UV-B on agricultural crops reported that biomass reduction under elevated UV-B radiation was observed only in 54 % of studies on various crops whereas 35 % of studies reported no such effects and a few studies demonstrated increases in crop dry matter accumulation. Nithia et al. (2005) reported differential response of UV-B on biomass production in radish and carrot, and find out that 20 % enhancement of UV-B above ambient caused a significant reduction in root and shoot biomass of carrot, while it caused a pronounced increment in radish. Urban et al. (2006) also reported decrease in above-ground biomass in *Calamagrostis villosa*, prevailing negative effect in response to enhanced UV-B radiation, while significant increase in biomass was reported in *C. arundinacea* prevailing positive effect. Table 8.2 presents a summary of research work conducted to evaluate the changes in biomass accumulation in different plants under sUV-B radiation.

As a concluding remark, In response to enhanced UV-B radiation, intraspecific variation in Biomass and yield responses has been determined in different crops.

8.11 UV-B and Reactive Oxygen Species

Exposure to UV-B leads to the generation of reactive oxygen species. In general, oxidative stress results from the disruption of cellular homeostasis between pro-oxidants and antioxidants due to over production of ROS from the excitation

Table 8.2 Effects of UV-B radiation on plant biomass in various plant species under different environmental conditions

Plants	UV-B treatment (KJ m^{-2} d^{-1})	Photosynthetic active radiation (μmol photons m^{-2} s^{-1})	Experimental Condition	Plant biomass	UV-B induced response	References	Country
Gossypium hirsutum	Ambient + UV_1 (+4.8 % of A)	Ambient	Field condition	Plant biomass	↓ (12 and 34 % at UV_1& UV_2)	Gao et al. (2003)	China
	+UV_2 (+9.5 % of A)			Cotton yield	↓ (31 and 72 % at UV_1& UV_2)		
Triticum aestivum	Ambient (8.6) +UV-B (+7.1)	Ambient	Field condition	Plant biomass	↓ (52.8 %–55.6 %)	Agrawal et al. (2004b)	India
				Grain yield	↓ (66.5 %–69.1 %)		
Amaranthus tricolor	Ambient + UV-B (+7.1)	Approx. 1,500	Field condition	Plant biomass	↓ (25.8 %)	Singh et al. (2009a)	India
Calamagrostis arundinacea & *C. villosa*	Ambient + UV-B (+25 % of A)	Ambient	Field condition	Above and below ground biomass	↓ (12 %) *C. villosa*	Urban et al. (2006)	Brno
					↑ (20 %) *C. arundinacea*		
					No effect		
Pseudotsuga menziesii	Ambient 2x and 3x (of A)	Range 600–1,800	Field condition	Total biomass	↓ at 2x (in Ist year)	Bassman et al. (2002)	USA
					↑ at 2x (in IInd year)		
Spinacea oleraceae	Ambient (8.6) +UV-B (+7.1)	1,100–1,200	Field condition	Plant biomass	↓ by 16.7 %	Mishra and Agrawal (2006)	India
				Net primary productivity	Decreased by approx. 23 %		
Ttiticum aestivum	Ambient + UV-B (+7.1)	1,100–1,200	Field condition	Plant biomass	↓ (max up to 55.6 %)	Agrawal et al. (2004b)	India
				Grain yield	↓ (69.1 %)		
Pisum sativum	Ambient + UV-B (+7.1)	1,100–1,200	Field condition	Plant biomass	↓ (18.9 %)	Agrawal and Mishra (2009)	India
				Grain yield	↓ (28.1 %)		

(continued)

Table 8.2 (continued)

Plants	UV-B treatment (KJ m^{-2} d^{-1})	Photosynthetic active radiation (μmol photons m^{-2} s^{-1})	Experimental Condition	Plant biomass	UV-B induced response	References	Country
Vigna radiata	Ambient + UV-B (+7.1)	Ambient	Field condition	Plant biomass	↓ (up to 66.8 %)	Agrawal et al. (2006)	India
				Seed yield	↓ (max. up to 71.6 %)		
Vegetable sp. (*Lectuca, Spinacea and Cucumis*)	Control (0.05)	550	Glass house	Shoot biomass	↓ in *P. vulgaris* (39 %), 30 % in *S. oleracea*, 26 and 31 % in *C. pepo*	Smith et al. (2000)	New Zealand
	UV-B (15.8)				Non significant reduction in *C. frutescens* and *S. melongena*		
Triticum aestivum	Ambient UV-B; T_1 and T_2 (11.4 and 5.8 % above Ambient)	Ambient	Field condition	Yield	↓ (24 and 13.3 % at T_1 and T_2)	Zheng et al. (2003)	China
				Dry matter accumulation	↓ (42.1 and 14 % at T_1 and T_2)		
Abelmoschus esculentus	Ambient + UV-B (+1.8)	1,200–1,500	Field condition	Plant biomass	↓ (39.6 %)	Kumari et al. (2009a)	India
Calamagrostis epigeios	–UV-B	800	Glass house	Plant biomass	↑ (15–33 %)	Tosserams and Rozema (1995)	The Netherlands
	UV-B (0, 6.7, 10, 14.9)			Shoot/root biomass	Not affected		

Fagopyrum tataricum	−UV-B Ambient + UV-B (+5.3, +8.5)	Ranged 87–93 & 82–90 of Ambient	Field condition	Total gross weight	↓ at both lower UV-B (up to 6 %) and elevated UV-B (17.8 %)	Yao et al. (2006)	China
				Seed yield & Above ground biomass	Decreased almost half		
				Harvest index	Not affected		
Jaborosa magellanica L.	Ambient + UV-B (+14.21)	1,200	Green house	Leaf fresh wt	↓ up to 88 %	Cuadra et al. (2004)	Chile
Raphanus sativus L. and *Daucus carota* L.	Ambient + UV-B (20 % above Ambient)	Ambient	Field condition	Shoot fresh mass	↑ up to twofold in radish.	Nithia et al. (2005)	India
				Root fresh mass	↓ up to 65 % in carrot.		
					↑ by 25 % in radish. ↓ up to 30 % in carrot		
Glycine max	Ambient + UV-B (+7.1)	1,010	Field condition	Plant biomass	↓ (up to 16.9 %)	Ambasht and Agrawal (2003a)	India
Picea asperata	Ambient + UV-B (30 % above A)	1,200	Open semi field	Plant biomass	↓ up to 171.3 % at 3 years. old plant, 44.7 increment at 6 years. old	Yao and Liu (2007)	China
Acorus calamus	Ambient + UV-B (+1.8, +3.6)	Ambient	Field condition	Plant biomass	↑ up to 45 % at lower sUVB dose, No significant variation at higher dose w.r.t. control	Kumari et al. (2009b)	India

(continued)

Table 8.2 (continued)

Plants	UV-B treatment (KJ m^{-2} d^{-1})	Photosynthetic active radiation (µmol photons m^{-2} s^{-1})	Experimental Condition	Plant biomass	UV-B induced response	References	Country
Picea asperata	Ambient (11.02) +UV-B (+3.3)	1,000 ± 50	Open semi field	Plant biomass	↓ (16.9 %)	Yao et al. (2008)	China
Triticum aestivum	Ambient + UV-B (30 % above A)	–	Field condition	Above ground biomass and yield	↓ (11–19) and↓ (12–20 %) at pre booting phase, no effect at later phase	Lizana et al. (2009)	Chile
Raphanus sativus	Ambient + UV-B (+7.2 kJ m^{-2} d^{-1})	1,100–1,500	Field condition	Biomass	↓ (approx. 30 %)	Singh et al. (2011)	India
Glycine max	Control (–UV-B) UV-B	–	Field condition	Crop yield	↓ soybean yield at ambient level UV-B	Mazza et al. (2013)	Argentina

Abbreviations and symbols used:
+UV-B (supplemental UV-B above ambient); –UV-B (UV-B excluded)
FC field condition, *OSF* Open semi field, *GLH* glass house, *GH* Green house. The ↓ and ↑ represent the decrease or increase of the parameter due to elevated levels of UV-B radiation, respectively, compared to their control values

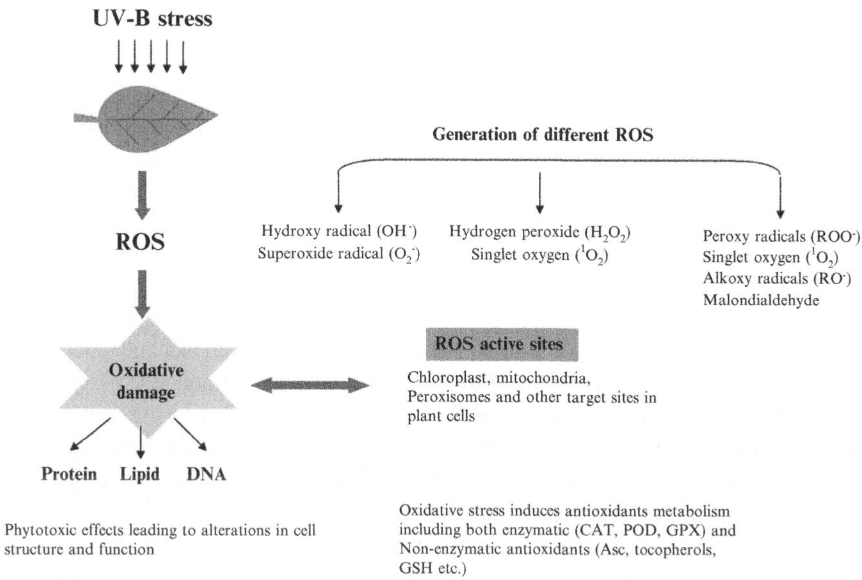

Fig. 8.1 UV-B induced reactive oxygen species production and oxidative stress in plant cell

of O2 to form singlet oxygen (${}^{\cdot}O_2$) and the transfer of 1, 2 or 3 electrons to O_2 to form superoxide ($O_2^{\cdot-}$), hydrogen peroxide (H_2O_2), and the hydroxyl radical ($HO^{\cdot-}$), respectively; the generating reactive oxygen species, leads to oxidative destruction of the cell components through oxidative damage of membrane lipids, nucleic acid, and protein (Shiu and Lee 2005). Intense UV-B radiation generate oxidative stress due to the over production of the reactive oxygen species caused by an unbalance within different cellular compartment of the plant cell. The main sources of reactive oxygen species in plants are chloroplast, mitochondria, peroxisomes, cytosols etc. (Foyer and Noctor 2005) (Fig. 8.1). In Chloroplast, photosystem I and II are major site for production of ${}^1O_2$ and $O_2^{\circ-}$. In mitochondria complex I, Ubiquinone and complex III of electron transport chain are the major site for generation of $O_2^{\circ-}$. In Chloroplast, UV-B radiation could accumulate reactive oxygen species, such as superoxide radicals O_2°-, singlet oxygen (O_2), hydrogen peroxide (H_2O_2), and hydroxyl radicals (·OH) during photosynthesis, which caused oxidative damage and photosynthesis reduction of plants. Eva (1999) reported that UV-B radiation induced oxidative damage of photosystem II and decreased electron transfer rate and thylakoid membrane stability. These reactive oxygen radicals to be formed in the plants damage the all major groups of bio-molecules like DNA, proteins, lipids, chlorophyll, membrane etc, by inducing lipid peroxidation, DNA mutation and protein denaturation in cells, the quantity of Rubisco enzyme and the expression of genes involved in photosynthesis (Mackerness 2000).

8.12 Damage to Nucleic Acid

DNA is a major and long studied target of UV-B damage because of its active absorption spectrum within UV range. UV-B absorption causes photo modification to DNA, produces multiple DNA photoproducts which may alter the nucleotide sequence and cause severe mutations during replication. The most common UV-B induced DNA photoproduct is the cyclobutane-type pyrimidine dimers and pyrimidine pyrimidone (6,4) photoproduct which formed between adjacent pyrimidines on the same strand (Britt 2004). CPD constitutes the majority of DNA lesion i.e. approximately 75 %, induced by UV exposure, whereas 6, 4 photoproduct accounts for remainder. DNA protein cross-links, DNA strand breaks and deletion or insertion of base pairs are also altered by DNA photoproduct formed at UV exposure. Such DNA damage can be lethal or mutagenic to organisms, disrupts the activity of DNA polymerase, interferes with gene transcription and can impede replication and transcription, results in mutation, developmental abnormality, senescence and cell death (Britt and May 2003). In another study, it is found that UV-B radiation cross-links RNA to particular ribosomal proteins with a concomitant decrease in translation concluding that RNA is also a major target of UV-B radiation (Casati and Walbot 2004). UV-B induced growth inhibition due to DNA photodamage has been tested in some plants (Ballaré et al. 2001; Teranishi et al. 2004). A greenhouse study on native herb *Gunnera magellanica* from southern Patagonia exposed to a gradient of UV-B radiation from zero to moderate UV-B fluxes, reduction in leaf expansion and increased density of cyclobutane pyrimidine dimer was observed with increase in UV-B radiation (Giordano et al. 2004).

8.13 Amino Acids

Aromatic amino acids particularly phenylalanine, tryptophan and tyrosine are highly vulnerable to UV-B stress because they absorb mainly in 280–320 spectral region. Histidine, cystine, cysteine can be also direct target of UV-B radiation. UV-B radiation causes modification and destruction of the form of both the amino acids and also in protein. UV-B radiation also induces photooxidation of tyrosine to 4, 4-dihydroxyphenylalanine as well as photochemical modification of tryptophan into N-formyl kynurenine. At high quantum efficiency cystine also undergo UV-B induced photolysis, causes split of its disulfide bond into reactive sulfhydryl group. This structural modification in amino acid and disulphide bond breakage can disrupt the tertiary structure of protein (Hollosy 2002). Murphy (1984) reported the alteration in integrity of sulfhydryl group of protein under UV radiation. Gerhardt et al. (1999) have reported the UV-B induced alteration on ribulose-1, 5-bisphosphate carboxylase/oxygenase i.e. Rubisco, into 66 kDa photoproduct formation due to tryptophan photolysis. A significant increase in amino acid contents under UV-B stress condition was reported in soybean plants by Ambasht and Agrawal (2003a). Increase in amino acids under UV-B exposure may be ascribed due to hydrolysis of protein or due to inhibition of protein synthesis (Table 8.3).

Table 8.3 Effects of UV-B on antioxidants, UV-B screening pigments and lipid peroxidation under different environmental conditions in various plant species

Plants	UV-B dose (kJ m^{-2} d^{-1})	Photosynthetic active radiation (μmol photons m^{-2} s^{-1})	Experimental condition	Biochemical parameters	UV-B induced responses	Reference	Country
Triticum aestivum & *Vigna radiata*	Ambient (8.6) sUV-B (+7.1)	1,100–1,200	Field condition	Asc. acid	↓ (max. 27.1 %) *V. radiata*	Agrawal and Rathore (2007)	India
					↑ (28.1–30.4 %) *T. aestivum*		
				Thiol	↑ (29.1–36.1 %) *V. radiata*		
Pisum sativum	Ambient (8.6) sUV-B (+7.1)	1,100–1,200	Field condition	CAT	↓ (22.4 %)	Agrawal and Mishra (2009)	India
				SOD, POD	↑ (17.4 &39.1 %)		
				Asc. Acid	↓ (16.9 %)		
				LPO	↑ (75.3 %)		
Helianthus annus	Control UV-B (15 & 30)	175	Growth chamber	LPO	↑ (15–25 %)	Costa et al. (2002)	Argentina
				APX & GR	No change		
				CAT & GPOD	↑ (20–36 %)		
				GSH & Ascorbate	↑ (sixfold in GSH & 53 % in Asc.)		
Glycine max	-UV-B Ambient	Near ambient	Field condition	Asc. acid DHA, ASC, GR & APX	↓ after treatment Significant ↑	Xu et al. (2008)	USA
				LPO	significant ↑		

(continued)

Table 8.3 (continued)

Plants	UV-B dose (kJ m^{-2} d^{-1})	Photosynthetic active radiation (μmol photons m^{-2} s^{-1})	Experimental condition	Biochemical parameters	UV-B induced responses	Reference	Country
Pisum sativum	Control (6) Low (+4.4), Moderate (+13.3)	160		Total phenol	↑ (13 and 45 %) at lUVB & mUV-B at 7th and ↓ (37 %) at hUV-B at 14th day	Katerova et al. (2009)	Poland
	High V-B (+26.5)			Proline	↑ by 19–29 % under low dose at 7th–14th day, ↓ by 40 % at 14th day		
				Thiol	↑ by 30–47 % at low and moderate UV-B dose, ↓ up to 80 % at high UV-B		
Pisum sativum L	Ambient sUV-B (+7.1)	Ambient	Field condition	Protein and Asc. acid	↓ (25.9 and 17.1 %)	Singh et al. (2009b)	India
				Phenol, Thiol & Proline	↑ (41.9, 40.4 and 32.7 % respectively)		
				LPO	↑ (71.9 %)		
Pisum sativum L	Control UV-B (0.72, 1.44 & 2.16)	150	Growth chamber	LPO and electrolyte leakage	↑ (max. up to 106 and 37 %)	Prasad et al. (2005)	India
				SOD, CAT, POD	Significant ↑		
Glycine max L.	Control (30)	300	Growth chamber	LPO	↑ (max. up to 60.6 %)	Galatro et al. (2001)	Argentina
	UV-B (60)			Asc. acid and Thiol	↑ (117 and 20.8 %)		
Cucumis sativus L.	Control + UV-B (2.6 mW/cm^{-2})	–	Growth chamber	SOD, APX, GR & POD	↑ significantly; SOD ↑ (up to 200 %)	Jain et al. (2004)	India

Zea mays L.	Control (PAR) UV-B (PAR + UVB)	400–700 nm	Growth chamber	Tot. ASC and proline α tocopherol	No significant change ↑ (18.2 %)	Carletti et al. (2003)	Italy
Spinacea oleraceae	Ambient (8.6) + UV-B (+7.1)	1,100–1,200	Field condition	CAT and POD	↓ CAT (54.8 %), ↑ POD (74.3 %)	Mishra and Agrawal (2006)	India
				LPO	↑ (74.3 %)		
				Asc. acid and proline	↓ in Asc. acid (26.5 %), proline (38.2 %)		
Ttiticum aestivum (2cv)	Ambient + UV-B (+7.1)	1,100–1,200	Field condition	CAT and POD	Significant ↓	Agrawal et al. (2004a)	India
				Phenol and Asc. acid	Significant ↑		
Helianthus annus	Control + UV-B (15 and 30)	175	Control environmental growth chamber	CAT and GPX	↑ CAT (20 %) and GPX (37 %)	Yannarelli et al. (2006)	Argentina
				APX	No significant variation		
Cassia auriculata	Control + UV-B (7.5 and 15.0)	160	Control environmental growth chamber	SOD, CAT, POD, Asc acid & Phenol	Significant ↑	Agrawal, (2007)	India
Abelmoschus esculentus	Ambient sUV-B (+1.8)	1,100–1,200	Field condition	SOD Ascorbic acid	Age wise variation. both ↓ and ↑ trend	Kumari et al. (2009a)	India
				Protein and Phenol	↑ at later age ↑ significant		
				LPO	↑ (194.5 %)		

(continued)

Table 8.3 (continued)

Plants	UV-B dose (kJ m^{-2} d^{-1})	Photosynthetic active radiation (μmol photons m^{-2} s^{-1})	Experimental condition	Biochemical parameters	UV-B induced responses	Reference	Country
Vigna unguiculata L. *Crotalaria juncea*	Control (0) UV-B (1.0, 1.4, 4.7 and 6.0)	–	Field condition	Total protein content	↓ (up to half)	Selvakumar (2008)	India
				LPO	↑ significant		
				SOD and CAT	↑ in SOD but ↓ in CAT		
				APX	↑ (max. 115 % at 6.4 kJ m^{-2} d^{-1})		
				Tot. ASC	↑ (max. 98 % and 84 % in *V. unguiculata* and *C. juncea*,)		
Cucumis sativus L.	Control + UV-B (0.2 W m^{-2} s^{-1})	–	Growth chamber	SOD and APX	↑ (approx. 4.5 fold)	Kondo and Kawashima (2000)	Japan
				GR	No significant effect		
				POD	Drastic ↑ (33.5 fold)		
Capsicum annuum L.	Control + UV-B (5.8 W m^{-2})	–	Green house	POD, APX, CAT, GR	↑ POD (35 %), CAT (23 %), APX (24 %) and GR (17 %)	Madhavian et al. (2008)	Iran

Helianthus annuus L.	Control + UV-B (8.6 W m^{-2})	–	Green house	MDA	↑ (up to threefold)	Cechin et al. (2008)	Brazil
Crepis capillaries L.	Control + UV-B (3, 9 Wm^{-2})	–	Green house	SOD	↑ (1.4 & 2 fold at 3 and 9 kJ m^{-2})	Raneeliene et al. (2005)	Lithuania
Daucus carota L.	Control + UV-B (141.4 mJ cm^{-2})	–	Green house	Tot. phenolics	↑ (3.2 and 6.6 fold)	Du et al. (2012)	USA
				Tot. antioxidant	↑ (3 times)		
Phaseolus vulgaris	Control + UV-B	–	Field condition	Protein	↑ (significantly)	Peykarestan and Seify (2012)	Iran
				POD	↓ (up to 2.5 times)		
Helianthus annuus	Control + UV-B (13 KJ m^{-2} d^{-1})	–	Growth chamber	MDA, GPX, APX	↑ MDA (max. 87 %), GPX (85 %), APX (61 %), CAT (78 %)	Hagh et al. (2012)	Iran
				CAT			

Abbreviations and symbols used:
The ↓ and ↑ represent the decrease or increase of the parameter due to elevated levels of UV-B radiation, respectively, compared to their control value

8.14 UV-B Induced Protein Oxidation

Proteins, due to a combination of their UV absorption characteristics and their abundance in cells, are primary targets of UV-mediated cellular damage. UV radiation can damage proteins by direct oxidation or by covalent binding of lipid peroxidation breakdown products, resulting in loss of protein function and/or enzymatic activity. The reactive oxygen species oxidative attack on proteins causes reversible and/or irreversible modifications, such as carbonylation, nitration, glycation, formation of adducts with lipid peroxidation products and protein protein cross linking. These modifications determine structural, functional and stability changes, leading to loss of function, fragmentation, unfolding/misfolding, protein aggregation and degradation (Perluigi et al. 2010).

8.15 Lipid Peroxidation

Reactive oxygen species attacks on lipids, especially those located on cellular and subcellular membranes, by acting on polyunsaturated fats through a peroxidation cascade, leading to the disorganization of the membrane, thereby altering its functions (Atmani et al. 2009). The peroxidation of lipid is considered as a most damaging process taken as a parameter to determine the level of membrane damage under UV-B induced peroxidative stress. Lipid peroxidation increased in both cellular and organelle take place when above –threshold reactive oxygen species levels are reached, thereby affecting both directly normal cellular functioning and aggravating the oxidative stress by through production of lipid derived radicals. Reactive oxygen species i.e. Hydroxyl radicals and singlet oxygen can react with lipids and form lipid peroxy radicals and hydroperoxide (Blokhina et al. 2003; Hollosy 2002). The peroxy radicals can abstract hydrogen from other unsaturated fatty acids, leading to a chain reaction of peroxidation. Products formed during Lipid peroxidation from polyunsaturated precursors such as ketones, Malondialdehyde, react with thiobarbituric acid to form complex product thiobarbituric acid reactive substances. The increase in thiobarbituric acid reactive substances content i.e. malondialdehyde equivalents is more precisely an indicator of general UV-induced oxidative damage (Costa et al. 2002). The overall effects of lipid peroxidation are to decrease membrane fluidity, increase the leakiness of membranes to all substances even those not normally crossed through it, which leads to the denaturations of membrane lipids due to consequential disruption of membrane functionality, increase in permeability and inactivation of proteins (Girotti 1985) and damage membrane proteins inactive receptors, enzymes and ion channels. It impairs the potassium exchange via ATPase reactions and calcium exclusion (Agrawal et al. 2009) through membrane leading to a rise of intercellular Ca^{2+} that would lead to alteration of all sorts of intracellular metabolism is reported and can induce enzyme activation and alter normal gene transcription. An analysis of data from numerous experiments on increase in level

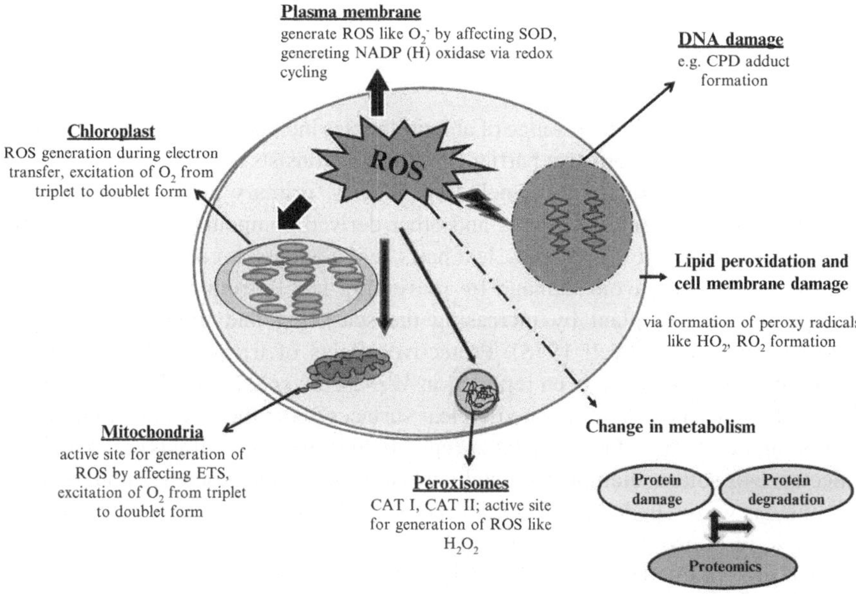

Fig. 8.2 Oxidative stress in different cellular compartment by UV-B induced ROS. *ROS* reactive oxygen species, *CPD* cyclobutane pyrimidine dimer, *SOD* superoxide dismutase, *CAT* catalase, *ETS* electron transport chain (Ref; Frohnmeyer and Staiger 2003; Agrawal et al. 2009; Gill and Tuteja 2010)

of lipid peroxidation in response to UV-B radiation is well documented in (Table 8.3) as reported in okra (Kumari et al. 2009a), soybean (Galatro et al. 2001), spinach (Mishra and Agrawal 2006) and pea (Prasad et al. 2005). Takeuchi et al. (1995) reported growth inhibition due to lipid peroxidation caused by UV-B exposure. But Giordano et al. (2004) reported no enhancement of lipid peroxidation in *Gunnera magellanica* under UV-B exposure 6.5 kJ m^{-2} d^{-1} fluence rate, while plant growth inhibited due to increased DNA damage (Fig. 8.2).

8.16 UV-B Response to Plant Defense Mechanism

Major defense mechanism in higher plants can be classified into categories: (i) optical screening of UV-B radiation by protective structural modifications in peripheral plant tissues and higher accumulation of secondary compounds (ii) an increased capacity of the plants to scavenge or detoxify activated oxygen species by antioxidants activity (iii) Direct Auto- repair of damaged DNA. Figure 8.3 well presented diagrammatically the major defense strategies in plants against UV-B stress

8.17 Optical Screening of UV-B Irradiation by Structural Modification

When plants subject to a wide range of abiotic stresses includes UV-B stress, cuticular wax layer provides a protective barrier (Fig. 8.3). It consists predominantly of long-chain hydrocarbon compounds, including alkanes, primary alcohols, aldehydes, secondary alcohols, ketones, esters and other derived compounds. Epidermal and cuticular wax structures, trichomes, leaf hairs, leaf bladders, is considered to be first line of defense that avoid damage by preventing UV-B from reaching sensitive mesophyll target of plant by increasing the scattering and reflectivity of light (Robberecht and Caldwell 1978). Protective effects of trichomes against UV-B radiation damage has also been reported in *Verbascum speciosum* (Manetas 2003). The configuration of wax deposited on leaf surfaces also influences radiation interception and certain surface structures (epicuticular wax crystalloids) increase the reflectance of solar radiation. Stomatal occlusion by the increased accumulation of epicuticular waxes under enhanced levels of UV-B radiation is also an important protective response. Stomata eventually covered by proliferating epidermal cells become less visible (Evans et al. 2001). An increase in the amount of cuticular wax as a result of UV-B irradiation has been reported for *Gossypium hirsutum*

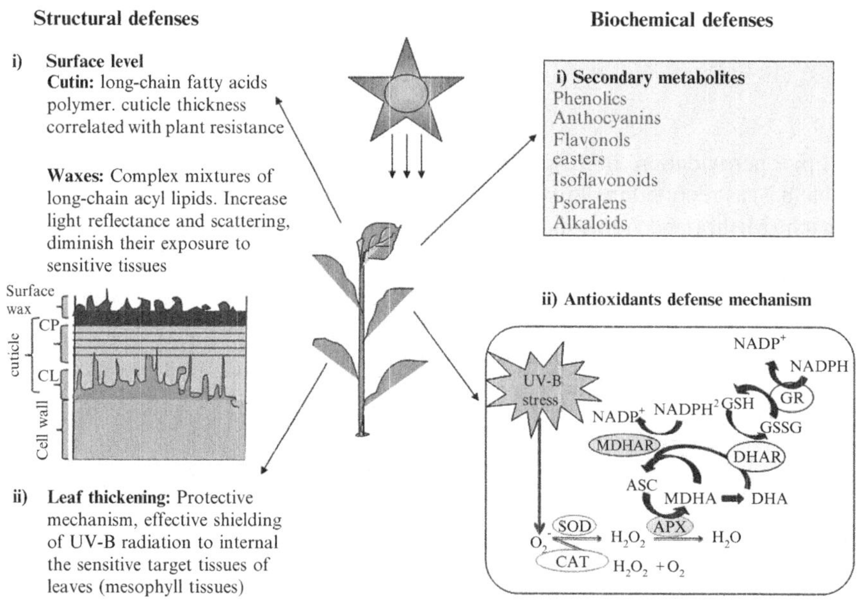

Fig. 8.3 Plant structural and biochemical defense responses against UV-B stress. *CP* cuticle proper, *CL* cuticular layer, *SOD* superoxide dismutase, *CAT* catalase, *APX* ascorbate peroxidase, *MDHAR* monodehydro reductase, *DHAR* dehydro reductase, *GR* glutathione reductase

(Kakani et al. 2003a), *Brassica napus* (Qaderi and Reid 2005), and *Pisum sativum,* (Gonzalez et al. 1996). Higher wax content observed by Kakani et al. (2003a) on adaxial leaf surface represents an important protection mechanism in response to UV-B (Table 8.1). Long et al. (2003) emphasized the importance of wax layers in the absorption of UV-B radiation by the maize epidermis.

Morphological changes such as leaf area reduction, leaf thickening, changing stage in stomatal opening and closing, etc. may also take place in an attempt to avoid damage. Architectural modifications in plants may lead to dense canopy to minimize the UV-B exposure on plant surface.

8.18 Induced Accumulation of UV-B Absorbing Compounds

One of the most important protective mechanisms in higher plants is the deposition of UV-absorbing phenolic compounds in epidermal tissues (Rozema et al. 2002). These compounds act as a sunscreen, reducing penetration of UV-B into the leaf. Genotypes lacking such protection suffer increased injury by UV-B. In fact, an increase in flavonoids, anthocyanins and other phenolics is often proposed as an adaptive mechanism to screen the potentially damaging UV-B radiation before its penetration to sensitive chromophore region of mesophyll tissue (Mazza et al. 2000). The presence of these compounds in epidermal and mesophyll tissues decrease the damage in the photosynthetic apparatus and in the DNA (Mazza et al. 2000) It has been documented that these compounds prevent the transmittance of up to 95–99 % of UV-B light to the inner mesophyll cells of the leaf (Robberecht and Caldwell 1978). It has been assumed that this is the primary plant defense i.e. phytochemical defenses to limit the penetration of UV-B within the mesophyll tissue. In addition to act as UV-B filter, it has been proposed that most of these also have antioxidant property to scavenge reactive oxygen species (Rozema et al. 2002). UV protection by phenolics compounds are well studied in several species of which several are cited: potato (*Solanum tuberosum* L.) (Santos et al. 2004), soyabean (*Glycine max* [L.] Merr.) (Mazza et al. 2000; Feng et al. 2003), broad bean (*Vicia faba* L.), Sweet flag (*Acorus calamus* L.) (Kumari et al. 2009c), cucumber (*Cucumis sativus* L.). Increase in UV-B absorbing compounds protected DNA in *Arabidopsis* (Fujibe et al. 2004) and reduced the sensitivity of PSII to UV-B in grape (Kolb et al. 2001).

8.19 Flavonoids

Flavonoids are important plant phenolics which act as UV-B screens, antioxidants and energy-dissipating agents (Rozema et al. 2002; Tattini et al. 2004). Flavonoids accumulate in epidermal vacuolar region, increased markedly under UV-B treatment. They have effective radical scavenging capabilities, and can contribute directly to enhance the photoprotection against UV-B (Ryan et al. 2002). Therefore, flavonoids can help to maintain the level of photosynthetic pigments and the normal photosynthetic activity (Day and Neale 2002). In *Brassica napus*, flavonoids reduce UV-B induced degradation of the D1 photosystem II reaction centre (Wilson and Greenberg 1993), and Ryan et al. (2002) in *Arabidopsis* reported that plants lacking flavonoids have enhanced UV-B injury and oxidative damage. A strong correlation between flavonoid accumulation and UV-B tolerance has been reported by various researchers (Tevini et al. 1983). It was verified in experiments with different phenylpropanoid mutants of *Arabidopsis*, that they were more sensitive to UV-B irradiation than the wild type (Kliebenstein 2004). Both field and *in-vitro* studies showed that certain flavonoids can protect DNA from UV-B induced damage (Mazza et al. 2000).

Flavonoids are synthesized by a complex multi-step phenylpropanoid biosynthetic pathway. Phenylalanine ammonia-lyase (PAL), which is the first enzyme in its biosynthetic pathway catalyse the deamination of phenylalanine to *trans-cinnamic* acid, which in turn leads to synthesis of an array of secondary metabolites. Increase in flavonoid concentration may relate with higher activity of key enzyme PAL and/or to higher rates of biosynthesis of this enzyme via transcriptional activation of genes of phenylpropanoid metabolism (Tevini and Teramura 1989; Rozema et al. 2002). Enhanced UV-B radiation has been found to increase flavonoid accumulation in *Crotalaria juncea* (Balakrishnan et al. 2005), *Glycine max* (Feng et al. 2003), *Triticum aestivum* (Ambasht and Agrawal 2003b; Rathore et al. 2003), *Spinacea oleracea* (Mishra and Agrawal 2006), in *Fagus sylvatica* (Láposi et al. 2005), *Pisum sativum* (Singh et al. 2009b). Tsormpatsidis et al. (2008) observed 20 % higher flavonoid content in their study with 'Lollo Rosso' lettuce plants grown under the UV transparent "UV280" film than plants under the UV blocking film "UV320". On the basis of UV-B exclusion experiment Kolb et al. (2003) have reported the stimulation of flavonoid biosynthesis under current level of UV-B. Activation of the expression of PAL and induction of flavonoids induced under UV-B stress has been reported by Kumari et al. (2009a, b) in *Acorus calamus* (sweet flag) and *Abelmoschus esculentus*. Sharma and Sharma (1999) have reported about four fold induction of PAL activity on exposure of rice seedlings to UV-B radiation.

Anthocyanins Anthocyanins are water-soluble pigments derived from flavonoids ***via*** the shikimic acid pathway. The characteristic red, blue and purple coloration seen in various tissues of a diverse assortment of plants is due to anthocyanins content. Many researchers have noted the presence of anthocyanins in the upper epidermal layer of leaves associate them with UV-B protection. UV-B protective effect of a polyacylated anthocyanin, in flower petals of the blue morning glory, *Ipomoea tricolor cv.* has been reported by Mori et al. (2005) and they mentioned that aromatic acyl residues of polyacylated anthocyanins absorb UV-B light overlapping the absorption spectrum of DNA; therefore, these pigments are expected to be an effective UV-screening pigment in plants. On basis of reports of several investigators it may be concludes that the accumulation of anthocyanins represents a multifunctional mechanism to; (i) directly reduce ROS through scavenging and possibly metal chelation, acting in conjunction with other antioxidants, and (ii) to shield photosynthetic processes experiencing excessive irradiances, thereby reducing the extent of photooxidation, photoinhibition, dissipating excess energy during intensive light (Gould 2002).

8.20 Defensive Role of Antioxidants Against UV-B Stress

In response of oxidative stress generated due to excess reactive oxygen species, enzymes involved in the elimination of free radicals are activated (Mackerness 2000). Major enzymatic antioxidants include superoxide dismutase (SOD), catalase (CAT), peroxidases (PODs), ascorbate peroxidase (APX), monodehydro-ascorbate reductase (MDHAR), dehydro-ascorbate reductase (DHAR), glutathione reductase (GR) and non enzymatic antioxidants are glutathione (GSH), ascorbic acid (AA), praline, carotenoids, tocopherols.

8.21 Enzymatic Defenses

8.21.1 Superoxide Dismutase and Catalase

Superoxide dismutase is the most effective intracellular enzymatic antioxidant; present in all subcellular compartments prone to reactive oxygen species induced oxidative stress. SOD classified as three different isozymes based on their metal co-factor group (1) Mn- SOD; found in mitochondria of eukaryotic cells and peroxisomes (2) Cu-Zn SOD; cytosolic and chloroplast of higher plant (3) Fe-SOD; (chloroplast compartment). SOD has been proposed to be important in plant stress tolerance. Inside a cell, the SOD constitutes the first line of defense against reactive oxygen species. It removes superoxide anion (O_2^-) by catalyzing its dismutation, being one molecule reduced to H_2O_2 and other oxidized to O_2. It removes O_2^- and hence decreased its risk to formation of toxic OH^- by Haber- Weiss reaction (Gill and Tuteja 2010). Although there are not known direct scavengers of single oxygen $1O_2$ or the hydroxyl radical $^{\cdot}OH^-$; SOD and is believed to function in their elimination by chemical reaction (Alvarez 2000). Significant increase in SOD activity under UV-B stress has been observed in various plants i.e. *Cucumis sativus* (Kondo and Kawashima 2000), *Triticum aestivum* (Yang et al. 2007; Alexieva et al. 2001), *Crotalarea juncea* (Balakrishnan et al. 2005), *Crepis capillaries* (Raneeliene et al. 2005). Agarwal (2007) reported that UV-B (7.5 and 15.0 KJ m^{-2}) irradiation showed significant increase in SOD activity in cassia seedlings.

Catalase, is tetrameric heme containing enzyme, involves in protection of cells against H_2O_2, dismutates into water and oxygen. CAT has one of the highest turnover rates among all enzymes. The enzyme is abundant in the peroxisomes of the leaves of C_3 plants, where it removes H_2O_2 produced during photorespiration by the conversion of glycolate into glyoxylate (Jayakumar et al. 2004). In germinating barley it accumulate abundantly in glyoxisomes of lipid bodies where it decomposes H_2O_2 formed during the oxidation of fatty acids (Fazeli et al. 2007). Though it is active scavenger of H_2O_2, the limited protective action of catalase is attributed to its poor affinity for its substrate, highly sensitive to light induced inactivation (Foyer et al. 1994) and feedback inhibition of its activity by high substrate/product (O_2 or H_2O_2) concentrations (Jaleel et al. 2009). The studies conducted on CAT in response to UV-B showed both increase and decrease in its activity after exposure. Decrease in CAT activity under UV-B treatments could be due to destruction of peroxisomes via peroxidation of lipids (Gao and Zhang 2008). The combined action of SOD and CAT abate the formation of the most toxic and highly reactive oxidant, the hydroxyl radical $^{\cdot}OH^{-}$; which can react in discriminately with all macromolecules.

8.21.2 Peroxidases

Peroxidase, CAT and APX appear to play an essential protective role in the reactive oxygen species scavenging process coordinately with SOD activity. They scavenge H_2O_2 generated primary through SOD action (Agrawal et al. 2009). Peroxidase uses multiple substrates mainly phenolics and catalyse its H_2O_2 dependent oxidation. It also involves in numerous important physiological roles, including lignin biosynthesis, auxin metabolism via indole-3-acetic acid oxidation, and defense against pathogens (Hollosy 2002). Acceleration in POD activity in response to elevated levels of UV-B is a common response to foliar tissues under field conditions. UV-B irradiation increased POD activity in soyabean (Ambasht and Agrawal 2003a), wheat and mungbean (Agrawal and Rathore 2007), spinach (Mishra and Agrawal 2006), Dragon spruce (Yao and Liu 2007).

8.21.3 Ascorbate Peroxidase and Glutathione Reductase

Ascorbate peroxidase is also one of important heme-containing enzyme involved in the detoxification of reactive oxygen species, protecting cells of higher plants against H_2O_2. It exists in multiple isoforms and mainly localized in specified cell compartments; the stroma (sAPX), thylakoid membrane (tAPX), microbodies (mAPX) and cytosolic (cAPX). APX is a potential enzyme of ascorbate-glutathione cycle, utilize ascorbate as electron donar. It has higher affinity for H_2O_2 than CAT and POD and it may play more crucial role in management of ROS under stress. APX utilizes ascorbate (AsA) to reduce H_2O_2 by the oxidation of AsA to the monodehydroascorbate (MDHA) radical, which can be reduced by photosynthetic electron flow through ferredoxin or reduced to AsA by NAD(P)H-dependent monodehydroascorbate reductase (MDHAR) and MDHA would be spontaneously disproportionated to AsA and dehydroascorbate (DHA) (Mittler 2002). DHA will then be reduced to generate AsA by dehydroascorbate reductase (DHAR) using glutathione as an electron donor; the glutathione disulphide formed after glutathione oxidization is reduced to glutathione by GR utilizing reducing equivalents from NAD(P)H (Fig. 8.3). Ascorbate peroxidase is distributed in at least four cellular compartments (chloroplast, mitochondria, cytoplasm and peroxisomes) and acts in photosynthetic organisms in combination with the so-called ascorbate–glutathione cycle (Asada 1992). The up-regulation of the APX and GR under UV-B stresses condition has been established in various plant species (Xu et al. 2008; Madhavian et al. 2008; Yao and Liu 2007).

8.22 Non Enzymatic Antioxidants

Low molecular weight antioxidants (non- enzymatic) such as α tocopherol, ascorbic acid glutathione, carotenoids, proline, etc. play a crucial role in plant defense as enzyme co-factor, cellular antioxidants act as redox buffers in plant cells (Foyer and Noctor 2005; Agrawal and Rathore 2007).

8.22.1 α -Tocopherol

α -tocopherol is lipophilic antioxidants, major and most active component of plant cell. In plants, α-tocopherol is found mainly in the envelope and the thylakoid membranes of chloroplasts. α-Tocopherols acts as powerful scavenger of free radicals of polyunsaturated fatty acid, having the ability to terminate chain reactions generated via lipid oxidation (Havaux et al. 2003). α -–Tocopherols interact with the polyunsaturated acyl groups of lipids, scavenges lipid peroxy radicals through the concerted action of other antioxidants, stabilize membranes, and scavenge and quench various reactive oxygen species (ROS) and lipid soluble byproducts of oxidative stress thereby, protects inner tissues against lipid peroxidation. α –tocopherols scavenges singlet oxygen most efficiently, can neutralize up to 120 singlet oxygen molecules by 1 molecule.

8.22.2 Ascorbic Acid (AsA)

AsA is one of the most extensively studied anti-oxidants occurs in all plant tissues, usually being higher in photosynthetic cells. AsA has effects on many physiological processes including the regulation of growth, differentiation and metabolism of plants. A fundamental role of AsA in the plant defense system is to protect metabolic processes against H_2O_2 and other toxic derivatives of oxygen. It may acts both as primary antioxidants reacts non-enzymatic with superoxide, hydrogen peroxide, and singlet oxygen or may also as secondary antioxidants as a cosubstrate of many enzymes, e.g. APX, to detoxifies H_2O_2. It influences many enzyme activities, and minimizes the damage caused by oxidative process through synergic function with other antioxidants. It reduces the oxidized form of α-tocopherol, or involve in the synthesis of zeaxanthin in the xanthophylls cycle. AsA are involved in the biosynthesis of plant hormones, including abscisic acid and gibberellic acid (Arrigoni and De Tullio 2002), in mitosis regulation and cell expansion (Noctor and Foyer 1998). Ascorbic acid also play defensive role against DNA damage. Protective effects of ascorbic acid (vitamin C) against DNA damage have also been widely documented (Konopacka et al. 1998).

8.22.3 Glutathione (GSH)

Glutathione is the major source of non-protein thiols in most plant cells. The GSH protects the cell against oxidative stress by reacting with peroxides and hydroperoxides (Deneke 2000). Glutathione with ascorbate are important constituents of ascorbate–glutathione cycle, which occur both in cytosol and chloroplasts providing protection against oxidative stress damage by dissociating toxic H_2O_2. The change in the ratio of its reduced (GSH) to oxidized (GSSG) form during the degradation of H_2O_2 is important in certain redox signaling pathways. It has been suggested that the GSH/GSSG ratio, indicative of the cellular redox balance, may be involved in ROS perception (Kumari et al. 2010). Levels of glutathione and ascorbate are up-regulated in response to UV-B (Gao and Zhang 2008).

8.22.4 Proline

Proline metabolism is a typical metabolism of the biochemical adaptation in living organisms subjected to stress conditions (Kumari et al. 2010). Proline is directly involved in ROS quenching and protection of cell membranes against sUV-B (Tripathi et al. 2011). It could function as a hydroxyl radical scavenger to prevent membrane damage. Proline also acts to stabilize the structural integrity of thylakoid membranous protein (Alia et al. 1997).

8.23 DNA Repair Induced in Plants Against UV-B Damage

To avoid the cytotoxic effects of UV-induced DNA damage, organisms use different types of DNA repair mechanisms including photorepair, nucleotide excision repair (also referred to as dark repair) and post-replication repair to maintain the genetic integrity of the plant (Frohnmeyer and Staiger 2003; Britt 2004). Photo repair is mediated by group of enzymes "DNA photolyases", and relies on radiant energy in the UVA/photosynthetically active radiation (PAR) bands as an energy source to convert the photodimers back to monomers. It is thought to be the major DNA repair pathway for CPDs and (6–4) photoproducts in higher plants (Britt 2004). Excision repair and post-replication repair are independent of radiant energy but are less efficient at repairing CPDs. Photo-repair of UV induced DNA lesions in plant cells have been reported in various plant species including cucumber (Takeuchi et al. 2007), *Arabidopsis thaliana* (Dany et al. 2001), rice (Hidema et al. 2001). Hidema et al. (2000) have demonstrated that UV-resistant rice cultivar Sasanishiki was better able to repair cyclobutane pyrimidine dimers through photorepair than UV-sensitive cultivar Norin 1. Iwamatsu et al. (2008) investigated the UV-B sensitivity in 12 rice strains and reported that increased UVB sensitivity depends on decrease in CPD photolyase activity resulting from spontaneously occurring mutations in the photolyase gene.

8.24 Conclusion

From the literature it is apparent that there has been a reduction in the amount of ozone in the stratosphere due to man-made CFCs resulting in an increase in UV-B radiation reaching the earth's surface. UV-B has many effects on plants; both photomorphogenic and damaging effects as well as genetic changes in plants. Much of the previous research was focused on damaging effect of above-ambient UV-B in an attempt to establish the consequences of stratospheric ozone depletion on agricultural important crops, it turned out the overall damaging effects of above-ambient UV-B plant growth and development in certain species; depends on sensitivity of plants, growth condition and exposure dose. Indeed, the focus is now beginning to be shifted toward effects of UV-B on native species plants as well as on medicinally important plants. UV-B effects are likely to play a significant role in aromatic plants in natural ecosystems in improving volatile yield and quality of herbal extracts, would be of great significance in understanding the application of UV-B for better cultivation and high potential for commercial exploitation of these herbal products for scientific and commercial purposes. Future UV-B research needs to integrate the different approach on molecular, chemical, physiological and biochemical information to understand these responses and the impacts of solar UV-B irradiance on the plant metabolism in different species at ecosystem level to gain a comprehensive understanding of the role of UV-B radiation for terrestrial life on earth.

Acknowledgement The author Rima Kumari acknowledges Dr. D.S. Kothari Post Doctoral fellowship from University Grant Commission for financial support. Thanks are also due to UGC, New Delhi for providing research grant "Precision stressing by UV-B radiation to improve the quality of *Coriander* and *Trigonella*, Ref. F. No. 41-389/2012.

References

Agarwal S (2007) Increased antioxidant activity in *Cassia* seedlings under UV-B radiation. Biol Plant 51:157–160

Agrawal SB, Mishra S (2009) Effects of supplemental ultraviolet-B and cadmium on growth, antioxidants and yield of *Pisum sativum* L. Ecotoxicol Environ Saf 72:610–618

Agrawal SB, Rathore D (2007) Changes in oxidative stress defense system in wheat (*Triticum aestivum* L.) and mung bean (*Vigna radiata* L.) cultivars grown with and without mineral nutrients and irradiated by supplemental ultraviolet-B. Environ Exp Bot 59:21–23

Agrawal SB, Rathore D, Singh A (2004a) Combined effects of enhanced UV-B radiation and additional nutrients on two cultivars of wheat (*Triticum aestivum* L.). Physiol Mol Biol Plant 10:99–108

Agrawal SB, Rathore D, Singh A (2004b) Effect of supplemental ultraviolet-B and mineral nutrients on growth, biomass allocation and yield of wheat (*Triticum aestivum* L.). Trop Ecol 45:315–325

Agrawal SB, Rathore D, Singh A (2006) Combined effects of enhanced ultraviolet-B radiation and mineral nutrients on growth, biomass accumulation and yield characteristics of two cultivars of *Vigna radiate* L. J Environ Biol 27:55–60

Agrawal SB, Singh S, Agrawal M (2009) UV-B induced changes in gene expression and antioxidants in plants, Chapter 3. Adv Bot Res 52:47–86

Alexieva V, Sergiev I, Mapelli S, Karanov E (2001) The effect of drought and ultraviolet radiation on growth and stress markers in pea and wheat. Plant Cell Environ 24:1337–1344

Alia B, Saradhi PP, Mohanty P (1997) Involvement of proline in protecting thylakoid membranes against free radical-induced photodamage. J Photochem Photobiol B 38:253–257

Allen DJ, Nogues S, Baker NR (1998) Ozone depletion and increased UV-B radiation: is there a threat to photosynthesis? J Exp Bot 49:1775–1788

Alvarez ME (2000) Salicylic acid in the machinery of hypersensitive cell death and disease resistance. Plant Mol Biol 44:429–442

Ambasht NK, Agrawal M (2003a) Interactive effects of ozone and ultraviolet-B radiation on physiological and biochemical characteristics of soybean plants. J Plant Biol 30:37–45

Ambasht NK, Agrawal M (2003b) Effects of enhanced UV-B radiation and tropospheric ozone on physiological and biochemical characteristics of field grown wheat. Biol Plant 47:625–628

Arrigoni O, De Tullio MC (2002) Ascorbic acid, much more than just an antioxidant. Biochim Biophys Acta 1569:1–9

Asada K (1992) Ascorbate peroxidase: a hydrogen peroxide-scavenging enzyme in plants. Physiol Plant 85:235–241

Atmani D, Chaher N, Atmani D, Berboucha M, Debbache N, Boudaoud H (2009) Flavonoids in human health: from structure to biological activity. Curr Nutr Food Sci 5:225–237

Balakrishnan V, Venkatesan K, Konganapuram E, Ravindrani C, Kulandaivelu G (2005) Protective mechanism in UV-B treated *Crotalaria juncea* L. seedlings. Plant Prot Sci 41:115–120

Ballaré CL, Scopel AL, Stapleton AE, Yanovsky MJ (1996) Solar ultraviolet-B radiation affects seedling emergence, DNA integrity, plant morphology, growth rate, and attractiveness to herbivore insects in Datura ferox. Plant Physiol 112:161–170

Ballaré CL, Rousseaux MC, Searles PS (2001) Impacts of solar ultraviolet- B radiation on terrestrial ecosystems of Tierra del Fuego (southern Argentina)—an overview of recent progress. J Photochem Photobiol B Biol 62:67–77

Barnes PW, Ballaré CL, Caldwell MM (1996) Photomorphogenic effect of UV-B radiation on plants: consequences for light competition. J Plant Physiol 148:15–20

Bassman JH, Edward GE, Robberecht R (2002) Long term exposure to enhanced UV-B radiation is not detrimental to growth and photosynthesis in Douglas-fir. New Phytol 154:107–120

Blokhina O, Virolainen E, Fagerstedt KV (2003) Antioxidants, oxidative damage and oxygen deprivation stress: a review. Ann Bot 91:179–194

Bonasoni P, Calzolari F, Cristofanelli P, Bonafe U, Evangelisti F, Van Dingenen R, Balkanski Y (2001) Ozone and aerosol correlation during Sahara dust transport episodes at Mount Climone during Minatroc project. In: 8th European symposium on the physico-chemical behavior of atmospheric pollutants, NASA Astrophysics Data report, Torino

Bornman JF (1989) Target sites of UV-B radiation in photosynthesis of higher plants. J Photochem Photobiol B Biol 4:145–158

Britt AB (2004) Repair of DNA damage induced by solar UV. Photosynth Res 81:105–112

Britt AB, May GD (2003) Re-engineering plant gene targeting. Trends Plant Sci 8:90–95

Brosche M, Strid A (2003) Molecular events following perception of ultraviolet-B radiation by plants. Physiol Plant 117:1–10

Caldwell MM, Flint SD (1994) Stratospheric ozone reduction, solar UV-B radiation and terrestrial ecosystems. Climate Change 28:375–394

Caldwell MM, Bornman JF, Ballaré CL, Flint SD, Kulandaivelu G (2007) Terrestrial ecosystems, increased solar ultraviolet radiation, and interactions with other climate change factors. Photochem Photobiol Sci 6:252–266

Carletti P, Masi A, Wonisch A, Grill D, Tausz M, Ferretti M (2003) Changes in antioxidant and pigment pool dimensions in UV-B irradiation maize seedlings. Environ Exp Bot 50:149–157

Casati P, Walbot V (2004) Cross linking of ribosomal proteins to RNA in maize ribosomes by UV-B and its effects on translation. Plant Physiol 136:3319–3332

Cechin I, Corniani N, de Fumis T (2008) Ultraviolet-B and water stress effects on growth, gas exchange and oxidative stress in sunflower plants. Radiat Environ Biophys 47:404–413

Cen Y, Bornman JF (1993) The effect of exposure to enhanced UV-B radiation on the penetration of monochromatic and polychromatic UV-B radiation in leaves of *Brassica napus*. Physiol Plant 87:249–255

Correia CM, Coutinho JF, Björn LO, Torres-Pereira JMG (2000) Ultraviolet-B radiation and nitrogen effects on growth and yield of maize under Mediterranean field conditions. Eur J Agron 12:117–125

Costa H, Gallego SM, Tomaro ML (2002) Effects of UV-B radiation on antioxidant defense system in sunflower cotyledons. Plant Sci 162:939–945

Cuadra P, Herrera R, Fajardo V (2004) Effects of UV-B radiation on the Patagonian *Jaborosa magellanica* Brisben. J Photochem Photobiol B Biol 76:61–68

Dany AL, Douki T, Triantaphylides C, Cadet J (2001) Repair of the main UV induced thymine dimeric lesions within *Arabidopsis thaliana* DNA: evidence for the major involvement of photoreactivation pathways. J Photochem Photobiol B 65:127–135

Day TA, Neale PJ (2002) Effects of UV-B radiation on terrestrial and aquatic primary producers. Annu Rev Ecol Syst 33:371–396

Deneke SM (2000) Thiol-based antioxidants. Curr Top Cell Regul 36:151–180

Dillenburg LR, Sullivan JH, Teramura AH (1995) Leaf expansion and development of photosynthetic capacity and pigments in Liquidambar styraciflua (Hamamelicaceae)—effects of UV-B radiation. Am J Bot 82:878–885

Du W-X, Avena-Bustillos RJ, Breksa AP III, McHugh T-H (2012) Effect of UV-B light and different cutting styles on antioxidant enhancement of commercial fresh-cut carrot products. Food Chem 134:1862–1869

Eva H (ed) (1999) Utilizing new adamantly spin traps in studying UV-B-induced oxidative damage of photosystem II. J Photochem Photobiol B: Biol 48:174–179

Evans NH, McAinsh MR, Hetherington AM (2001) Calcium oscillations in higher plants. Curr Opin Plant Biol 4:415–420

Executive summary (2003) Environmental effects of ozone depletion and its interactions with climate change: 2002 assessment. Photochem Photobiol Sci 2:1–4

Farooq M, Shankar U, Ray RS, Agrawal N, Verma K, Hans RK (2005) Morphological and metabolic alterations in duckweed *Spirodela polyrhiza* on long term low level chronic exposure. Ecotoxicol Environ Saf 62:408–414

Fazeli F, Ghorbanli M, Niknam V (2007) Effect of drought on biomass, protein content, lipid peroxidation and antioxidant enzymes in two sesame cultivars. Biol Plant 51:98–103

Feng H, An L, Chen T, Qiang W, Xu S, Zhang M, Wang X, Cheng G (2003) The effect of enhanced ultraviolet-B radiation on growth, photosynthesis and stable carbon isotope composition (^{13}C) of two soybean cultivars (*Glycine max*) under field conditions. Environ Exp Bot 49:1–8

Foyer CH, Noctor G (2005) Oxidant and antioxidant signaling in plants, a re-evaluation of the concept of oxidative stress in a physiological context. Plant Cell Environ 28:1056–1071

Foyer CH, Lelandais M, Kunert KJ (1994) Photooxidative stress in plants. Physiol Plant 92:696–717

Frohnmeyer H, Staiger D (2003) Ultraviolet-B radiation mediated responses in plants, balancing damage and protection. Plant Physiol 133:1420–1428

Fujibe T, Saji H, Arakawa K, Yabe N, Takeuchi Y, Yamamoto K (2004) A methyl viologen-resistant mutant of *Arabidopsis*, which is allelic to ozone-sensitive *rcd1*, is tolerant to supplemental ultraviolet- B irradiation. Plant Physiol 134:275–285

Fukuda S, Satoh A, Kasahara H, Matsuyama H, Takeuchi Y (2008) Effects of ultraviolet-B irradiation on the cuticular wax of cucumber (*Cucumis sativus*) cotyledons. J Plant Res 121:179–189

Galatro A, Simontacchi M, Puntarulo S (2001) Free radical generation and antioxidant content in chloroplasts from soybean leaves exposed to ultraviolet-B. Physiol Plant 113:564–570

Gao Q, Zhang L (2008) Ultraviolet-B-induced oxidative stress and antioxidant defense system responses in ascorbate-deficient vtc1 mutants of *Arabidopsis thaliana*. J Plant Physiol 165:138–148

Gao W, Zheng Y, Slusser JR, Heisler GM (2003) Impact of enhanced ultraviolet-B radiation on cotton growth, development, yield and qualities under field conditions. Agric For Meteorol 120:214–248

Gerhardt KE, Wilson MI, Greenberg BM (1999) Tryptophan photolysis leads to a UVB-induced 66 kDa photoproduct of ribulose-1,5- bisphosphate carboxylase/oxygenase (Rubisco) *in vitro* and *in vivo*. Photochem Photobiol 70:49–56

Gerhardt KE, Wilson MI, Greenberg BM (2005) Ultraviolet wavelength dependence of photomorphological and photosynthetic responses in *Brassica napus* and *Arabidopsis thaliana*. Photochem Photobiol 81:1061–1068

Gilbert M, Pors Y, Grover K, Weingart I, Skotnica J, Grimm B, Seidlitz HK, Langebartales C, Wilhelm C (2009) Intra- and inter-specific differences of 10 barley and 10 tomato cultivars in response to short-time UV-B radiation: a study analysing thermoluminescence, fluorescence, gas-exchange and biochemical parameters. Environ Pollut 157:1603–1612

Gill SS, Tuteja N (2010) Reactive oxygen species and antioxidant machinery in abiotic stress tolerance in crop plants. Plant Physiol Biochem 48:909–930

Giordano CV, Galatro A, Puntarulo S, Ballaré CL (2004) The inhibitory effects of UV-B radiation (280–315 nm) on *Gunnera magellanica* growth correlate with increased DNA damage but not with oxidative damage to lipids. Plant Cell Environ 27:1415–1423

Girotti AW (1985) Mechanisms of lipid peroxidation. Free Radic Biol Med 1:87–95

Gonzalez R, Paul ND, Percy K, Ambrose M, McLaughlin CK, Barnes JD, Areses M, Wellburn AR (1996) Responses to ultraviolet-B radiation (280–315 nm) of pea (*Pisum sativum*) lines differing in leaf surface wax. Physiol Plant 98:852–860

Gonzalez JA, Rosa M, Parrado MF, Hilal M, Prado FE (2009) Morphological and physiological responses of two varieties of a highland species. J Photochem Photobiol B 96:144–151

Gould KS (2002) The colour red. N Z J Geogr 60:59–63

Hagh AG, Khara J, Darvishzadeh R (2012) Effect of UV-B radiation on activity of antioxidant enzymes in four sunflower cultivars. Int J Agric Res Rev 2:528–534

Havaux M, Lutz C, Grimm B (2003) Chloroplast membrane stability in chlP transgenic tobacco plants deficient in tocopherols. Plant Physiol 132:300–310

He J, Huang L-K, Whitecross MI (1994) Chloroplast ultrastructure changes in *Pisum sativum* associated with supplementary ultraviolet (UV-B) radiation. Plant Cell Environ 17:771–775

Hidema J, Kumagai T, Sutherland BM (2000) UV radiation-sensitive Norin 1 rice contains defective cyclobutane pyrimidine dimer photolyase. Plant Cell 12:1569–1578

Hidema J, Song IK, Kumagai T (2001) Relationship between ultraviolet-B sensitivity and cyclobutane pyrimidine dimer photorepair in rice. J Radiat Res 42:295–303

Hofmann RW, Campbell BD (2011) Response of *Trifolium repens* to UV-B radiation: morphological links to plant productivity and water availability. Plant Biol 2011(13):896–901

Hollosy F (2002) Effect of ultraviolet radiation on plant cell. Micron 33:179–197

Hopkins L, Bond MA, Tobin AK (2002) Ultraviolet-B radiation reduces the rates of cell division and elongation in the primary leaf of wheat (*Triticum aestivum* L cv Maris Huntsman). Plant Cell Environ 25:617–624

Iwamatsu Y, Aoki C, Takahashi M, Teranishi M, Ding Y, Sun C, Kumagai T, Hidema J (2008) UV-B sensitivity and cyclobutane pyrimidine dimer (CPD) photolyase genotypes in cultivated and wild rice species. Photochem Photobiol Sci 7:311–320

Jain K, Kataria S, Guruprasad KN (2004) Effect of UV-B radiation on antioxidant enzymes and its modulation by benzoquinone and *a*-tocopherol in cucumber cotyledons. Curr Sci 87:87–90

Jaleel CA, Riadh K, Gopi R, Manivannan P, Ines J, Al-Juburi HJ, Zhao C-X, Hong-Bo S, Panneerselvam R (2009) Antioxidant defense responses: physiological plasticity in higher plants under abiotic constraints. Acta Physiol Plant 31:427–436

Jansen MAK, Gaba V, Greenberg BM (1998) Higher plant and UV-B radiation: balancing damage, repair and acclimation. Trends Plant Sci 3:131–135

Jayakumar M, Amudha P, Kulandaivelu G (2004) Effect of low doses of UV-A and UV-B radiation on photosynthetic activities in *Phaseolus mungo* L. J Plant Biol 47:105–110

Jenkins GI (2009) Signal transduction in responses to UV-B radiation. Annu Rev Plant Biol 60:407–431

Jordan BR (1996) The effects of ultraviolet-B radiation on plants: a molecular perspective. Adv Bot Res 22:97–162

Jordan BR (2002) Molecular response of plant cells to UV-B stress. Funct Plant Biol 29:909–916

Kakani VG, Reddy KR, Zhao D, Mohammed AR (2003a) Effect of ultraviolet-B radiation on cotton (*Gossypium hirsutum* L.) morphology and anatomy. Ann Bot 91:817–827

Kakani VG, Reddy KR, Zhao D, Sailaja K (2003b) Field crop responses to ultraviolet-B radiation: a review. Agric For Meteorol 120:191–218

Katerova Z, Ivanov S, Mapelli S, Alexieva V (2009) Phenols, proline and low-molecular thiol levels in pea (*Pisum sativum*) plants respond differently toward prolonged exposure to ultraviolet-B and ultraviolet-C radiations. Acta Physiol Plant 31:111–117

Keil M, Jackson DR, Hort MC (2006) The January 2006 low ozone event over the UK. Atmos Chem Phys Discuss 6:8457–8483

Kliebenstein DK (2004) Secondary metabolites and plant/environment interactions: a view through *Arabidopsis thaliana* tinged glasses. Plant Cell Environ 27:675–684

Kolb CA, Käser MA, Kopecký J, Zotz G, Riederer M, Pfündel EE (2001) Effects of natural intensities of visible and ultraviolet radiation on epidermal ultraviolet screening and photosynthesis in grape leaves. Plant Physiol 127:863–875

Kolb CA, Kopecky J, Riederer M, Pfundel EE (2003) UV screening by phenolics in berries of grapevine (*Vitis vinifera*). Funct Plant Biol 30:1177–1186

Kondo N, Kawashima M (2000) Enhancement of the tolerance to oxidative stress in cucumber (*Cucumis sativus* L.). J Plant Res 113:311–317

Konopacka M, Widel M, Rzeszowska-Wolny M (1998) Modifying effect of vitamins C, E and beta-carotene against gamma-ray-induced DNA damage in mouse cells. Mutat Res 417:85–94

Kostina E, Wulff A, Julkunen-Tiitto R (2001) Growth, structure, stomatal responses and secondary metabolites of birch seedlings (*Betula pendula*) under elevated UV-B radiation in the field. Trees 15:483–491

Kumari R, Agrawal SB (2010) Supplemental UV-B induced changes in leaf morphology, physiology and secondary metabolites of an Indian aromatic plant *Cymbopogon citratus* (D.C.) Staph under natural field conditions. Int J Environ Stud 67:655–675

Kumari R, Singh S, Agrawal SB (2010) Responses of ultraviolet-B induced antioxidant defense system in a medicinal plant Acorus calamus L. J Environ Biol 31:907–911

Kumari R, Singh S, Agrawal SB (2009a) Combined effects of Psoralens and ultraviolet-B on growth, pigmentation and biochemical parameters of *Abelmoschus esculentus* L. Ecotoxicol Environ Saf 72:1129–1136

Kumari R, Singh S, Agrawal SB (2009b) Responses of supplemental ultraviolet-B radiation on growth and physiology of *Acorus calamus* L. (Sweet flag). Acta Biol Cracov Ser Bot 52:19–27

Kumari R, Agrawal SB, Singh S, Dubey NK (2009c) Supplemental ultraviolet-B induced changes in essential oil composition and total phenolics of *Acorus calamus* L. (Sweet flag). Ecotoxicol Environ Saf 72:2013–2019

Landry LG, Chapple CC, Last RL (1995) Arabidopsis mutants lacking phenolic sunscreens exhibit enhanced ultraviolet- B injury and oxidative damage. Plant Physiol 109:1159–1166

Láposi R, Veres S, Mile O, Mészáros I (2005) Effects of supplemental UV-B radiation on the photosynthesis—physiological properties and flavonoid content of beech seedlings (*Fagus sylvatica* L.) in outdoor conditions. Acta Biol Szeged 49:151–153

Laposi R, Veres S, Lakatos G, Olah V, Fieldsend A, Meszaros I (2009) Responses of leaf traits of European beech (*Fagus sylvatica* L.) saplings to supplemental UV-B radiation and UV-B exclusion. Agric For Meterol 149:745–755

Li YA, Zu YQ, Chen HY, Chen JJ, Yang JL, Hu ZD (2000) Intraspecific responses in crop growth and yield of 20 wheat cultivars to enhanced ultraviolet- B radiation under field conditions. Field Crop Res 67:25–33

Liakoura V, Bornman JF, Karabourniotis G (2003) The ability of abaxial and adaxial epidermis of sun and shade leaves to attenuate UV-A and UV-B radiation in relation to the UV absorbing capacity of whole leaf methanolic extracts. Physiol Plant 117:33–43

Liu L-X, Xu SM, Woo KC (2005) Solar UV-B radiation on growth, photosynthesis and xanthophyll cycle in tropical acacias and eucalyptus. Environ Exp Bot 54:121–130

Lizana XC, Hess S, Calderini DF (2009) Crop phenology modifies wheat responses to increased UV-B radiation. Agric For Meteorol 149:1964–1974

Long LM, Patel HP, Wendy CC, Stapleton AE (2003) The maize epicuticular wax layer provides UV protection. Funct Plant Biol 35:75–81

Mackerness A-H-S (2000) Plant responses to ultraviolet-B (UV-B: 280–320 nm) stress: what are the key regulators? Plant Growth Regul 37:27–39

Madhavian K, Ghorbanli M, Kalantari KM (2008) The effects of ultraviolet radiation on some antioxidant components and enzymes in *Capsicum annuum* L. Turk J Bot 32:129–134

Madronich S, McKenzie RL, Caldwell MM, Bjorn LO (1995) Changes in ultraviolet reaching the earth's surface. Ambio 24:143–152

Manetas Y (2003) The importance of being hairy: the adverse effects of hair removal on stem photosynthesis of *Verbascum speciosum* are due to solar UV-B radiation. New Phytol 158:503–508

Mazza CA, Boccalandro HE, Giordano CV, Battista D, Scopel AL, Ballaré CL (2000) Functional significance and induction by solar radiation of ultraviolet-absorbing sunscreens in field-grown soybean crops. Plant Physiol 122:117–126

Mazza CA, Gimenezb PI, Kantolicb AG, Ballaré CL (2013) A beneficial effects of solar UV-B radiation on soybean yield mediated by reduced insect herbivory under field conditions. Physiol Plant 147:307–315. doi:10.1111/j.1399-3054.2012.01661.x

McKenzie R, Connor B, Bodeker G (1999) Increased summertime UV radiation in New Zealand in response to ozone loss. Science 285:1709–1711

McKenzie RL, Aucamp PJ, Bais AF, Bjorn LO, Ilyas M, Madronich S (2011) Ozone depletion and climate change: impacts on UV radiation. Photochem Photobiol Sci 10:182–198

Mishra S, Agrawal SB (2006) Interactive effects between supplemental UV-B radiation and heavy metals on growth and biochemical characteristics of *Spinacia oleracea* L. Braz J Plant Physiol 18:1–8

Mittler R (2002) Oxidative stress, antioxidants and stress tolerance. Trends Plant Sci 7:405–410

Monsi M (1968) Mathematical models of plant communities. In: Eckardt F (ed) Function of terrestrial ecosystem at the primary production level. UNESCO, Paris, pp 131–150

Mori M, Yoshida K, Ishigaki Y et al (2005) UV-B protective effect of a polyacylated anthocyanin, HBA, in flower petals of the blue morning glory. *Ipomoea tricolor* cv. Heavenly Blue. Bioorg Med Chem 13:2015–2020

Murphy TM (1984) Effect of sulfhydryl reagents on K+ efflux from rose cells, relationship to ultraviolet-stimulated efflux. Plant Physiol 75:138–141

Nara A, Takeuchi Y (2002) Ethylene evolution from tobacco leaves irradiated with UV-B. J Plant Res 115:247–253

Nithia SMJ, Shanthi N, Kulandaivelu G (2005) Different responses to UV-B enhanced solar radiation in radish and carrot. Photosynthetica 43:307–311

Noctor G, Foyer CH (1998) Ascorbate and glutathione: keeping active oxygen under control. Annu Rev Plant Physiol Plant Mol Biol 49:249–279

Oravecz A, Baumann A, Mate Z, Brzezinska A, Molinier J, Oakeley EJ, Adam E, Schafer E, Nagy F, Ulm R (2006) Constitutively photomorphogenic1 is required for the UV-B response in *Arabidopsis*. Plant Cell 18:1975–1990

Perluigi M, Domenico FD, Blarzino C, Foppoli C, Cini C, Giorgi A, Grillo C, Federico De Marco, Butterfield DA, Schininà ME, Coccia R (2010) Effects of UVB-induced oxidative stress on protein expression and specific protein oxidation in normal human epithelial keratinocytes: a proteomic approach. Protein Sci 8:13

Peykarestan B, Seify M (2012) UV irradiation effects on seed germination and growth, protein content, peroxidase and protease activity in redbean. Int Res J Appl Basic Sci 3:92–102

Prasad SM, Srivastava G, Mishra V, Dwivedi R, Zeeshan M (2005) Active oxygen species generation, oxidative damage and antioxidative defense system in *Pisum sativum* exposed to UV-B irradiation. Physiol Mol Biol Plant 11:303–311

Qaderi MM, Reid DM (2005) Growth and physiological responses of canola (*Brassica napus*) to UV-B and CO_2 under controlled environment conditions. Physiol Plant 125:247–259

Raneeliene V, Vyoniauskiene R, Janeys Z, Dlekyte K (2005) Action of UV-B on *Crepis apillaris* (L.) Wallr plants in controlled environmental conditions. Biologija 3:74–80

Rathore D, Agrawal SB, Singh A (2003) Influence of supplemental UV-B radiation and mineral nutrients on biomass, pigments and yield of two cultivars of wheat. Int J Biotron 32:1–15

Reddy KR, Kakani VG, Zhao D, Koti S, Gao W (2004) Interactive effects of ultraviolet-B-radiation and temperature on cotton physiology, growth development and hyperspectral reflectance. Photochem Photobiol 79:416–427

Robberecht R, Caldwell MM (1978) Leaf epidermal transmission of ultraviolet radiation and its implications for plant sensitivity to ultraviolet-radiation induced injury. Oecologia 32:277–287

Rozema J, Bjorn LO, Bornman JF, Gaberscik A, Häder D-P, Trost T, Germ M, Klisch M, Gröniger A, Sinha RP, Lebert M, He Y-Y, Buffoni-Hall R, de Bakker NVJ, van de Staaij J, Meijkamp BB (2002) The role of UV-B radiation in aquatic and terrestrial ecosystems – an experimental and functional analysis of the evolution of UV-absorbing compound. J Photochem Photobiol B Biol 66:2–12

Ryan KG, Hunt JE (2005) The effects of UVB radiation on temperate southern hemisphere forests. Environ Pollut 137:415–427

Ryan KG, Swinny EE, Markham KR, Winefield C (2002) Flavonoid gene expression and UV photoprotection in transgenic and mutant Petunia leaves. Phytochem 59:23–32

Sahoo A, Sarkar S, Singh RP, Kafatos M, Summers ME (2005) Declining trend of total ozone column over the northern parts of India. Int J Remote Sens 26:3433–3440

Sakalauskaite J, Viskelis P, Dambrauskiene E, Sakalauskiene S, Samuoliene G, Brazaityte A, Duchovskis P, Urbonavicien˙e D (2013) The effects of different UV-B radiation intensities on morphological and biochemical characteristics in *Ocimum basilicum L.* J Sci Food Agric 93:1266–1271. doi:10.1002/jsfa.5879

Santos I, Almeida JM, Salema R (1993) Plants of *Zea mays* L. developed under enhanced UV-B radiation. I. Some ultrastructural and biochemical aspects. J Plant Physiol 141:450–456

Santos I, Fidalgo F, Almeida JM, Salema R (2004) Biochemical and ultrastructural changes in leaves of potato plants grown under supplementary UV-B radiation. Plant Sci 167:925–935

Sasaki M, Takeshita S, Oyanagi T, Miyake Y, Sakata T (2002) Increasing trend of biologically active solar ultraviolet-B irradiance in mid-latitude Japan in the 1990s. Opt Eng 41:3062–3069

Searles PS, Flint SD, Caldwell MM (2001) A meta-analysis of plant field studies simulating stratospheric ozone depletion. Oecologia 127:1–10

Selvakumar V (2008) Ultraviolet-B radiation (280–315 nm) invoked antioxidant defence systems in *Vigna unguiculata* (L.) Walp. and *Crotalaria juncea* L. Photosynthetica 46:98–106

Sharma AD, Sharma R (1999) Purification and characterization of UV-B induced phenylalanine ammonia-lyase from rice seedlings. Phytochemistry 49:618–626

Shi S-B, Zhu W-Y, Li H-M, Zhou D-W, Han F, Zhao X-Q, Tang Y-H (2004) Photosynthesis of *Saussurea superba* and *Gentiana straminea* is not reduced after long-term enhancement of UV-B radiation. Environ Exp Bot 51:75–83

Shiu C-T, Lee T-M (2005) Ultraviolet-B-induced oxidative stress and responses of the ascorbate–glutathione cycle in a marine macroalga *Ulva fasciata*. J Exp Bot 56:2851–2865

Singh RP, Sarkar S, Singh A (2002) Effect of El Nino on inter-annual variability of ozone during the period 1978–2000 over the Indian sub-continent and China. Int J Remote Sens 23:2449–2456

Singh S, Kumari R, Agrawal M, Agrawal SB (2009a) Modification of growth and yield responses of *Amaranthus tricolor* L. to sUV-B under varying mineral nutrient supply. Sci Hortic 120:173–180

Singh S, Mishra S, Kumari R, Agrawal SB (2009b) Response of supplemental ultraviolet-B and nickel on pigments, metabolites, and antioxidants of *Pisum sativum* L. cv Arkel. J Environ Biol 30:677–684

Singh S, Kumari R, Agarawal M, Agrawal SB (2011) Modification in growth, biomass and yield of radish under supplemental UV-B at different NPK levels. Ecotoxicol Environ Saf 74:897–903

Smith JL, Burritt DJ, Bannister P (2000) Shoot dry weight, chlorophyll and UV-B absorbing compounds as indicators of plant sensitivity to UV-B radiation. Ann Bot 86:1057–1063

Staxen I, Bornman JF (1994) A morphological and cytological study of *Petunia hybrida* exposed to UV-B radiation. Physiol Plant 91:735–740

Stratmann JW, Stelmach BA, Weller EW, Ryan CA (2000) UVB/UVA radiation activates a 48 kDa myelin basic protein kinase and potentiates wound signaling in tomato leaves. Photochem Photobiol 71:116–123

Takeuchi Y, Fukomuto R, Kasahara H, Sakaki T (1995) Peroxidation of lipid and growth inhibition induced by UV-B irradiation. Plant Cell Rep 14:566–570

Takeuchi Y, Inoue T, Takemura K, Hada M, Takahashi S, Ioki M, Nakajima N, Kondo N (2007) Induction and inhibition of cyclobutane pyrimidine dimer photolyase in etiolated cucumber (*Cucumis sativus*) cotyledons after ultraviolet irradiation depends on wavelength. J Plant Res 120:365–374

Tattini M, Guidi L, Morassi-Bonzi L, Pinelli P, Remorini D, Degl Innocenti E, Giordano C, Massai R, Agati G (2004) Differential accumulation of flavonoids and hydroxycinnamates in leaves of *Ligustrum vulgare* under excess light and drought stress. New Phytol 163:547–561

Teranishi M, Iwamatsu Y, Hidema J, Kumagai T (2004) Ultraviolet-B sensitivities in Japanese lowland rice cultivars: cyclobutane pyrimidine dimer photolyase activity and gene mutation. Plant Cell Physiol 45:1845–1856

Tevini M, Teramura AH (1989) UV-B effects in terrestrial plants. Photochem Photobiol 50:479–487

Tevini M, Iwanzik W, Teramura AH (1983) Effects of UV-B radiation on plants during mild water stress, II Effects on growth, protein and flavonoid content. Z Pflanzenphysiol 110:459–467

Tosserams M, Rozema J (1995) Effects of ultraviolet-B radiation (UV-B) on growth and physiology of the dune grassland species *Calamagrostis epigeios*. Environ Pollut 89:209–214

Tosserams M, Magendans GWH, Rozema J (1997) Differential effects of elevated ultraviolet-B radiation on plant species from a dune grassland ecosystem. Plant Ecol 128:266–283

Tripathi R, Sarkar A, Rai SP, Agrawal SB (2011) Supplemental ultraviolet-B and ozone, impact on antioxidants, proteome and genome of linseed (*Linum usitatissimum* L. cv Padmimi). Plant Biol 13:93–104

Tsormpatsidis E, Henbest RGC, Davis FJ, Battey NH, Hadley P, Wagstaffe A (2008) UV irradiance as a major influence on growth, development and secondary products of commercial importance in Lollo Rosso lettuce 'Revolution' grown under polyethylene films. Environ Exp Bot 63:232–239

Turtola S, Sallas L, Holopainen JK, Jalkunen-Tiitto R, Kainulainen P (2006) Long term exposure to enhanced UV-B radiation has no significant effect on growth or secondary compounds of outdoor-grown Scots pine and Norway spruce seedlings. Environ Exp Bot 56:80–86

UNEP, United Nations Environment Programme (2005) Environmental Effects Assessment Panel. Photochem Photobiol Sci 4:177–184

Urban O, Tuma I, Holub P, Marek MV (2006) Photosynthesis and growth response of *Calamagrostis arundinacea* and *C. villosa* to enhanced UV-B radiation. Photosynthetica 44:215–220

Weatherhead B, Tanskanen A, Stevermer A (2005) Ozone and ultraviolet radiation. In: Symon C, Arris L, Heal B (eds) Arctic climate impact assessment scientific report. Cambridge University Press, New York, pp 151–182

Wilson MI, Greenberg BM (1993) Specificity and photomorphogenic nature of ultraviolet-B induced cotyledon curling in *Brassica napus* L. Plant Physiol 102:671–677

WMO (World Meteorological Organization), Scientific Assessment of Ozone Depletion (2007) Global ozone research and monitoring project – report no. 50, Geneva, Switzerland, 572 pp

Wulff A, Ahonen J, Kärenlampi L (1996) Cell ultrastructural evidence of accelerated ageing of Norway spruce needles in industrial areas. New Phytol 133:553–561

Xu C, Natarajan S, Sullivan JH (2008) Impact of solar ultraviolet-B radiation on the antioxidant defense system in soybean lines differing in flavonoid contents. Environ Exp Bot 63:39–48

Yang S-H, Wang Li-J, Li S-H, Duan W, Loescher W, Liang Z-C (2007) The effects of UV-B radiation on photosynthesis in relation to Photosystem II photochemistry, thermal dissipation and antioxidant defenses in winter wheat (*Triticum aestivum* L.) seedlings at different growth temperatures. Funct Plant Biol 34:907–917

Yannarelli GG, Gallego SM, Tomaro ML (2006) Effect of UV-B radiation on the activity and isoforms of enzymes with peroxidase activity in sunflower cotyledons. Environ Exp Bot 56:174–181

Yao X, Liu Q (2007) Changes in photosynthesis and antioxidant defenses of *Picea asperata* seedlings to enhanced ultraviolet-B and to nitrogen supply. Physiol Plant 129:364–374

Yao Y, Xuan ZG, Li Y, He Y, Helena K, Li C (2006) Effects of ultraviolet-B radiation on crop growth, development, yield and leaf pigment concentration of tartary buckwheat (*Fagopyrum tataricum*) under field conditions. Eur J Agron 25:215–222

Yao Y, Xuan Z, He Y, Lutts S, Korpelainen H, Li C (2007) Principal component analysis of intraspecific responses of tartary buckwheat to UV-B radiation under field conditions. Environ Exp Bot 61:237–245

Yao YN, Yang YQ, Li Y, Lutts S (2008) Intraspecific responses of *Fagopyrum esculentum* to enhanced ultraviolet B radiation. Plant Growth Regul 56:297–306

Zepp RG, Erickson Iii DJ, Paul ND, Sulzberger B (2011) Effects of solar UV radiation an climate change on biogeochemical cycling: interactions and feedbacks. Photochem Photobiol Sci 10:261–279

Zheng Y, Gao W, Slusser JR, Grant RH, Wang C (2003) Yield and yield formation of field winter wheat in response to supplemental solar ultraviolet-B radiation. Agric For Meteorol 120:279–283

Chapter 9
Soil Quality and Plant-Microbe Interactions in the Rhizosphere

Debarati Bhaduri, Sharmistha Pal, T.J. Purakayastha, K. Chakraborty, R.S. Yadav, and Mohd. Sayeed Akhtar

Abstract The rhizosphere is a microenvironment contrastingly different from non-rhizosphere soil. The high microbial activity in the rhizosphere leads to better cycling and availability of nutrients and improves chemical soil quality indicators. The biological soil quality indicators are improved in rhizosphere due to enhanced microbial activities either in terms of microbial biomass carbon, dehydrogenase activity, activities of various hydrolytic enzymes e.g., phosphatases, sulfatases, proteases, amidases and glucosidase responsible for breakdown of organically bound nutrients in soil. Over the last two decades lot of interests developed on soil quality research due to degradation in soil quality by anthropogenic activity. Besides being a rich source of carbon and energy for the heterotrophic organisms, plant roots exudates secrete a variety of carbonaceous materials that can act as a binding agent to increase stability of soil aggregates. The adoption of proper soil, crop, nutrient and organic manure management strategies has the direct impact on the soil quality. However, it may also be affected indirectly by the plant microbe interactions. This article focuses on induced changes in the rhizosphere ecology by plant microbe interactions and also to integrate these changes to soil quality indicator of unified soil in the sustainable agriculture soil management practices.

D. Bhaduri (✉) • K. Chakraborty • R.S. Yadav
ICAR-Directorate of Groundnut Research, Junagadh 362001, Gujarat, India
e-mail: debarati.ssiari@gmail.com; koushik_iari@rediffmail.com; yadavrs2002@gmail.com

S. Pal
ICAR-Indian Institute of Soil and Water Conservation, Research Centre, Chandigarh 160019, India
e-mail: sharmistha.ars@gmail.com

T.J. Purakayastha
Division of Soil Science and Agricultural Chemistry, ICAR-Indian Agricultural Research Institute, New Delhi 110012, India
e-mail: tpurakayastha@gmail.com

M.S. Akhtar (✉)
Institute of Tropical Agriculture, Universiti Putra Malaysia, 43400 UPM Serdang, Darul Ehsan, Selangor, Malaysia
e-mail: sayeedbot@yahoo.co.in

E. Lichtfouse (ed.), *Sustainable Agriculture Reviews*,
Sustainable Agriculture Reviews 17, DOI 10.1007/978-3-319-16742-8_9

Keywords Carbon dynamics • Rhizo-depositions • Soil-plant-microbe relations • Soil quality indicators

9.1 Introduction

Soil is gradually renewable over the human time scale of decades to centuries. Soil is fragile and easily degraded when misused and mismanaged. Soil is unequally distributed among geographical regions of the world because of limitation of climate and slope gradient, and is fixed in location and cannot be transported (Lal 2013). The toll of soil erosion in natural ecosystems occurs at the steady rate and cumulative impacts cumulative effects on the soil quality. Worldwide, erosion rates range from a low of 0.001–2 t/ha/year on relatively flat land with grass and/or forest cover to rates ranging from 1 to 5 t/ha/year on mountainous regions with normal vegetative cover (Pimentel and Kounang 1998). According to Crosson (1997) approximately 24 billion tons of fertile soils are lost annually from world agricultural system. However, Lal and Stewart (1990) and Wen (1998) reported that 6.6 billion tons of soils per year are lost in India and 5.5 billion tons are lost annually in China. The term soil quality received considerable scientific attention during the 1990s (Lal 1993, 1997, 1999; Doran and Parkin 1994; Doran et al. 1996; Carter et al. 1997; Bezdicek et al. 1996; Karlen et al. 1997). The interest is quite logical as the soil always remains at receiving end of all irresponsible anthropogenic activities. Soil performs various functions and one of the major functions is to act as a supporting medium for plants or food crops feeding the human and animal population. The condition of our soils ultimately determines human health by serving as a major medium for food and fibre production and a primary interface with the environment, influencing the quality of the air we breathe and water we drink. Thus, there is a clear linkage between soil quality and human and environmental health. As such the health of our soil resources is a primary indicator of the sustainability of our land management practices (Doran et al. 2002).

Plant-microbial interactions is a matter of discussion among scientific fraternity for quite a few time, as it directly linked with microbiologists, agronomists, soil scientists, botanists and pathologists. Plant-microbial interaction can influence the crop production as an applied aspect other than relating with issue of basic science research. But how the plant-microbial interaction impacts various soil processes and therefore changes in soil quality as a cumulative effect have not been much highlighted earlier. As a complex functional state, soil quality on the other hand, cannot be measured directly, but soil quality may be inferred from management-induced changes in soil properties known as soil quality indicators (Bouma 2002). Such indicators are measurable soil attributes that influence the capacity of soil to perform crop production or environmental functions and are sensitive to change in land use, management and conservation practices. Traditionally, soil quality research focused primarily on chemical and physical properties because of their simple analytical methods (Larson and Pierce 1991). Moreover, it has been suggested by Islam and Weil (2000) that soil biological properties can serve as early and sensitive indicators of agro-ecosystems in response to soil management practices. Integrated soil quality indices based on a combination of soil properties provide a better indication of soil quality than individual parameters.

Karlen et al. (1994) developed a soil quality index through 'Conceptual Framework model' to quantify parameters such as chemical, biological and physical properties that interact in a complex way to give a soil its quality. Further, Andrews et al. (2004) designed a "Soil Management Assessment Framework" for soil quality assessment, where indicators were selected based on the management goals, associated soil functions and other site-specific factors. A valid soil quality index would help to interpret data from different soil measurements and show whether management and land use are having the desired results for productivity, environmental protection, and health (Granatstein and Bezdicek 1992).

In recent years, soil quality research has focused on the linkages among the following: management practices and systems; observable soil characteristics; and soil processes and performance of soil functions (Lewandowski and Zumwinkle 1999). Choosing the appropriate soil attributes to include in an index must include consideration of soil function and management goals that are site specific and user-oriented and must focus on sustainability rather than just crop yields. These indices would be useful in ascertaining the fragility of soil and for understanding how improved management might strengthen its resilience (Chaudhury et al. 2005).

9.2 Role of Plant Roots in Modifying Rhizosphere Environment

Soil organic matter is one of the key components in maintaining soil fertility, soil quality, and agricultural sustainability. The rhizosphere has strongly influenced by plant roots and also has a significant role in regulation of soil organic matter decomposition and nutrient cycling (Dijkstra et al. 2013). The presence of live roots significantly controls the rate of original soil organic matter decomposition through the rhizosphere effect. Soil food webs are mainly based on three primary carbon sources: root exudates, litter or residues, and soil organic matter. These carbon sources vary in their availability and accessibility to soil organisms, and can thus; increase the carbon flow and biodiversity within the food web. Processes that are shared both by roots and associated microorganisms are commonly referred as rhizosphere processes, such as: exudation of soluble compounds, water uptake, nutrient mobilization by roots and microorganisms, rhizosphere-mediated soil organic matter decomposition, and the subsequent release of CO_2 through respiration. In this way, rhizosphere processes serve as are major gateways for nutrients and water. Roots leak carbohydrates, amino acids, organic acids, and a number of other complex compounds that feed the microorganisms in the rhizosphere. Each plant species leaks its own unique signature of compounds into the rhizosphere. The more diverse the plant community above ground the more diverse the community below ground.

At the global scale, rhizosphere processes utilize approximately 50 % of the energy fixed by photosynthesis in terrestrial ecosystems, contribute roughly 50 % of the total CO_2 emitted from terrestrial ecosystems, and mediate virtually all aspects of nutrient cycling. Therefore, plant roots and their rhizosphere interactions are at the center of many ecosystem processes. However, the linkage between rhizosphere processes and soil organic matter decomposition is not well understood though it is often large in magnitude. The study of Dijkstra and Cheng (2007) has indicated that the rate of soil organic matter decomposition can be accelerated by as much as 380 % or inhibited by as much as 50 % by the presence of live roots. On the other hand, microbial growth is limited by supply of fresh organic carbon that directly linked with plant biomass production. Plants have a clear cut effect to modify the functional and structural composition of microbes present in rhizosphere (Dey et al. 2012). But this effect of rhizosphere priming varies with plant phenology, seasonal and temporal variations and soil mineral nutrients applied (Cheng et al. 2003). With help of improved knowledge and up-to-date technologies for studying the rhizosphere organization will facilitate greater understanding in this aspect.

9.3 Role of Microbes in Rhizosphere Zone

Microorganisms are the tiniest forms of life; still their role towards ecosystem sustainability cannot be ignored. Being present in numerous genus and species in soil and even a lot as unexplored carrying useful metabolic processes are influenced by several ecological factors like moisture, temperature, organic matter, rhizosphere structure of plant, and plant diversity etc. They are also described to be affected by some crop management practices like tillage, irrigation, nutrient, organic and green manuring, mulching and few others. Microorganisms play a key role in nutrient cycling and energy flow. Potentiality of biological nitrogen fixation as low-cost source of nitrogen and phosphorus solubilization by phosphate solubilizing bacteria and phosphorus mobilization by arbuscular mycorrhizal fungi is already well-known among scientific community. Comparing the rhizosphere effect on microbes, the rhizosphere/bulk soil ratios are found to be ranged from 2–20, 10–20 and 5–10 for bacteria, fungi and actinomycetes, respectively (Dey et al. 2012). Soil microorganisms are also recognized as potential indicator of some ecological stresses such as heat, moisture, salts, heavy metals, pesticides and contaminants. Microbial abundance and diversity was found to be reduced under soil perturbations and pollution. Even their changing metabolism and functions are also significant in the present context of climate change. Probably their sensitivity towards such imposed condition proved them as effective soil quality indicators. However, it is important to screen those useful microbial indicators that can vary from one ecosystem to another like agriculture, pasture and forests. Microbes present in the rhizosphere thus have a prominent role to play as sensitive ecological indicator. Though, these microbial indicators have some merits and demerits and should be selected on the basis of measurement protocols, reproducibility and easiness of interpretation that can immensely represent soil quality and health.

9.4 Rhizosphere Ecology

The term 'rhizosphere' was coined by Hiltner (1904) to describe the narrow zone of intense bacterial activity around legume roots, via root stimulation of any soil inhabiting organism. Currently, it is described as the zone of soil surrounding the root which is affected by it. As the definition has changed from the specific to the more general, so the width of the rhizosphere itself has also broadened, from a very narrow zone extending at most 1–2 mm from the root surface, to one extending several centimetres in the case of nutrient or water depletion profiles, or several tens of centimetres for volatile compounds released by the root. Perhaps the only recognizable characteristic of the original definition is that the rhizosphere is operationally described in terms of a concentration gradient between the root surface and the bulk soil.

9.4.1 The Rhizosphere Effect

As seeds germinate and roots grow through the soil, the release of organic material provides the driving force for the development of active microbial populations in a zone that includes plant root and surrounding soil in a few mm of thickness. This phenomenon is referred as the rhizosphere effect (van Veen et al. 2007). Broadly, there are three distinct components recognized in the rhizosphere; the rhizosphere *per se* soil, the rhizoplane, and the root itself. The rhizosphere is thus the zone of soil influenced by roots through the release of substrates that affect microbial activity. The rhizoplane is the root surface, including the strongly adhering root particles. The rhizosphere effect can thus be viewed as the creation of a dynamic environment where microbes can develop and interact.

9.4.2 The Rhizosphere as a Battle Field

The number and diversity of microorganisms are related to the quantity and quality of the exudates but also to the outcome of the microbial interactions that occur in the rhizosphere. Soil biota e.g. bacteria, fungi, micro-fauna and the plant root are themselves embedded in food webs and thus interactions with consumers or predators in the microbial as well as macro and mesofaunal world are important to understand rhizosphere processes. A high number of soil microbes attained properties enabling them to interact more efficiently with roots and withstand the quite challenging conditions of rhizosphere life. The rhizosphere inhabiting microorganisms competing each other for water, nutrients and space and sometimes improve their competitiveness by developing an intimate association with plant. This process can be regarded as an on-going process of micro-evolution in low-nutrient environments, which are quite common in natural ecosystems (Schloter et al. 2000).

9.4.3 Alteration of Soil Characteristics in Rhizosphere

As a consequence of normal growth and development, a large range of organic and inorganic substances are secreted by roots into the soil, which inevitably leads to changes in its biochemical and physical properties (Walker et al. 2003). Various functions have been attributed to root cap exudation including the maintenance of root-soil contact, lubrication of the root tip, protection of roots from desiccation, stabilization of soil micro-aggregates, and selective adsorption and storage of ions (Hawes et al. 2000). Root mucilage is a reasonably studied root exudate that is believed to alter the surrounding soil as it is secreted from continuously growing root cap cells (Sims et al. 2000). Young (1995) found that rhizosheath soil was significantly wetter than bulk soil and suggested that exudates within the rhizosheath increase the water holding capacity of the soil. Furthermore, it has recently been proposed that in dry soil, the source of water to hydrate and expand exudates is the root itself. The root exudation plays a major role in maintaining root-soil contact in the rhizosphere by modifying the biochemical and physical properties of the rhizosphere and contributing to root growth and plant survival. However, the exact fate of exuded compounds in the rhizosphere, and the nature of their reactions in the soil, remains poorly understood. The rhizosphere inhabiting microorganisms compete for water, nutrients and space and sometimes improve their competitiveness by developing an intimate association with plant (Hartmann et al. 2009). These microorganisms play important roles in the growth and ecological fitness of their host.

9.4.4 Root Exudation and Nutrient Availability

Root exudation is the release of organic compounds from living plant roots into the surrounding soil; it is a ubiquitous phenomenon. Roots release compounds via at least two potential mechanisms, and the rates of exudation *sensu stricto* vary widely among species and environmental conditions. At the moment of exudation and thereafter, the organic materials are subject to microbial attack, and thus cannot be enriched and separated from the roots in the natural environments. In this context, root exudation has been quantified by measuring the production of labelled CO_2 in the rhizosphere of ^{14}C-labelled plants. It has been estimated that approximately 40 % of the total amount of carbohydrates produced by photosynthesis is released into the soil surrounding roots (Akhtar and Siddiqui 2008). Root exudates are mainly composed of water soluble sugars, organic acids, and amino acids, but also contain hormones, vitamins, amino compounds, phenolics and sugar phosphate esters. Release of these low molecular weight compounds is a passive process along the steep concentration gradient which usually exists between the cytoplasm of intact root cells (millimolar range) and the external soil solution (micromolar range).

It is assumed that both the qualitative and quantitative compositions of root exudates are affected by various environmental factors, including pH, soil type, oxygen status, light intensity, soil temperature, nutrient availability and the presence of microorganisms. The composition of root exudates is also largely dependent on the plant species and plant developmental stages (Jaeger et al. 1999). It is estimated at the maturity that the rhizodeposition of N amounted to 20 % of the total plant nitrogen (Jensen 1996). Low concentrations of some nutrients such as K^+, Na^+ and Mg^{++} readily stimulate the activity of major enzymes of the glycolytic pathway, namely phosphofructokinase and pyruvate kinase, which together regulate glycolysis in plant cells (Plaxton 1996). Individual micronutrients are similarly important components of major enzymes, which regulate all biological processes. So low nutrient availability can create constraint to plant growth in many environments of the world, especially the tropics where soils are extremely deficient in these oligoelement nutrients (Pinton et al. 2007). Some species typically exude organic acid anions in response to P and Fe deficiency or phytosiderophores due to Fe and Zn deficiency (Haynes 1990). Root exudates, on the other way provides nutritional source to several bacteria inhabiting in rhizosphere that takes part in nutrient cycling in soil (Jaeger et al. 1999).

9.5 Rhizosphere Effect on Soil Quality Indicators

Soil is a dynamic, living, natural body that is vital to the function of terrestrial ecosystems and represents a unique balance between physical, chemical and biological factors. Quality of soil as distinct from health is largely defined by the ability of soil to perform various intrinsic and extrinsic functions. Hence number of definitions has drawn attraction over the time. Quality is represented by a suite of physical, chemical and biological properties that together (i) provide a medium for plant growth; (ii) regulate and partition water flow and storage in the environment; (iii) and serve as an environmental buffer in the formation and destruction of environmentally hazardous compounds (Larson and Pierce 1991). Soil qualities can be defined as inherent attributes of soil that are inferred from soil characteristics of indirect observations e.g. compatibility, erodibility and fertility. The ability of soil to support crop growth which includes factors such as degree of tilth, aggregation, organic matter content, soil depth, water holding capacity, infiltration rate, pH changes, nutrient capacity and so forth (Grassini et al. 2010; Piedallu et al. 2011).

The soil quality has been defined by scientists as the "fitness for use" (Doran et al. 2002), and by others as the "capacity of a soil to function" (Doran and Parkin 1994; Karlen et al. 1997). According to Arshad and Coen (1992) soil quality is "the sustaining capability of a soil to accept, store and recycle water, and energy for production of crops at optimum levels while preserving a healthy environment".

Karlen et al. (1992) defined soil quality as “the ability of the soil to serve as a natural medium for the growth of plants that sustain human and animal life”. Soil quality refers to its ability to sustain productivity and maintain environmental quality (Lal 1993). “The capacity of a soil to function within boundaries to sustain biological productivity, maintain environmental quality and promote plant and animal health”, was the definition of soil quality put forth by Doran and Parkin (1994). Lal and Stewart (1995) described soil quality as the inherent attribute of soil and to characteristics and processes that determined the soil’s capacity to produce economic goods and services and regulate the environment. Sims et al. (1997) proposed a non-polluted soil criterion for soil quality, that they refer soil quality as the clean state of soil. According to Holden and Friestone (1997) soil quality is the degree to which the physical, chemical and biological characteristics of the soil to attenuate environmental pollution. The soil quality definition given by Karlen et al. (1997) mentioned as “The capacity of a specific kind of soil to function within natural or managed ecosystem boundaries, to sustain plant and animal productivity, maintain or enhance water and air quality, and support human health and habitation”. It can be conceptualized as a three-legged stool, the function of which requires an integration of three major components-sustained biological productivity, environmental quality and plant and animal health. Sojka and Upchurch (1999) defined soil quality in terms of distinct management and environmental condition specific to one soil under explicit circumstance for a given use.

Soil quality traditionally has focused on, and has been equated with agricultural system productivity. Crop yield is an important indicator of system productivity that is, in part, dependent upon soil quality. Crop yield can serve as a bioassay for several interacting factors such as soil, water, air, disease, germplasm, and management. Crop yield alone, however, is an incomplete measure of system productivity. Soil quality may better represent system productivity and function. Soils have biological, chemical and physical properties that interact in a complex way to give a soil its quality or capacity to function. Indicators of soil quality should give measures of the capacity of the soil to perform these functions (Larson and Pierce 1991; Doran and Parkin 1994). The soil quality indicators could be used to measure changes caused by soil and crop management practices. Efforts to quantify soil quality are increasing throughout the world and several physical and chemical indicators have been identified (Karlen and Stott 1994). However it is very closely linked to soil biological properties also. Most of the indicators are governed by rhizosphere either by direct or indirect way.

The package of indicators used for assessing soil quality, can vary from location to location depending on the kind of land eg. Rangeland, wasteland, agricultural land or land use, soil function and soil forming factors (Arshad and Coen 1992; Helkamp et al. 1995). Henceforth, indicators should be easily measured and measurement should be reproducible (Gregorich et al. 1994). The proper approach in defining soil quality and health must be holistic, not reductionistic. However, it would be unrealistic to use all ecosystems or soil attributes as indicators, so a minimum dataset consisting of a core set of soil attributes encompassing chemical, physical and biological soil properties are selected for soil quality assessment (Larson and Pierce 1991).

9.5.1 *Rhizosphere Effect and Physical Soil Quality Indicators*

Soil physical characteristics are a necessary part of soil quality assessment because they cannot be easily improved (Karlen and Cambardella 1996; Wagnet and Hutson 1997). Larson and Pierce (1991) summarized the physical indicators of soil quality as those properties which influence crop productivity by:

1. Whether a soil can accommodate unobstructed root growth and provide pore space of sufficient size and continuity for root penetration and expansion
2. The extent to which the soil matrix will resist deformation
3. The capacity of soil for water supply and aeration.

Physical conditions of a soil have direct and indirect effects on crop production and environmental quality (Fig. 9.1). These include parameters that are relatively static (standard soil profile characteristics, rooting depth, morphological features, texture etc.) and sensitive parameters (bulk density, aggregate stability, penetration resistance etc.).

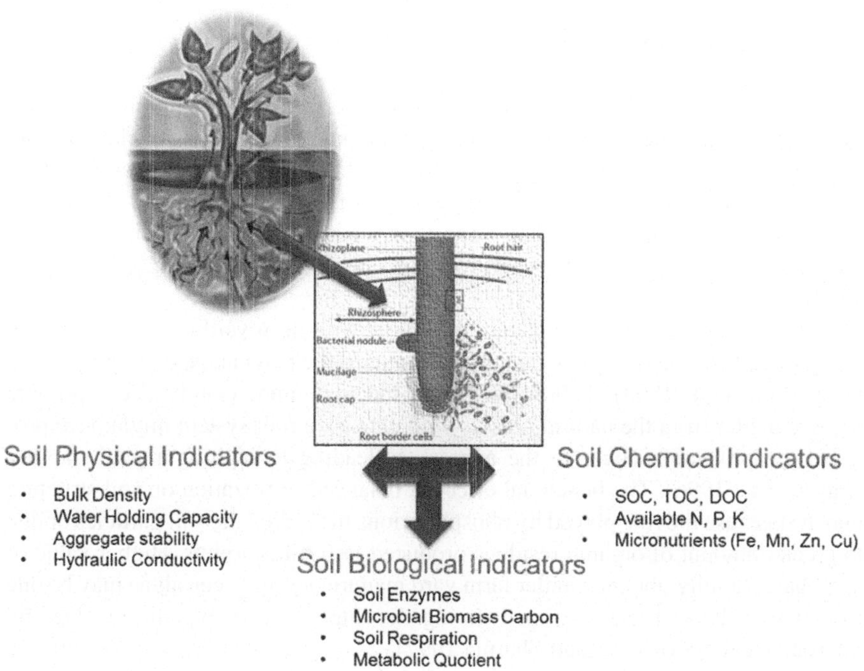

Fig. 9.1 Effect of soil quality indicators on the rhizospheric soil

9.5.1.1 Bulk Density

Bulk density is considered as one of the most important dynamic property of soil. It varies with structural condition of the soil that is altered by cultivation, compression, animal grazing, agricultural machinery, weather and others. Soil bulk density also can serve as an indicator of soil compaction and relative restriction to root growth (Arshad et al. 1996).

Bulk density is reported to be influenced by various manuring and fertilization practices. Nambiar (1994) reported that incorporation of farm yard manure at the rate 10–15 t ha^{-1} annually for many years brought about a slight reduction in bulk density in almost all the soils indicating improvement in soil physical properties. The effect of 10 years of rice-rice rotation showed that the lowest bulk density (1.26 Mg m^{-3}) was recorded with paddy straw incorporation along with 50 % NPK, hence incorporation of organic matter along with fertilizer had significantly increased the porosity (Oguike et al. 2006). The bulk density (0–15 cm) ranged from 1.39 Mg m^{-3} for 100 % NPK plus farm yard manure to 1.47 Mg m^{-3} for control under maize-wheat crop rotation of LTFE in inceptisol (Masto et al. 2007). Puddling is another important management practice which significantly influences the soil bulk density. It is reported that bulk density increases when the puddled soils undergo desiccation in a lowland-upland e.g. rice-wheat situation because of soil shrinkage (Sharma et al. 2005).

9.5.1.2 Aggregate Stability

Aspects of soil structure are quantitatively characterized by determining the stability of aggregates. The amount and quantity of organic matter, types of clay, wetting and drying, freezing and thawing, type and amount of electrolytes affecting colloidal dispersion, biological activity, cropping and tillage system affect soil aggregates. Aggregate stability is an important measure of soil quality for crop establishment, water infiltration and resistance to erosion and compaction (Beare and Bruce 1993).

Either the application of application of NPK with farm yard manure or sewage compost and continuous green manuring increased the percentage of aggregate stability (Sur et al. 1993). Fallow plots had the maximum readily decomposable organic matter from the natural grasses with extensive root system might have provided the protection cover to the aggregates leading to high aggregate stability (Rawat et al. 1996). The beneficial effect of balanced fertilization on soil structure may be because the role played by phosphate ions in binding the soil particles or due to greater amount of organic residues produced in fertilized plots. Higher value of aggregate stability obtained under farm yard manure or blue green algae may be due to certain polysaccharides formed during decomposition of organic residues by microbial activity (Mishra and Sharma 1997).

9.5.1.3 Hydraulic Conductivity

The rate at which water flows through soil affects many properties of the soil, including infiltration, drainage, nutrient flow within the soil, and soil erosion. Hydraulic conductivity values provide a potential measure for comparing the impact of different management practices on water flow, be it either cropping system, or tillage and nutrient management. Paddy caused greater reduction in hydraulic conductivity than wheat, which may be attributed to reorientation of particles into impermeable crust at the surface and the deposition of finer particles in lower layers forming a flow restricting layer in puddled rice cultivation. Hence tillage for rice caused a significant reduction in saturated hydraulic conductivity (Sharma et al. 2005). While hydraulic conductivity in residue-covered no-till soil increased due to improved macroporosity (Blanco-Canqui and Lal 2007). Application of NPK fertilizers (Schjonning et al. 2005), organic materials in combination with fertilizers (Oguike et al. 2006), and combined use of farm yard manure and blue green algae (Mishra and Sharma 1997) reported to increase soil's hydraulic conductivity.

9.5.1.4 Water Holding Capacity

Water holding capacity is the amount of water retained by the soil after it has been saturated. Water holding capacity depends on the number and size of pores that are primarily governed by the texture, structure, organic matter and mineralogy of the soil. Incorporation or substitution of nitrogen by organic materials increased the maximum water holding capacity than the chemical fertilizer alone (Reddy et al. 1999). However, the maximum water holding capacity in no-tilled soil was due to change in soil organic carbon content (Masto 2004).

9.5.2 Rhizosphere Effect and Carbon Dynamics

Carbon dynamics in soil is directly related with rhizosphere zone. Carbon mineralization and subsequent release of CO_2 is governed either by respiration activity of root tissues or microbial cells present in the root vicinity (Melillo et al. 2002). On the other hand, Carbon sequestration implies to trap carbon in the soil so that the carbon loss from soil can be prevented (Lal 2008). Deposition of root tissues along with leaf litters and stems of plant contribute here.

9.5.2.1 Soil Organic Carbon

Soil organic matter has long been considered the key quality factor of soil. Soil organic matter is a source and a sink of plant nutrients and is important in maintaining soil tilth, improving aeration and infiltration of water, promoting water retention, reducing erosion and controlling the efficacy and fate of applied pesticides (Gregorich et al. 1994). Amongst the various attributes organic matter content is the most important determinant of soil quality. Therefore, the evaluation of soil quality should be taken into account the multifarious beneficial function of organic matter (Gill et al. 2006).

Changes in soil organic carbon over a long-term period showed an observable trend. The annual addition of farm yard manure had increased initially the soil organic carbon content rapidly and then grows slowly as a steady state. Johnston (1994) reported that organic carbon in agricultural soils is found to be constant for about 100 years in both unmanured and manured plots supplied with NPK fertilizers. Application of farm yard manure Application of farm yard manure visibly enhanced the soil organic carbon content in various cropping systems (Iqbal et al. 2012).

9.5.2.2 Microbial Biomass Carbon

The microbial biomass is the living component of soil organic matter (Jenkinson and Ladd 1981) excluding the macrofauna and plant root. In many models of organic matter formation microbial biomass is used as a precursor to more stable fraction of organic matter (Parton et al. 1987). About 95 % of the total soil organic matter is non-living, therefore, relatively stable or resistant to change, decade may be required to observe in measurable changes in soil organic matter. It provides an indication of soil ability to store and recycle nutrients and energy and also serves as a sensitive indicator of change and future trends in organic matter levels and equilibria (Gregorich et al. 1994). Microbial biomass carbon is a relatively small (approximately 1–4 % of total soil organic carbon), labile fraction that quickly responds to C availability and also strongly influenced by management practices and system perturbations (Smith and Paul 1990).

No-tillage has been shown an improvement in microbial biomass carbon as compared to conventional tillage site (Purakayastha et al. 2008). Whereas many researches have cited that reported that the application of farm yard manure leads to higher content of microbial biomass carbon (Bucher 1999; Rudrappa et al. 2006) with the increasing microbial biomass level per unit soil organic matter and caused a shift in organic matter equilibrium. Other than farm yard manure, organic amendments also supply readily decomposable organic matter in addition to increasing root biomass and root exudates due to increased crop growth, so further improve the microbial biomass carbon content in soil (Goyal et al. 1999; Manjaiah et al. 2000).

9.5.2.3 Soil Respiration

Soil respiration is the production of CO_2 or consumption of O_2 as a result of the metabolic processes of living organisms at the rhizosphere and in the soil. Soil respiration has been used as a useful measure of soil activity. However all the microbial biomass in the soil is not in active stage and with the increase in microbial biomass the dormant cells also increase. The term 'dormant' is based on the assumption that under field condition energy limitations allow only small parts of the total potentially active population to be active at any given time. The majority exist as vegetative cell with lower metabolic activity (Heim et al. 2002).

Soil respiration is a well-established parameter to monitor decomposition (Birkas et al. 2008), but soil respiration is also highly variable and can show wide natural fluctuation depending on substrate availability, moisture and temperature. The great variability in soil respiration means that this measure taken alone is very difficult to interpret in terms of soil health (Brooks 1995). For valid comparison between soils, respiration measurements must normally be made under controlled laboratory conditions. Again, interpretation of respiration in terms of soil health or quality is problematic. Rapid decomposition of organic matter is not necessarily desirable because stable organic matter has an important role in soil physical and chemical characteristics. On the other hand, the decomposition of organic residues to release plant nutrients at times of plant demand is a desirable character. Influence of organic matter added to soil on carbon mineralization has increased the soil respiration ability (Rudrappa et al. 2006). However, Jassal et al. (2005) stated that seasonal pattern also plays a key role in determining CO_2 evolution rates from soil.

9.5.2.4 Metabolic Quotient

Microbial metabolic quotient (qCO_2) also known as respiratory quotient, is the ratio of carbon mineralized to microbial biomass carbon. As per the reported literature, the average soil respiratory activity increased with increasing levels of farm yard manure application but respiratory activity as a percent of microbial biomass carbon decreased with increasing levels of manure application (Lupwayi et al. 1998).

A more stable microbial population and a greater microbial efficiency in utilizing carbon substrates lead to lower the metabolic quotient (Purakayastha et al. 2008), which is desirable for carbon sequestration, often found low under the agricultural sites (Smith 2002) than the pasture and uncultivated lands. On the other way, High microbial quotient has been reported with long-term N or recent cattle manure applications and low with recent N applications (Fauci and Dick 1994). NPK plus organic treatments was most efficient in preserving C in soil as exhibited lower quotient (Rudrappa et al. 2006; Majumder et al. 2008).

Effect of fertilizer and manure applications on metabolic quotient depends on the nutrient status of the soil, with applications enhancing qCO_2 in non-nutrient stressed situations, while the opposite true under nutrient stress so the difficulty persists in ascribing differences in metabolic quotient to disturbance and stress limit the usefulness of metabolic quotient as a diagnostic indicator of soil quality (Wardle and Ghani 1995). However, it clearly does have potential when used in conjunction with other biochemical and microbial analyses.

9.5.3 *Rhizosphere Effect and Soil Nutrient Dynamics*

Measurement of nutrient cycling and concentration of elements that may be potential contaminants or of those needed for plant growth and development often determined by other chemical indicators of soil such as pH, salinity, organic carbon and cation exchange capacity. Karlen et al. (1992) suggested that total and available plant nutrients, and nutrient cycling rates should be included in chemical soil quality assessment. Larson and Pierce (1991) chose those chemical properties that either inhibits root growth or that effect nutrient supply due to the quantity present or the availability. Reganold and Palmer (1995) used chemical parameters related to nutrient availability as measures of soil quality including CEC, total N and P, pH and extractable P, S, K, Mg, Ca.

9.5.3.1 Available Nitrogen

Nitrogen being a primary nutrient element to crop plants controls their vegetative and reproductive growth. It supplies majorly from soil available nitrogen either organic or inorganic forms, constantly converts to readily-available forms (NH_4^+, NO_3^-) through certain biochemical interventions, mostly carried out by native microbial population in soil. Rhizosphere also contributes in supplying nitrogen via biological nitrogen fixation by nodule formation in leguminous crops. The regular incorporation of farm yard manure in soil enhanced nitrogen build up remarkably (Jenkinson 1991), however that varies among soil types (Nambiar 1994), but optimal to super optimal dose of fertilizer nitrogen showed little and gradual improvement in the soil available nitrogen.

9.5.3.2 Available Phosphorus

Phosphorus content of most of the soils is usually and only a very small proportion of the total P is present in available form. In contrast to certain inorganic N compounds which are not stable in soil and are subjected to volatilization and leaching, P is relatively stable and is not subject to such losses. The low solubility of P is partly responsible for the relative low availability of this nutrient to the plant (Mengel and Kirkby 1982). Continuous cropping without fertilization and manuring decreased the available P content in soil (Srinivasarao et al. 2013) whereas application of P-fertilizers had shown to increase the content of available P even with continuous cropping (Kumar and Yadav 2005).

9.5.3.3 Available Potassium

Although not much pronounced and studied, soil available potassium has the role to maintain soil quality. As this element present in the soil in a single cationic form (K^+) and transforms from non-exchangeable to exchangeable and soluble forms for maintaining equilibrium, no other biochemical factors plays any role in K-dynamics. Root uptake of K occurs mostly by diffusion mechanism. Balanced application of farm yard manure, farm yard manure plus NP fertilizer, or NPK fertilizer, increased the effectiveness of the K applied (Blake et al. 1999). Long-term usage of organic amendments such as groundnut shells and farm yard manure along with inorganic fertilizers is reported to enrich both the exchangeable and nonexchangeable-K forms in K-deficient Alfisol (Srinivasarao et al. 2010, 2013).

9.5.3.4 Available Micronutrients

Intensive cropping with high yielding varieties fertilized with high analysis fertilizers catalyzed the rapid depletion of available micronutrients. Enhanced removal of the micronutrients caused by the realization of higher yields without their supplementation has already made some of these nutrients as yield-limiting.

Continuous application of chemical fertilizers was accompanied by a drop in the available Zn (Nambiar 1994; Mathur 1997; Kumar and Yadav 2005) and Cu content (Singh and Nambiar 1986). But incorporation of farm yard manure along with NPK either maintained the available Zn at the initial level or raised it due to enhanced mobilization of the available micronutrients in the surface and sub-surface soils (Bellaki and Badanur 1997). However, Mishra et al. (2006) reported that application of green manure of Sesbania with farm yard manure and wheat straw is advantageous for sustained supply of Zn to rice wheat cropping system soils. On the other hand, continuous use of NPK resulted in an increase in the acidity which in turn increased the available micronutrient status in soil (Setia and Sharma 2004) as a result available Fe content was ascribed to the enhanced solubilization of Fe induced by the intensification of soil acidity (Nambiar 1994).

9.5.4 Biological and Biochemical Changes in Rhizosphere

Soil quality does not just depend on the physical and chemical properties of the soil but it is very closely linked to its biological properties also. Biological properties include size and diversity of the microbial, macro and microfaunal biomass, enzyme quantities and activities, mineralizable C, N, P, S etc., respiration and soil organic matter content (Fig. 9.2). Biological processes provide the resilience and buffering capacity to ameliorate stress (Karlen et al. 1992).

Soil biota is considered an important and labile fraction of soil organic matter involved in energy and nutrient cycling. It has been well established that the more dynamic characteristics such as microbial biomass, soil enzyme activity and soil respiration respond more quickly to changes in crop management practices or environmental conditions than do characteristics such as total soil organic matter (Dick 1992; Doran et al. 1996).

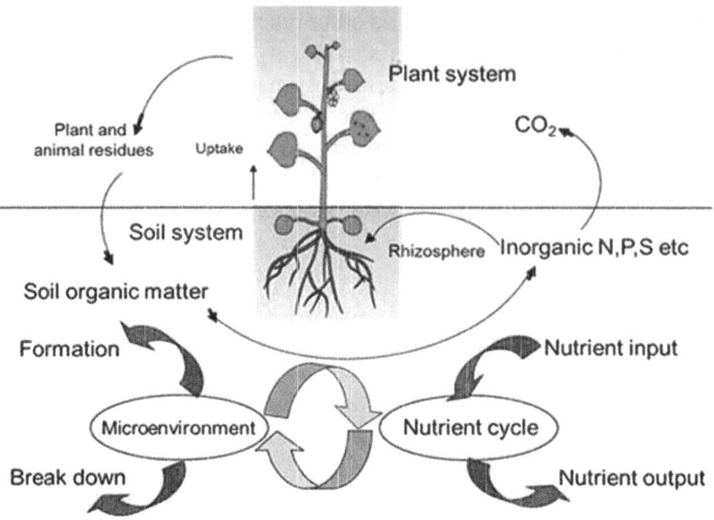

Fig. 9.2 Role of microbial biomass in the cycling of nutrients in rhizosphere

9.5.4.1 Potentially Mineralizable Nitrogen

Potentially mineralizable nitrogen is one of the most important potential indicators for use to assess the function of ecological habitat and biodiversity (Cardoso et al. 2013). Reduction in the soil's natural ability to provide nitrogen, which they attribute to interactions between organic matter and soil microbes, may even hamper the crop yield (Cassman and Pingali 1995). The effect of green manure in improving the potentially mineralizable N was more pronounced (Kang et al. 2005). Thus, the potential mineralizable nitrogen has been selected as soil biological indicator to assess to soil quality (Doran and Parkin 1994; Hseu et al. 1999).

9.5.4.2 Enzymes Activity

Enzymes are important soil components involved in the dynamics of soil nutrient transformation. Enzyme activity in the soil environment is considered to be a major contributor of overall soil microbial activity (Frankenberger and Dick 1983) and, also to soil quality (Dick 1994). For the mineralization of the organic substrate to occur, both the synthesis and the activity of a specific enzyme complex are needed. These later processes may be linked to the presence of countless factors directly implicated in the mechanism of enzyme synthesis and secretion (Martens et al. 1992).

Dehydrogenase

Dehydrogenases reflect the total range of oxidative activity of soil microflora and consequently may be associated to be a good indicator of microbiological activity (Watts et al. 2010). Dehydrogenase was highly sensitive to the inhibitory effects associated with large fertilizer addition (Manjaiah and Singh 2001; Masto et al. 2007) suggested an optimum and balanced application of nutrients that led to significant increase in dehydrogenase activity. Whereas, the dehydrogenase activity was stimulated by the application of organic manures (Goyal et al. 1999; Kang et al. 2005), even the effect is much pronounced in saline and alkaline soils (Rao and Pathak 1996).

Phosphatase

Phosphatases are a group of enzymes that catalyze hydrolysis of esters and anhydrides of phosphoric acid, so they are meant to play critical roles in P cycles in soil ecosystems (Speir and Ross 1978). Phosphomonoesterases have been the most studied soil enzymes as they have shown activity both under acidic and alkaline conditions, according to its optimal pH, and because they act on low molecular P-compounds including nucleotides, sugar phosphates and polyphosphates (Makoi and Ndakidemi 2008). Due to the relative importance of this enzyme in soil organic-P mineralization and plant nutrition, their assay in soil assumes more importance. As the alkaline phosphatases are derived from microorganisms only, the acid phosphatases are secreted both by plant roots and microorganisms in rhizosphere (Chhonkar et al. 2002). Being such an important enzyme in soil system phosphatase activity in temperate grassland, are found to have a strong correlation with other soil properties such as pH, total N, organic P and clay content (Turner and Haygarth 2005).

β–Glucosidase

It has been established as a key soil quality indicator due to its importance in catalytic reactions on cellulose degradation, releasing glucose as a source of energy that maintains metabolically active microbial biomass in soil (Dick 1996). At the same time, its role in energy flow in the soil that directly related to labile-C and with the ability to stabilize soil organic matter. As a free enzyme in soil solution, it normally has a time-limited activity, so rapidly degrades and denatures, and its activity has been detected in soil, fungi and plants (Martinez-Salgado et al. 2010).

Urease

Urease enzymes are mainly responsible for urea hydrolysis which breaks into CO_2 and NH_3 as a result nitrogen losses take place by NH_3 volatilization and soil pH increase. As urea is supplied to soil as a major source of nitrogen fertilizers but due to the substantial nitrogen lost to the atmosphere through volatilisation, and a large share of the process is mediated by the urease enzyme thus it often proves very inefficient fertilizer practice. Soil urease originates mainly from plants and microorganisms found as both intra- and extra-cellular enzymes. However, mostly as an extracellular enzyme it represents up to 63 % of total activity in soil. (Martinez-Salgado et al. 2010); but is rapidly degraded in soil by proteolytic enzymes. Urease is widely reported to be a soil quality indicator, as its activity increases with organic fertilization and decreases with soil tillage (Saviozzi et al. 2001), also showing its influence by other factors such as cropping history, management practices, heavy metals, organic matter content, soil depth, and some environmental factors like temperature and pH (Yang et al. 2006).

9.6 Soil Quality Management *vis-a-vis* Rhizosphere Interferences

A better understanding of the basic principles of the rhizosphere ecology, including the function and diversity of inhabiting microorganisms is on the way but further knowledge is necessary to optimize soil microbial technology to the benefit of plant-growth and health in the natural environment. In sum, this can constitute overwhelming evidence indicating that an ever exploitation of plant growth promoting rhizobacteria can be a true success story in sustainable agriculture. As a consequence, current production methods in agriculture, *e.g.*, the improper use of chemical pesticides and fertilizers creating a long list of environmental and health problems will reduce.

9.6.1 Plant-Bacteria Interactions in the Rhizosphere

Microorganisms present in the rhizosphere play important roles in ecological fitness of their plant host. Plant-microbe interactions may thus be considered beneficial, neutral, or harmful to the plant, depending on the specific microorganisms and plants involved and on the prevailing environmental conditions.

9.6.1.1 Pathogenic Interactions

Roots exudates can attract beneficial organisms, but they can also be equally attractive to pathogenic populations, that many express virulence on only a limited number of host species. Plant diseases play a direct role in the destruction of natural resources in agriculture. Soil-borne pathogens cause important losses, fungi being the most aggressive. The extent of their harmful effects ranges from mild symptoms to catastrophes where large fields planted with agricultural crops are destroyed. Thus, they are major and chronic threats to food production and ecosystem stability worldwide. Common and well investigated bacterial agents include *Erwinia carotovora*, *Pseudomonas*, *Ralstonia* spp., *Streptomyces scabies*. The fungal and oomycete phytopathogens include members of *Fusarium*, *Phytophthora*, *Pythium*, *Rhizopus*, *Rhizoctonia* and *Verticillium* (Tournas and Katsoudas 2005) and amongst the forest pathogens most of them are filamentous fungi *Heterobasidion* and *Armillariella* (Asiegbu et al. 2005), and *Phytophthora* spp. (Rizzo et al. 2005).

9.6.1.2 Beneficial Microorganisms and Modes of Action

Plant-beneficial microbial interactions can be roughly divided into three categories. First, those microorganisms that, in association with plants, is responsible for its nutrition (i.e. microorganisms that can increase the supply of mineral nutrients to the plant). In this case, while most may not directly interact with the plant, their effects on soil biotic and abiotic parameters certainly have an impact on plant growth. Second, there is a group of microorganisms that stimulate plant growth indirectly by preventing the growth or activity of pathogens. Such microorganisms are referred to as biocontrol agents, and they have been well documented. A third group involves those microorganisms responsible for direct growth promotion, for example, by production of phytohormones. There has been a large body of literature describing potential uses of plant associated bacteria as agents stimulating plant growth and managing soil and plant fitness (Welbaum et al. 2004).

9.6.2 Plant Growth Promotion

9.6.2.1 Phytostimulation

Phytostimulation may directly enhance plant growth. In the processes of plant growth, phytohormones (e.g. production of indole-3-acetic acid, auxins, cytokinins, and gibberellins) play an important role. These hormones can be synthesized by the plant themselves but also by their associated microorganisms such as *Azospirillum* spp., besides having nitrogen-fixing ability (Steenhoudt and Vanderleyden 2000). Species of *Pseudomonas* and *Bacillus* can produce as yet not well characterized phytohormones or growth regulators that cause crops to have greater amounts of fine roots which have the effect of increasing the absorptive surface of plant roots for uptake of water and nutrients. The phytohormones they produce include indole-acetic acid, cytokinins, gibberellins and inhibitors of ethylene production. Indole-3-acetic acid is a phytohormone which is known to be involved in root initiation, cell division, and cell enlargement commonly produced by plant growth promoting rhizobacteria (Barazani and Friedman 2001). Auxins are quantitatively the most abundant phytohormones secreted by *Azospirillum* (Bloemberg and Lugtenberg 2001).

Furthermore, plant-associated bacteria can influence the hormonal balance of the plant. Ethylene is an important example to show that the balance is most important for the effect of hormones: at low levels, it can promote plant growth in several plant species including *Arabidopsis thaliana*, while it is normally considered as an inhibitor of plant growth and known as a senescence hormone.

9.6.2.2 Biofertilization

There are several plant growth promoting rhizobacterial inoculants currently commercialized that seem to promote growth through at least one mechanism; suppression of plant disease (termed bioprotectants), phytohormone production (termed biostimulants), or improved nutrient acquisition (termed biofertilizers). The mode of action of these biofertilizers act directly by help in the nutrient uptake or indirectly influenced the root growth and morphology (Vessey 2003). The most prominent example of bacterial nitrogen fixation is the symbiotic association between rhizobia and its legume host plants (Table 9.1). In this association bacteria metabolize root exudates (carbohydrates) and in turn provide nitrogen to the plant for amino acid synthesis. The ability to fix nitrogen also occurs in free-living bacteria like *Azospirillum*, *Burkholderia*, and *Stenotrophomonas* (Dobbelare et al. 2003). Biofertilization accounts for approximately 65 % of the nitrogen supply to crops worldwide (Bloemberg and Lugtenberg 2001). Another nutrient is sulfate, which can be provided to the plant via oxidation by bacteria (Banerjee and Yesmin 2002).

9.6.3 Microbial Co-operation in the Rhizosphere-Interactions for Improving Soil Quality

Soil structure is one of the most influential factors controlling soil quality. A well-aggregated soil structure ensures appropriate soil tilth, water infiltration rates, aeration, root penetrability and organic matter accumulation leading to better soil quality. The contribution of microbial co-operation in the rhizosphere to the formation and stabilization of soil aggregates has been demonstrated frequently (Miller and Jastrow 2000). Firstly, soil particles are bound together by bacterial products and by hyphae of saprophytic and arbuscular mycorrhizal fungi, into stable micro-aggregates (ranges in between 2 and 20 μm in diameter). These are bound by microbial products into larger micro-aggregates (ranges in between 20 and 250 μm in diameter), with bacterial polysaccharides acting as binding agents. Micro-aggregates are then bound into macro-aggregates (higher than 250 μm in diameter), with bacterial polysaccharides acting as binding agents and arbuscular mycorrhizal mycelia increasing the size of macro-aggregates. The effect of the arbuscular mycorrhizal fungi in co-operation with other microbes in the formation of water-stable soil aggregates is evident in different ecological situations (Requena et al. 2001), and the

Table 9.1 Multitrophic effect of rhizobacteria on the various host plants

Rhizobacteria	Agricultural crop	References
Bacillus cereus UW 85	Grain legumes	Vessey and Buss (2002)
P. fluorescens CHA0	*Arabidopsis* sp.	Iavicoli et al. (2003)
Bacillus subtilis, *B. amyloliquefaciens* IN 937, *Enterobacter cloaca*	*Arabidopsis* sp.	Ryu et al. (2003)
P. putida KD	Tomato and cucumber	Rezzonoco et al. (2005)
Pseudomonas fluorescens PCL1606	Avocado	Cazorla et al. (2006)
Bradyrhizobium and rhizobacteria	Mungbean	Shaharoona et al. (2006)
Pseudomonas brassicacearum, *P. marginali*, *P. oryzihabitans*, *P. putida*, *Alcaligenes xylosoxidans*	Indian mustard and rape	Belimov et al. (2007)
P. fluorescens WCS 365	Tomato	Kamilova et al. (2007)
Collimonas fungivorans	Tomato	Kamilova et al. (2008)
Agrobacterium amazonense	Rice	Rodrigues et al. (2008)
Bacillus subtilis FB17	*Arabidopsis thaliana*	Rudrappa et al. (2008)
Pseudomonas BA-8, *Bacillus* OSU-142, *Bacillus* M-3	Strawberry	Pirlak and Kose (2009)
Bacillus cepacia strain OSU-7	Stored potatoes	Recep et al. (2009)
Comamonas acidovorans	Kiwi	Erturk et al. (2010)

involvement of glomalin, a glycoprotein produced by the external hyphae of arbuscular mycorrhizal fungi, has been demonstrated (Wright and Upadhyaya 1998). The co-operation of microbial symbionts inoculated in the rhizosphere of target indigenous species of plants is a successful biotechnological tool to aid the recovery of desertified ecosystems.

9.7 Future Research Directions

Various bacteria and fungi colonising in the rhizosphere, carry out numerous interactive chemical processes that influence nutrition of plants, thus benefit plant growth and health, and also function-based soil quality. Thus the significance of maintaining microbial diversity in rhizosphere ecology and need-based use of microbial inoculates must take into consideration for better future scenario. It will help in achieving effective biotechnological applications, known as 'rhizosphere technology'. In this connection, application of selected microbial consortia as plant inoculates to benefit plant production systems has gained attention. Either to maintain sustainable plant productivity or environmental quality, basic and strategic research should be carried on to improve our existing knowledge on microbial interactions in the root–soil environments. Here lies the scope of molecular and biotechnological interventions. New generation genetically-modified and environmentally-friendly microbial inoculates can be used to protect plants from disease and to promote plant growth. As future research guide, several key questions still need to be answered. Firstly, it has been reported that less than 5 % of all soil bacteria and Archaea (Curtis et al. 2002) and less than 5 % of all soil fungi (Hawksworth 2001) are culturable. Rest huge diverse population of microbes cannot be cultured. Hence, the function of these non-culturable microbes in rhizosphere ecosystems is challenging to test how these microbes alter, or respond to their environment, is less or poorly under knowledge. Secondly, there are number of unexplored species of root-associated soil microbes, those have definite roles on plant nutrition and productivity but they are yet to be identified. Finally, global climate change definitely has an influence on microbial diversity and community structure. But the changing pattern and its extent is another researchable option.

9.8 Conclusion

Degradation of land and deterioration of soil fertility is a concern for quite a long time around the globe. All such alarming issues and their possible management strategies are addressed by soil quality. Soil quality, a function-based holistic representation of soil system, is often directed by plant-microbial processes in rhizosphere. Although both the topics are not often discussed and intends to be interconnected, still it needs to be highlighted in the present scenario. Plant and microbes, as both are integral parts of rhizosphere contribute in nutrient mobilization, mediated by enzymatic interfaces. Nevertheless, some physical attributes such as aggregate stability, porosity, water retention and mobility are partially determined by the rhizosphere that altogether balances soil physical environment. But, the root system, as traditionally only believed to provide anchorage and uptake of water and nutrients, is also a chemical factory that simultaneously performing plentiful underground interactions. These happen due to mutualistic associations with beneficial microbes such as rhizobia, mycorrhizae, endophytes and plant-growth promoting rhizobacteria and parasitic interactions with other plants, pathogenic microbes e.g. fungi, bacteria, and invertebrate herbivores. Roots discharge several types of chemicals at a substantial amount that combat pathogenic microorganisms and attract beneficial ones. Every single soil quality indicator responds to changing equation of plant-microbial interactions that visibly altered by crop management practices like tillage, fertilizer and organic nutrient, amendments etc. in agricultural systems. Now-a-days research focus has been shifted to unravel the core processes that affect rhizosphere ecology. Rhizosphere ecology is depicted as highly heterogeneous microenvironments that varies with crop, species, cropping system and even with the crop cultivars, attributed by a combination of the physical architecture of the soil matrix, coupled with the rhizo-deposits, protons and other ions, gases e.g. CO_2, O_2, CH_4 etc. and that finally determines the role of roots as sinks for water and nutrients. Methodological advances will target to resolve the future research priorities for comprehensive understanding of the simultaneous processes in the rhizosphere.

References

Akhtar MS, Siddiqui ZA (2008) Biocontrol of a root-rot disease complex of chickpea by *Glomus intraradices, Rhizobium* sp. and *Pseudomonas straita*. Crop Prot 27:401–417. doi:10.1016/j.cropro.2007.07.009

Andrews SS, Karlen DL, Cambardella CA (2004) The soil management assessment framework: a quantitative soil quality evaluation method. Soil Sci Soc Am J 68:1945–1962

Arshad MA, Coen GM (1992) Characterization of soil quality. Am J Altern Agric 7:25–31. doi:10.1017/S0889189300004410

Arshad MC, Lowery B, Grossman B (1996) Physical tests for monitoring soil quality. In: Doran JW, Jones AJ (eds) Methods for assessing soil quality. Soil Science Society of America, Inc., Madison, pp 123–141

Asiegbu FO, Nahalkova J, Li G (2005) Pathogen-inducible cDNAs from the interaction of the root rot fungus *Heterobasidion annosum* with Scots pine (*Pinus sylvestris* L.). Plant Sci 168:365–372. doi:10.1016/j.plantsci.2004.08.010

Banerjee M, Yesmin L (2002) Sulfur-oxidizing plant growth promoting *Rhizobacteria* for enhanced canola performance. US7491535 B2

Barazani O, Friedman J (2001) Allelopathic bacteria and their impact on higher plants. Crit Rev Microbiol 27:41–55. doi:10.1080/20014091096693

Beare MH, Bruce RR (1993) A comparison of methods for measuring water stable aggregates: implications for determining environmental effects on soil structure. Geoderma 56:87–104. doi:10.1016/0016-7061(93)90102-Q

Belimov AA, Dodd IC, Safronova VI, Hontzeas N, Davies WJ (2007) *Pseudomonas brassicacearum* strain Am3 containing 1-aminocyclopropane-1-carboxylate deaminase can show both pathogenic and growth-promoting properties in its interaction with tomato. J Exp Bot 58:1485–1495. doi:10.1093/jxb/erm010

Bellaki MA, Badanur VP (1997) Long term effect of integrated nutrient management on properties of vertisol under dryland agriculture. J Indian Soc Soil Sci 45:438–442

Bezdicek DF, Papendick RI, Lal R (1996) Importance of soil quality to health and sustainable land management. In: Doran JW, Jones AJ (eds) Methods for assessing soil quality. Soil Science Society of America, Inc., Madison, pp 1–7

Birkas M, Stingli A, Szemok A, Kalmar T, Bottlik L (2008) Soil condition and plant interrelations in dry years. Cereal Res Commun 36S:15–18

Blake L, Mercik S, Koerschens M, Goulding KWT, Stempen S, Weigel A, Poulton PR, Powlson DS (1999) Potassium content in soil, uptake in plants and the potassium balance in three European long-term field experiments. Plant Soil 216:1–14. doi:10.1023/A:1004730023746

Blanco-Canqui H, Lal R (2007) Soil quality and crop response harvesting corn residues for biofuel production. Geoderma 141:355–362

Bloemberg GV, Lugtenberg BJJ (2001) Molecular basis of plant growth promotion and biocontrol by rhizobacteria. Curr Opin Plant Biol 4:343–350. doi:10.1016/S1369-5266(00)00183-7

Bouma J (2002) Land quality indicator of sustainable land management across scales. Agric Ecosyst Environ 88:129–136. doi:10.1016/S0167-8809(01)00248-1

Brooks PC (1995) The use of microbial parameters in monitoring soil pollution by heavy metals. Biol Fertil Soils 19:269–279. doi:10.1007/BF00336094

Bucher AE (1999) Evaluating soil management using indicators of soil quality. M.Sc. thesis, Pennsylvania State University, University Park

Cardoso EJBN, Vasconcellos RLF, Bini D, Miyauchi MYH, Santos CA, Alves PRL, Paula AM, Nakatani AS, Pereira JM, Nogueira MA (2013) Soil health: looking for suitable indicators. What should be considered to assess the effects of use and management on soil health. Sci Agric 70:274–289. doi:10.1590/S0103-90162013000400009

Carter MR, Gregorich EG, Anderson DW, Doran JW, Janzen HH, Pierce FJ (1997) Concepts of soil quality and their significance. In: Gregorich EG, Carter MR (eds) Soil quality for crop production and ecosystem health. Elsevier, Amsterdam, pp 1–19

Cassman KG, Pingali PL (1995) Extrapolating trends from long term experiments to farmers fields: the case of irrigated rice systems in Asia. In: Barnett V, Payne R, Steiner R (eds) Agricultural sustainability: economic, environmental and statistical consideration. Wiley Press, London, pp 63–86

Cazorla FM, Duckett SB, Bergstrom FT, Noreen S, Odik R (2006) Biocontrol of avocado dematophora root rot by the antagonistic *Pseudomonas fluorescens* PCL 1606 correlates with the production 2-hexyl-5-propyl resorcinol. Mol Plant Microbe Interact 19:418–428. doi:10.1094/MPMI –19-0418

Chaudhury J, Mandal UK, Sharma KL, Ghosh H, Mandal B (2005) Assessing soil quality under long-term rice-based cropping system. Commun Soil Sci Plant Anal 36:1141–1161. doi:10.1081/CSS-200056885

Cheng W, Johnson DW, Fu S (2003) Rhizosphere effects on decomposition: controls of plant species, phenology, and fertilization. Soil Sci Soc Am J 67:1418–1427. doi:10.2136/sssaj2003.1418

Chhonkar PK, Bhadraray S, Patra AK, Purakayastha TJ (2002) Practical manual on soil biology and biochemistry. Indian Agricultural Research Institute, New Delhi

Crosson P (1997) Will erosion threaten agricultural productivity. Environ Sci Policy Sustain Dev 39:4–31. doi:10.1080/00139159709604756

Curtis TP, Sloan WT, Scannell JW (2002) Estimating prokaryotic diversity and its limits. Proc Natl Acad Sci 99:10494–10499. doi:10.1073/pnas.142680199

Dey R, Pal KK, Tilak KVBR (2012) Influence of soil and plant types on diversity of rhizobacteria. Proc Natl Acad Sci India Sect B Biol Sci 82:341–352

Dick RP (1992) A review: long-term effects of agricultural systems on soils biochemical and microbial parameters. Agric Ecosyst Environ 40:25–36. doi:10.1016/0167-8809(92) 90081-L

Dick RP (1994) Soil enzymes activities as indicators of soil quality. In: Doran JW, Coleman DC, Bezdicek DE, Stewart BA (eds) Defining soil quality for a sustainable environment. Soil Science Society of America, Inc., Madison, pp 104–124

Dick R (1996) Soil enzyme activities as integrative indicators of soil health. In: Doran JW, Jones AJ (eds) Methods for assessing soil quality. Soil Science Society of America, Inc., Madison, pp 121–156

Dijkstra FA, Cheng W (2007) Interactions between soil and tree roots accelerate long-term soil carbon decomposition. Ecol Lett 10:1046–1053. doi:10.1111/j.1461-0248.2007. 01095.x

Dijkstra FA, Carrillo Y, Pendall E, Morgan JA (2013) Rhizosphere priming: a nutrient prospective. Front Microbiol 4:216. doi:10.3389/fmicb.2013.00216

Dobbelare S, Vanderleyden J, Okon Y (2003) Plant growth promoting effects of diazotrophs in the rhizosphere. Crit Rev Plant Sci 22:107–149. doi:10.1080/713610853

Doran JW, Parkin TB (1994) Defining and assessing soil quality. In: Doran JW, Coleman DC, Bezdicek DF, Stewart BA (eds) Defining soil quality for a sustainable environment. Soil Science Society of America, Inc., Madison, pp 3–21

Doran JW, Sarrantonio M, Lieberg MA (1996) Soil health and sustainability. In: Sparks DL (ed) Advances in agronomy, vol 56. Academic, San Diego, pp 1–54

Doran JW, Stamatiadis S, Haberern J (2002) Soil health as an indicator of sustainable management. Agric Ecosyst Environ 88:107–110. doi:10.1016/S0167-8809(01)00250-X

Erturk Y, Ercisli S, Haznedar A, Cakmakci R (2010) Effects of plant growth promoting rhizobacteria (PGPR) on rooting and root growth of kiwifruit (*Actinidia deliciosa*) stem cuttings. Biol Res 43:91–98. doi:10.4067/S0716-97602010000100011

Fauci MF, Dick RP (1994) Soil microbial dynamics short and long term effects of inorganic and organic nitrogen. Soil Sci Soc Am J 58:801–806. doi:10.2136/sssaj1994.03615995005800030023x

Frankenberger WT, Dick WA (1983) Relationships between enzyme activities and microbial growth and activity indices in soil. Soil Sci Soc Am J 47:945–951

Gill S, Abid M, Ahmad Z, Azam F (2006) Organic amendment accelerates nitrification in soil. Soil Environ 25:35–39

Goyal S, Chander K, Mundra MC, Kapoor KK (1999) Influence of inorganic fertilizers and organic amendments on soil organic matter and soil microbial properties under tropical conditions. Biol Fertil Soils 29:196–200. doi:10.1007/s003740050544

Granatstein D, Bezdicek DF (1992) The need for a soil quality index: local and regional perspectives. Am J Altern Agric 7:12–16. doi:10.1017/S0889189300004380

Grassini P, Jinsheng Y, Hubbard KG, Cassman KG (2010) Soil water recharge in a semi-arid temperate climate of the Central U.S. Great Plains. Agric Water Manag 97:1063–1069. doi:10.1016/j.agwat.2010.02.019

Gregorich EG, Monreal CM, Carter MR, Angers DA, Ellert BH (1994) Towards a minimum data set to assess soil organic matter quality in agricultural soils. Can J Soil Sci 74:367–385. doi:10.4141/cjss94-051

Hartmann A, Schmid M, van Tuinen D, Berg G (2009) Plant-driven selection of microbes. Plant Soil 321:235–257. doi:10.1007/s11104-008-9814-y

Hawes MC, Gunawardena U, Miyasaka S, Zhao X (2000) The role of root border cells in plant defense. Trends Plant Sci 5:128–133. doi:10.1016/S1360-1385(00)01556-9

Hawksworth DL (2001) The magnitude of fungal diversity: the 1.5 million species estimate revisited. Mycol Res 105:1422–1432. doi:10.1017/S0953756201004725

Haynes RJ (1990) Active ion uptake and maintenance of cation-anion balance: a critical examination of their role in regulating rhizosphere pH. Plant Soil 126:247–264. doi:10.1007/BF00012828

Heim S, Mar Lleo MD, Bonato B, Guzman CA, Canepari P (2002) The viable but nonculturable state and starvation are different stress responses of *Enterococcus faecalis*, as determined by proteome analysis. J Bacteriol 184:6739–6745. doi:10.1128/JB.184.23.6739–6745.2002

Helkamp AS, Bay JM, Easterling KN, Hess GS, Mcquaid BF, Munster MJ, Nether DA, Olson GL, Sidik K, Stenfanski LA, Tooley MB, Campbell CL (1995) Environmental monitoring and assessment program: agricultural lands pilot field program report- 1993. U.S. Environmental Protection Agency, Washington, DC, p 64

Hiltner L (1904) Uber neuere Erfahrungen und probleme auf dem gebiet der bodenbakteriologie und unter besonderer berucksichtigung der grundungung und brache. Arbeiten der Deutschen Landwirtschafts Gesellschaft 98:59–78

Holden PA, Firestone MK (1997) Soil microorganisms in soil cleanup: how can we improve our understanding. J Environ Qual 26:32–40. doi:10.2134/jeq1997.00472425002600010006x

Hseu ZY, Chen ZS, Tsai CC (1999) Selected indicators and conceptual framework for assessment methods of soil quality in arable soils of Taiwan. Soil Environ 2:77–88

Iavicoli A, Boutet E, Buchala A, Metraux JP (2003) Induced systemic resistance in *Arabidopsis thaliana* in response to root inoculation with *Pseudomonas fluorescens* CHA0. Mol Plant Microbe Interact 16:851–858. doi:10.1094/MPMI.2003.16.10.851

Iqbal M, Khan AG, Hassan AU, Raza MW, Amjad M (2012) Soil physical health indices, soil organic carbon, nitrate contents and maize growth as influenced by dairy manure and nitrogen rates. Int J Agric Biol 14:20–28. doi:11-284/MFA/2012/14–1–20–28

Islam KR, Weil RR (2000) Soil quality indicator properties in mid-Atlantic soils as influenced by conservation management. J Soil Water Conserv Ankeny 55:69–78

Jaeger CH, Lindow SE, Miller W, Clark E, Firestone MK (1999) Mapping of sugar and amino acid availability in soil around roots with bacterial sensors of sucrose and tryptophan. Appl Environ Microbiol 65:2685–2690

Jassal R, Black A, Novak M, Morgenstern K, Nesic Z, Gaumont-Guay D (2005) Relationship between soil CO_2 concentrations and forest-floor CO_2 effluxes. Agric For Meteorol 130:176–192. doi:10.1016/j.agrformet.2005.03.005

Jenkinson DS (1991) The Rothamsted long term experiments: are they still in use. Agron J 83:2–10

Jenkinson DS, Ladd JN (1981) Microbial biomass in soil: measurement and turnover. In: Paul EA, Ladd JN (eds) Soil biochemistry, vol 5. Marcel Dekker, New York, pp 415–471

Jensen ES (1996) Rhizodeposition of N by pea and barley and its effects on soil N dynamics. Soil Biol Biochem 28:65–71. doi:10.1016/0038-0717(95)00116-6

Johnston AE (1994) The Rothamsted classical experiment. In: Leigh RA, Johnston AE (eds) Long-term experiments in agricultural and ecological sciences. CAB International, Wallingford, pp 9–38

Kamilova F, Leveau JHJ, Lugtenberg BJJ (2007) *Collimonas fungivorans*, an unpredicted in vitro but efficient in vivo biocontrol agent for the suppression of tomato foot and root rot. Environ Microbiol 9:1597–1603. doi:10.1111/j.1462-2920.2007.01348.x

Kamilova F, Lamers G, Lugtenberg B (2008) Biocontrol strain *Pseudomonas fluorescens* WCS365 inhibits germination of *Fusarium oxysporum* spores in tomato root exudate as well as subsequent formation of new spores. Environ Microbiol 10:2455–2461. doi:10.1111/j.1462-2920.2008.01638.x

Kang GS, Beri V, Sidhu BS, Rupela OP (2005) A new index to assess soil quality and sustainability of wheat-based cropping systems. Biol Fertil Soils 41:389–398. doi:10.1007/s00374-005-0857-4

Karlen DL, Cambardella CA (1996) Conservation strategies for improving soil quality and organic matter storage. In: Advances in soil science. CRC Press, Boca Raton

Karlen DL, Stott DE (1994) A framework for evaluating physical and chemical indicators of soil quality. In: Doran JW, Coleman DC, Bezdicek DF, Stewart BA (eds) Defining soil quality for a sustainable environment. Soil Science Society of America, Inc., Madison, pp 53–72

Karlen DL, Eash NS, Unger PW (1992) Soil and crop management effects on soil quality indicators. Am J Altern Agric 7:48–55. doi:10.1017/S0889189300004458

Karlen DL, Wollenhaupt NC, Erbach DC, Berry EC, Swan JB, Eash NS, Jordahl JL (1994) Crop residue effects on soil quality following 10-years of no-till corn. Soil Tillage Res 31:149–167. doi:10.1016/0167-1987(94)90077-9

Karlen DL, Mausbach JW, Doran JW, Cline RG, Harris RF, Schuman GE (1997) Soil quality: a concept, definition and framework for evaluation. Soil Sci Soc Am J 61:4–10. doi:10.2136/sssaj1997.03615995006100010001x

Kumar A, Yadav DS (2005) Influence of continuous cropping and fertilization on nutrient availability and productivity of an alluvial soil. J Indian Soc Soil Sci 53:194–198

Lal R (1993) Tillage effects on soil degradation, soil resilience, soil quality, and sustainability-introduction. Soil Tillage Res 27:1–8. doi:10.1016/0167-1987(93)90059-X

Lal R (1997) Degradation and resilience of soils. Philos Trans R Soc Lond B 352:997–1010. doi:10.1098/rstb.1997.0078

Lal R (1999) Soil quality and soil erosion, 1st edn. CRC Press, Boca Raton

Lal R (2008) Carbon sequestration. Philos Trans R Soc B 363:815–830. doi:10.1098/rstb.2007.2185

Lal R (2013) Cropping systems and soil quality. J Crop Prod 8:33–52. doi:10.1300/J144v08n01_03

Lal R, Stewart BA (1990) Soil degradation. Springer, New York

Lal R, Stewart BA (1995) Need for long-term experiments in sustainable use of soil resources. In: Lal R, Stewart BA (eds) Soil management: experimental basis for sustainability and environmental quality. CRC Press, Boca Raton, pp 537–545

Larson WE, Pierce FJ (1991) Conservation and enhancement of soil quality. Int Board Soil Res Manag 2:175–203

Lewandowski A, Zumwinkle M (1999) Assessing the soil system: a review of soil quality literature. In: Fish A (ed) Energy and sustainable agriculture program. Biocentric Inc., St. Paul, Minnesota

Lupwayi NZ, Rice WA, Clayton GW (1998) Soil microbial biomass and carbon dioxide flux under wheat as influenced by tillage and crop rotation. Can J Soil Sci 79:273–280. doi:10.4141/S98-052

Majumder B, Mandal B, Bandyopadhyay PK, Gangopadhyay A, Mani PK, Kundu AL, Mazumdar D (2008) Organic amendments influence soil organic carbon pools and rice-wheat productivity. Soil Sci Soc Am J 72:775–785. doi:10.2136/sssaj2006.0378

Makoi J, Ndakidemi P (2008) Selected soil enzymes: examples of their potential roles in the ecosystem. Afr J Biotechnol 7:181–191. doi:10.5897/AJB07.590

Manjaiah KM, Singh D (2001) Soil organic matter and biological properties after 26 years of maize–wheat–cowpea cropping as affected by manure and fertilization in a Cambisol in semiarid region of India. Agric Ecosyst Environ 86:155–162. doi:10.1016/S0167-8809(00)00280-2

Manjaiah KM, Voroney RP, Sen U (2000) Soil organic carbon stocks, storage profile and microbial biomass under different crop management systems in a tropical agricultural ecosystem. Biol Fertil Soils 31:273–278. doi:10.1007/s003740000248

Martens DA, Johanson JB, Frankenberger WT Jr (1992) Production and persistence of soil enzymes with repeated addition of organic residues. Soil Sci 153:53–61

Martinez-Salgado MM, Gutierrez-Romero V, Jannsens M, Ortega-Blu R (2010) Biological soil quality indicators: a review. In: Mendez-Vilas A (ed) Current research, technology and education topics in applied microbiology and microbial biotechnology. Formatex, Badajoz, Spain, pp 319–328

Masto RE (2004) Soil quality assessment in maize-wheat-cowpea cropping system under long-term fertilizer use. PhD thesis, Indian Agricultural Research Institute, Pusa, New Delhi

Masto RE, Chhonkar PK, Singh D, Patra AK (2007) Soil quality response to long-term nutrient and crop management on a semi-arid Inceptisol. Agric Ecosyst Environ 118:130–142. doi:10.1016/j.agee.2006.05.008

Mathur GM (1997) Effect of long-term application of fertilizers and manures on soil properties and yield under cotton-wheat rotation in north-west Rajasthan. J Indian Soc Soil Sci 45:288–292

Melillo JM, Steudler PA, Aber JD, Newkirk K, Lux H, Bowles FP, Catricala C, Magill A, Ahrens T, Morrisseau S (2002) Soil warming and carbon-cycle feedbacks to the climate system. Science 298:2173–2176. doi:10.1126/science.1074153

Mengel K, Kirkby EA (1982) Principles of plant nutrition. International Potash Institute, Bern

Miller RM, Jastrow JD (2000) Mycorrhizal fungi influence soil structure. In: Kapulnik Y, Douds DD Jr (eds) Arbuscular mycorrhizas: physiology and functions. Kluwer Academic Publishers, Dordrecht, pp 3–18

Mishra BK, Sharma RB (1997) Effect of fertilizer alone and in combination with manure on physical properties and productivity of entisol under rice-based cropping system. J Indian Soc Soil Sci 45:84–88

Mishra BN, Prasad R, Gangaiah B, Shivakumar BG (2006) Organic manures for increased productivity and sustained supply of micronutrients Zn and Cu in a rice-wheat cropping system: innovations for long-term and lasting maintenance and enhancement of agricultural resources, production and environmental quality. J Sustain Agric 28:55–66

Nambiar KKM (1994) Soil fertility and crop productivity under long term fertilizer use in India. Indian Council of Agricultural Research, New Delhi, pp 234–267

Oguike PC, Chukwu GO, Njoku NC (2006) Physico-chemical properties of a Haplic Acrisol in Southeastern Nigeria amended with rice mill waste and NPK fertilizer. Afr J Biotechnol 5:1058–1061

Parton WJ, Schimel DS, Cole CV, Ojima DS (1987) Analysis of factors controlling soil organic matter levels in Great Plains grasslands. Soil Sci Soc Am J 51:1173–1179. doi:10.2136/sssaj1987.03615995005100050015x

Piedallu C, Jean-Claude G, Bruand A, Seynave I (2011) Mapping soil water holding capacity over large areas to predict potential production of forest stands. Geoderma 160:355–366. doi:10.1016/j.geoderma.2010.10.004

Pimentel D, Kounang N (1998) Ecology of soil erosion in ecosystems. Ecosystem 1:416–426. doi:10.1007/s100219900035

Pinton R, Veranini Z, Nannipieri P (2007) The rhizosphere: biochemistry and organic substances at the soil-plant interface. Taylor and Francis, New York

Pirlak M, Kose M (2009) Effects of plant growth promoting rhizobacteria on yield and some fruit properties of strawberry. J Plant Nutr 32:1173–1184. doi:10.1080/0190416090 2943197

Plaxton WC (1996) The organization and regulation of plant glycolysis. Annu Rev Plant Physiol Plant Mol Biol 47:185–214. doi:10.1146/annurev.arplant.47.1.185

Purakayastha TJ, Huggins DR, Smith JL (2008) Carbon sequestration in native prairie, perennial grass, no-till, and cultivated palouse silt loam. Soil Sci Soc Am J 72:534–540. doi:10.2136/sssaj2005.0369

Rao DLN, Pathak H (1996) Ameliorative influence of organic matter on biological activity in salt affected soils. Arid Soil Res Rehabil 10:311–319. doi:10.1080/15324989 609381446

Rawat MS, Tripathi RP, Nand R (1996) Long-term effect of puddling and fertilizer use in rice-wheat-cowpea sequence on structural properties of soil. J Indian Soc Soil Sci 44:364–368

Recep K, Fikrettin S, Erkol D, Cafer E (2009) Biological control of the potato dry rot caused by *Fusarium* species using PGPR strains. Biol Control 50:194–198. doi:10.1016/j.biocontrol.2009.04.004

Reddy MN, Sitaramayya M, Narayanaswamy S, Sairam A, Krishnakanth G (1999) Productivity and soil fertility changes under continuous fertilization and rice (*Oryza sativa*)- rice cropping system. Indian J Agric Sci 69:395–398

Reganold JP, Palmer AS (1995) Significance of gravimetric versus volumetric measurements of soil quality under biodynamic, conventional and continuous grass management. J Soil Water Conserv 50:298–305

Requena N, Perez-Solis E, Azcon-Aguilar C, Jeffries P, Barea JM (2001) Management of indigenous plant-microbe symbioses aids restoration of desertified ecosystems. Appl Environ Microbiol 67:495–498. doi:10.1128/AEM.67.2.495-498.2001

Rezzonoco F, Binder C, Defago G, Moenne-Loccoz Y (2005) The type III secretion system of biocontrol *Pseudomonas fluorescens* KD targets the phytopathogenic chromista *Pythium ultimum* and promotes cucumber protection. Mol Plant Microbe Interact 9:991–1001. doi.org/10.1094/MPMI-18-0991

Rizzo DM, Garbelotto M, Hansen EA (2005) *Phytophthora ramorum*: integrative research and management of an emerging pathogen in California and Oregon forests. Annu Rev Phytopathol 43:309–335. doi:10.1146/annurev.phyto.42.040803.140418

Rodrigues EP, Rodrigues LS, de Oliveira ALM, Baldani VLD, dos Santos Teixeira KR, Urquiaga S, Reis VM (2008) *Azospirillum amazonense* inoculation: effects on growth, yield and N_2 fixation of rice (*Oryza sativa* L.). Plant Soil 302:249–261. doi:10.1007/s11104-007-9476-1

Rudrappa L, Purakayastha TJ, Singh D, Bhadraray S (2006) Long-term manuring and fertilization effects on soil organic carbon pools in a Typic Haplustept of semi-arid sub-tropical India. Soil Tillage Res 88:180–192. doi:10.1016/j.still.2005.05.008

Rudrappa T, Czymmek KJ, Pare PW, Bais HP (2008) Root-secreted malic acid recruits beneficial soil bacteria. Plant Physiol 148:1547–1556. doi:10.1104/pp.108.127613

Ryu CM, Farag MA, Hu CH, Reddy MS, Wei HX, Pare PW, Kloepper JW (2003) Bacterial volatiles promote growth in Arabidopsis. Proc Natl Acad Sci 100:4927–4932. doi:10.1073/pnas.0730845100

Saviozzi A, Levi-Minzi R, Cardelli R, Riffaldi R (2001) A comparison of soil quality in adjacent cultivated, forest and native grassland soils. Plant Soil 233:251–259. doi:10.1023/A:1010526209076

Schjonning P, Iversen BV, Munkholm LJ, Labouriau R, Jacobsen OH (2005) Pore characteristics and hydraulic properties of a sandy loam supplied for a century with either animal manure or mineral fertilizers. Soil Use Manag 21:265–275. doi:10.1111/j.1475-2743.2005.tb00398.x

Schloter M, Lebuhn M, Heulin T, Hartmann A (2000) Ecology and evolution of bacterial microdiversity. FEMS Microbiol Rev 24:647–660. doi:10.1111/j.1574-6976.2000.tb00564.x/pdf

Setia RK, Sharma KN (2004) Effect of continuous cropping and long-term differential fertilization on profile stratification of DTPA-extractable micronutrients. J Food Agric Environ 2:260–265

Shaharoona B, Arshad M, Zahir ZA (2006) Effect of plant growth promoting rhizobacteria containing ACC-deaminase on maize (*Zea mays* L.) growth under axenic conditions and on nodulation in mungbean (*Vigna radiata* L.). J Appl Microbiol 42:155–159. doi:10.1111/j.1472-765X.2005.01827.x

Sharma P, Tripathi RP, Singh S (2005) Tillage effects on soil physical properties and performance of rice-wheat-cropping system under shallow water table conditions of Tarai, Northern India. Eur J Agron 23:327–335. doi:10.1016/j.eja.2005.01.003

Sims JT, Cunningham SD, Summer ME (1997) Assessing soil quality for environmental purposes: roles and challenges for soil scientists. J Enviorn Qual 26:20–25. doi:10.2134/jeq1997.00472425002600010004x

Sims IM, Middleton K, Lane AG, Cairns AJ, Bacic A (2000) Characterisation of extracellular polysaccharides from suspension cultures of members of the Poaceae. Planta 210:261–268. doi:10.1007/PL00008133

Singh GB, Nambiar KKM (1986) Crop productivity and fertility under intensive use of chemical fertilizers in long-term field experiments. Indian J Agron 31:115–127

Smith JL (2002) Soil quality: the role of microorganisms. In: Bitton G (ed) Encyclopedia of environmental microbiology. Wiley, New York, pp 2944–2957

Smith JL, Paul EA (1990) The significance of soil microbial biomass estimation. In: Bollag J, Stotzky G (eds) Soil biochemistry, vol 6. Marcel Dekker, New York, pp 357–396

Sojka RE, Upchurch DR (1999) Reservations regarding the soil quality concept. Soil Sci Soc Am J 63:1039–1054

Speir TW, Ross DJ (1978) Soil phosphatase and sulphatase. In: Burns RG (ed) Soil enzymes. Academic, London, pp 197–250

Srinivasarao C, Vittal KPR, Kundu S, Gajbhiye PN, Vijayasankarbabu M (2010) Continuous cropping, fertilization, and organic manure application effects on Potassium in an Alfisol under arid conditions. Commun Soil Sci Plant Anal 41:783–796. doi:10.1080/00103620903565993

Srinivasarao C, Kundu S, Venkateswarlu B, Lal R, Singh AK, Balaguravaiah G, Vijayasankarbabu M, Vittal KPR, Reddy S, Manideep VR (2013) Long-term effects of fertilization and manuring on groundnut yield and nutrient balance of Alfisols under rainfed farming in India. Nutr Cycl Agroecosyst. doi:10.1007/s10705-013-9575-8

Steenhoudt O, Vanderleyden J (2000) *Azospirillum*, a free-living nitrogen-fixing bacterium closely associated with grasses: genetic, biochemical and ecological aspects. FEMS Microbiol Rev 24:487–506. doi:10.1016/S0168-6445(00)00036-X

Sur HS, Sidhu AS, Singh R, Aggarwal GC, Sandhu KS (1993) Long-term effect of green manuring on soil physical properties and production potential in green manure-maize-wheat sequence. Ann Agric Res 14:125–131

Tournas VH, Katsoudas E (2005) Mould and yeast flora in fresh berries, grapes and citrus fruits. Int J Food Microbiol 105:11–17. doi:10.1016/j.ijfoodmicro.2005.05.002

Turner B, Haygarth P (2005) Phosphatase activity in temperate pasture soils: potential regulation of labile organic phosphorous turnover by phosphodiesterase activity. Sci Total Environ 344:27–36. doi:10.1016/j.scitotenv.2005.02.003

van Veen JA, Morgan JAW, Whipps JM (2007) Methodological approaches to the study of rhizosphere carbon flow and microbial population dynamics. In: Pinton A, Varanini Z, Nannipieri P (eds) The rhizosphere: biochemistry and organic substances at the soil-plant interface. CRC Press, Boca Raton, pp 371–399. doi:10.1201/9781420005585.ch13

Vessey JK (2003) Plant growth promoting rhizobacteria as biofertilizers. Plant Soil 255:571–586. doi:10.1023/A:1026037216893

Vessey JK, Buss TJ (2002) *Bacillus cereus* UW85 inoculation effects on growth, nodulation, and N accumulation in grain legumes: controlled environmental studies. Can J Plant Sci 82:283–290. doi:10.4141/P01-047

Wagnet RJ, Hutson JL (1997) Soil quality and its dependence on dynamic physical processes. J Environ Qual 26:41–48. doi:10.2134/jeq1997.00472425002600010007x

Walker TS, Bais HP, Grotewold E, Vivanco JM (2003) Root exudation and rhizosphere biology. Plant Physiol 132:44–51. doi:10.1104/pp. 102.019661

Wardle DA, Ghani A (1995) A critique of the microbial metabolic quotient (qCO_2) as a bioindicator of disturbance and ecosystem development. Soil Biol Biochem 27:1601–1610. doi:10.1016/0038-0717(95)00093-T

Watts DB, Allen TH, Feng Y, Prior SA (2010) Soil microbial community dynamics as influenced by composted dairy manure, soil properties, and landscape position. Soil Sci 175:474–486

Welbaum GE, Sturz AV, Dong Z, Nowak J (2004) Fertilizing soil microorganisms to improve productivity of agroecosystems. Crit Rev Plant Sci 23:175–193. doi:10.1080/07352680490433295

Wen D (1998) Agriculture in China: water and energy resources. In: Tso T, Tuan F, Faust M (eds) Agriculture in China: 1949–2030. IDEALS, Beltsville, pp 479–497. http://conservancy.umn.edu/handle/58873. Accessed on 1st Aug 2013

Wright SF, Upadhyaya A (1998) A survey of soils for aggregate stability and glomalin, a glycoprotein produced by hyphae of arbuscular mycorrhizal fungi. Plant Soil 198:97–107. doi:10.1023/A:1004347701584

Yang YZ, Liu S, Zheng D, Feng S (2006) Effects of cadium, zinc and lead on soil enzyme activities. J Environ Sci 18:1135–1141. doi:10.1016/S1001-0742(06)60051-X

Young IM (1995) Variation in moisture contents between bulk soil and the rhizosheath of wheat (*Triticum aestivum* L. cv. Wembley). New Phytol 130:135–139. doi:10.1111/j.1469-8137.1995.tb01823.x

Chapter 10
Vetiver Production for Small Farmers in India

E.V.S. Prakasa Rao, Srinivas Akshata, C.T. Gopinath, N.S. Ravindra, Aparna Hebbar, and Nagendra Prasad

Abstract Vetiver (*Vetiveria zizanioides*) is a perennial grass of Poaceae family, native to India. Vetiver production systems in Western Ghat region in India support livelihoods of small farmers. Vetiver systems have diverse economic and ecological uses. Vetiver is a major industrial crop and is grown for its essential oil that is used extensively world over in flavor and fragrance industries. Vetiver is also used in manufacture of handy-crafts, thatching houses, and organic compost production. Vetiver has been traditionally incorporated in the cropping systems of the region. Vetiver has a variety of environmental applications such as soil erosion control, phytoremediation, carbon sequestration etc. which are reviewed. Large scale cultivation of vetiver for essential oil production was initiated by using improved agronomic and field distillation methods covering an area of 250 ha in coastal Karnataka, a Western Ghats region, India, a region characterized by tropical climate with a well defined rainy season. The soils of the region are lateritic, of low pH below 5.0, and low to medium in fertility. Cultivation by adopting improved agronomy and field distillation was successful. The field root yield was estimated at 2.5 t ha^{-1}. The roots on steam distillation by conventional and improved methods produced 0.6–0.8 % and 1.0–1.2 % essential oil respectively. Fifty to seventy-five percent more net returns were obtained from improved agronomic and distillation methods in vetiver in comparison to traditional crops such as paddy, areca nut and cashew. This article shows that economic gains from vetiver oil production can help livelihoods of farmers while helping to maintain ecological sustainability in this region.

Keywords Vetiver oil • Flavor and fragrance industry • Large scale production • Improved agronomy and distillation • Economic returns • Environmental protection • Sustainable livelihoods

---

E.V.S. Prakasa Rao (✉)
Climate and Environmental Modelling Programme, CSIR Centre for Mathematical Modelling and Computer Simulation (CMMACS), NAL Belur Campus, Bangalore 560 037, India
e-mail: evsprakasarao@gmail.com

S. Akshata • C.T. Gopinath • N.S. Ravindra • A. Hebbar • N. Prasad
Central Institute of Medicinal and Aromatic Plants (CIMAP), Research Centre, Bangalore 56 065, India

E. Lichtfouse (ed.), *Sustainable Agriculture Reviews*,
Sustainable Agriculture Reviews 17, DOI 10.1007/978-3-319-16742-8_10

10.1 Introduction

Sustainable agricultural systems are vital in Western Ghats regions of India which is one of most ecologically sensitive regions of the world. The local farmers who are resource poor have to adopt cropping systems which provide food, economic returns and at the same time provide ecological protection to the region. Vetiver (*Vetiveria zizanioides*), a native of India, known for its perfumery and medicinal value since ancient times (Lavania 2003b), has been a traditional crop of the region which provides economic returns to the farmers and ecological value. In Uttara Kannada and Dakshina Kannada districts in western Ghats region of south India (Fig. 10.1) vetiver is traditionally cultivated for its roots and essential oil in around 3,000 ha using conventional agronomic and distillation practices resulting in poor yields and low economic returns (Prakasa Rao et al. 2008). Vetiver is cultivated as a secondary crop, with major crops being paddy, areca nut and cashew. Vetiver has been a crop of choice for providing economic returns to farmers. The variety of uses of vetiver has been reviewed briefly and results of the study aimed to improve essential oil production and economic returns have been reported in following sections.

10.2 Economic and Environmental Uses of Vetiver

10.2.1 Essential Oil

Vetiver, a member of the family Poaceae commonly known as the *Khas-Khas*, *Khas* or *Khus* in India, is a perennial grass with thick fibrous adventitious roots which are aromatic and highly valued. They can grow up to 1.5 m high and form clumps as wide as 1 m. The oil extracted from its roots by hydro/steam distillation is one of the finest oriental perfumes with a persistent fragrance. The oil is used in the flavor and fragrance industries for the manufacture of soaps, cosmetics, perfumery, agarbatti (incense sticks), soft drinks etc. In blended perfumes, vetiver oil acts as an excellent fixative for volatile compounds. It is known for its cooling properties and hence used in aromatherapy.

The current worldwide production of vetiver oil is 250 t per year. This production volume is made up of a dozen different varieties of vetiver oil: Haiti (100 tonnes), Indonesia (80 tonnes), China (20 t), India (20 t), Brazil (15 t), Dominican Republic (12 t), Vietnam (3 t), El Salvador (2 t), Madagascar (2 t), Nepal (0.5 t), Reunion (Bourbon) (0.5 t) and Ghana (0.4 t) (Thwaites 2010).

Vetiver oil, with its heavy, woody and earthy character, is one of the perfumer's most basic and traditional materials. Chemical composition of vetiver oil is highly complex mixture of more than 150 sesquiterpenoid constituents of which vetiverols, their carbonyl compounds and esters, are the main constituents and their relative abundance normally establishes the oil quality. Three carbonyl compounds α–vetivone, β-vetivone and khusimol are considered the primary odour influencing compounds and are considered to be fingerprints of the oil in the perfumery

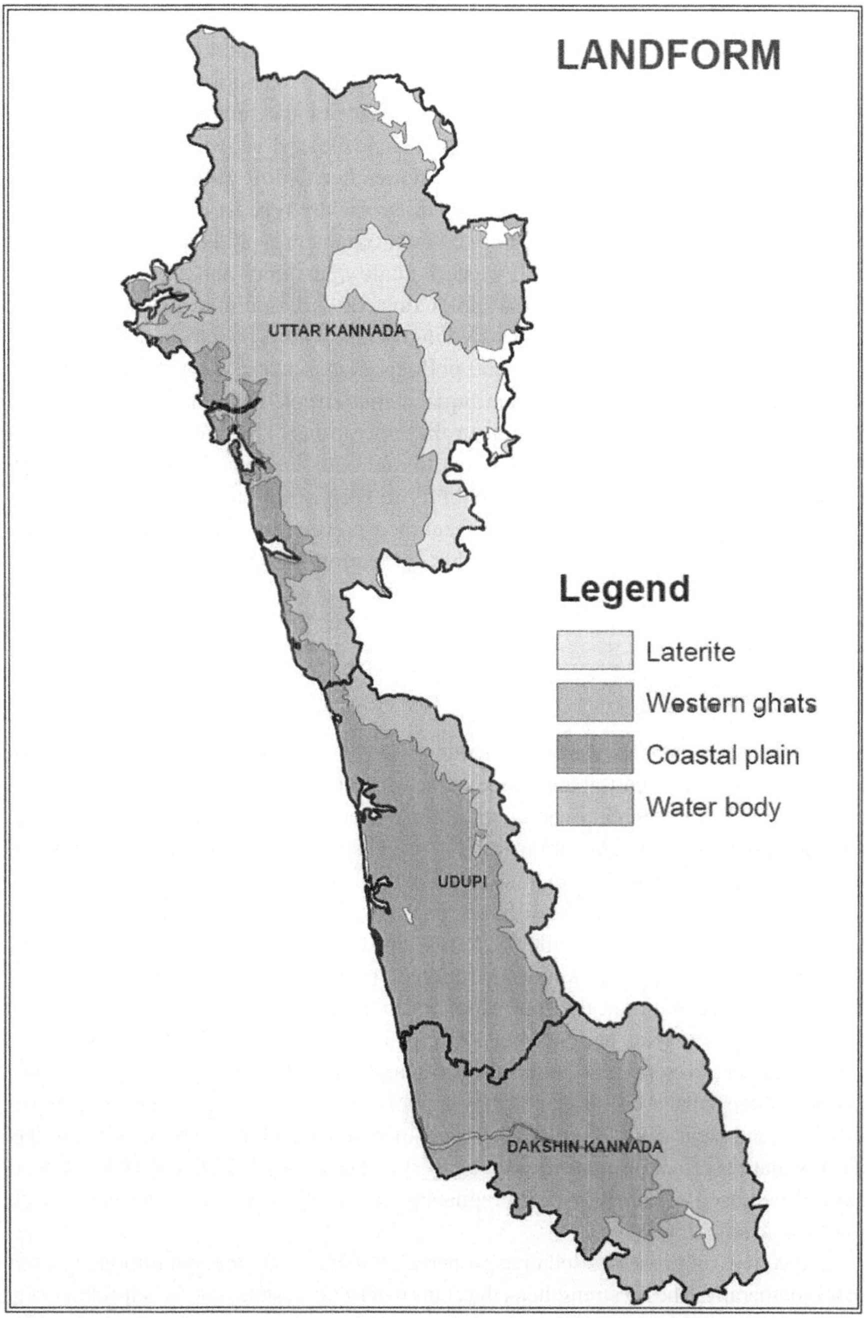

Fig. 10.1 Land forms in Uttara and Dakshina Kannada districts in Western Ghats area, India. The study area is ecologically sensitive region with complex landforms ranging from coastal to forest areas with lateritic soil formation

industries (Demole et al. 1995) even though they do not possess the typical odor characteristics associated with vetiver. Vetiver oil is used as part of the woody notes for luxury perfumes. The oils of vetiver, patchouli and sandalwood in combination with a jasmine and gardenia complex, is the base of the famous Crêpe de Chine note. In addition to its importance in classical perfumery, vetiver oil is also used as a base for many modern men's colognes (Guenther Ohloff 1994). The oil and its constituents are used extensively for blending oriental type of perfumes and floral compounds as well as in other cosmetic and aromatherapy applications. Vetiver oil is a main ingredient in 36 % of all western quality perfumes and 20 % of all men's fragrances (eg. Channel no. 5, Miss Dior, Cravache and Shalimar) (Thwaites 2010).

Vetiver oil is an extremely expensive ingredient with a precarious supply situation subject to the forces of nature and politics. This material is so crucial to industry, in fact, that in the wake of earthquake that struck Haiti in 2010, fragrance industry contributed to the earth quake relief operations (Thwaites 2010). The main reason for this increasing demand is its unique odor, for which it is used in both flavor and fragrance industries and inability to reconstitute this oil. The oil is preferred for 'base note' in flavors and fragrance production. Therefore, production systems have to be evolved for large scale cultivation and distillation of vetiver oil.

10.2.2 Medicinal Uses

Vetiver has been used for various ailments. Ayurvedic literature mentioned that plant is used as digestive, carminative stomachic, constipating, haematinic, expectorant, antispasmodic, antiasthmatic, antigout, anthelmintic, antimicrobial and diuretic. The roots are used for cooling the brain and also used in the treatment of ulcers. In addition to these, the plant is used for anemia, amenorrhea and dysmenorrheal (Jain 1991).

The tribals in India use the different parts of vetiver for many of their ailments such as a mouth ulcer, boils, epilepsy, burns, snake bite, scorpion stings, rheumatism, fever, headache, etc.; decoction of the roots has been used as tonic for weakness; the leaf juice as anthelmintic; the root vapor for malarial fever; the root ash is given to patients for acidity (Rao and Suseela 2000; Jain 1991; Singh and Maheshwari 1983).

Vetiver oil owes several beauty benefits and emotional effects. It balances the activity of the sebaceous oil glands, has deodorizing properties, and helps normalize oily skin and clear acne. It replenishes moisture in dry and dehydrated skin and has a rejuvenation effect on mature skin, as well as cuts/wounds/irritated and inflamed skin. When used regularly during pregnancy, vetiver oil reportedly prevents stretch marks (Lavania 2003a).

The vetiver oil possesses sedative property and hence it has been traditionally used in aromatherapy. The oil strengthens the central nervous system, and is helpful in overcoming depression, insomnia, anxiety, stress, tension and nervousness (Wilson 1995). It may be added to sports oil blends and massaged into muscles before and after sports. Vetiver oil is particularly useful for jet lag and for grounding and clarity while travelling (Shealy 1998). Researchers are exploring the therapeutic potential of this plant as it has more therapeutic properties which are not known (Bharath Bhushan et al. 2013).

10.2.3 Soil Erosion Control

Vetiver being a perennial grass and having deep and robust root system binds soil firmly. This characteristic has been used to control soil erosion, especially in sloppy lands and vulnerable conditions (Grimshaw 2008). Vetiver has been used world over for soil and water conservation in farm land, for land stabilisation: road, railway batter and river, for environmental protection: water and land pollution, and for combating climate changes (Truong 2011). In India on cropping land with 1.7 % slope, vetiver contour hedges reduced runoff (as percentage of rainfall) from 23.3 % (control) to 15.5 %, soil loss from 14.4 t/ha to 3.9 t/ha and sorghum yield increased from 2.52 t/ha to 2.88 t/ha over a 4 year period. The yield increase was attributed to mainly *in situ* soil and water conservation over the entire toposequence under the vetiver hedge system (Truong 1993). Relative to control plots, average reductions of 69 % in runoff and 76 % in soil loss were recorded from vetiver plots (Rao et al. 1992).

Results from Nigeria showed that soil physical and chemical conditions were ameliorated behind the vetiver strip for a distance of 20 m. Crop yields were increased by a range of 11–26 % for cowpea and by about 50 % for maize under vetiver management. Soil loss and runoff water at the end of 20 m runoff plots were 70 % and 130 % higher respectively in non-vetiver plots than vetiver plots. Vetiver strips increased soil moisture storage by a range of 1.9–50.1 % at various soil depths. Eroded soils on non-vetiver plots were consistently richer in nutrient contents than on vetiver plots. Nitrogen use efficiency was enhanced by about 40 %. This work demonstrates the usefulness of vetiver grass as a soil and water conservation measure in the Nigerian environment (Babalola et al. 2003).

In Brazil, vetiver could rehabilitate and maintain slopes affected by landslides even after 4 years of planting (Eboli et al. 2011) and also helped coastal erosion control (Luiz Lucena and Paula Leão 2011). Vetiver is used for large road stabilization projects in China, Malaysia, Thailand, Argentina, Venezuela (Smyle 2011).

In Thailand., vetiver is used in farmers' fields for soil and water conservation (Panklang 2011) and in Vietnam for sand dune protection, river bank erosion control, coastal erosion control, road batter stabilization (Van and Truong 2011) and in mountain slopes to prevent erosion (Quang et al. 2014). Vetiver is used in infrastructure stabilization in countries like Madagascar, S. Africa, Mozambique, Uganda, DR Congo, Guinea, Swaziland and for high absorption of heavy metals, phosphates and nitrates (Nöffke 2011). In India, vetiver cultivation in sloppy lands help to reduce run-off losses and control soil erosion (Fig. 10.2, Prakasa Rao et al. 2008).

10.2.4 Amelioration of Problem Soils and Phyto-Remediation

Vetiver tolerates extreme climatic variation such as prolonged drought, flood, submergence and extreme temperature from −14 to 55 °C, frost, salt and other adverse soil conditions such as soil acidity, salinity, sodicity and acid sulfate conditions (Van et al. 2008).

Fig. 10.2 Cultivation of vetiver on hill slopes in Western Ghats, India. Vetiver provides livelihood support besides protecting the soil from erosion

Vetiver is one such plant which is having extensive rooting system and can uptake heavy metals and been exemplified to rise in polluted soils (Yang et al. 2003; Chiu et al. 2006). The metal accumulating ability of this plant, coupled with metal tolerance and high shoot biomass, makes this plant ideal for extraction (Randloff et al. 1995; Truong and Baker 1998; Chen 2000). Vetiver could tolerate high Pb concentrations in soil (10,750 mg kg^{-1}) and had very good growth performance (Rotkittikhun et al. 2007). Uses of vetiver grass in relation to petroleum-contaminated soils are promising for amelioration of slightly polluted sites, to allow other species to get established and for erosion control (Brandt et al. 2006). Truong and Truong (2011) reported that vetiver has a very high capacity for uptake of N and P and can tolerate high levels of Cr, Cd, Cu, Pb and Zn. They have also reported phytoremediation of a contaminated land near an explosive factory in Australia where the soil is highly contaminated with nitrate and NH_3: soil total N up to 5 400 mg/kg, soil total NH_3 up to 1,220 mg/kg, water total N up to 18,300 mg/kg, water total NH_3 up to 12,300 mg/ kg.

A hydroponical experiment has shown that 1 mg/ml of cadmium increased chlorophyll content, root activity, and biomass to 2.2 % when compared to the control. The high level of Cd accumulation was seen in 30 mg/L with the cadmium accumulation of 93 and 2,232 mg/kg Cd in shoot and root respectively, but the cadmium treatment increased the catalase and the peroxidase activity in vetiver plant (Aibibu et al. 2010; Andra et al. 2009). Truong (1999) reported that the distribution of heavy metals in vetiver plant can be divided into three groups: very little (1–5 %) of arsenic, cadmium, chromium and mercury absorbed were translocated to the shoots, moderate proportion of (16–33 %) of copper, lead, nickel and selenium were translocated to the top and zinc (40 %) was almost evenly distributed between shoot and root. However, Yang et al. (2003), Roongtanakiat et al. (2007) and Singh et al. (2007) concluded that heavy metal accumulations were found higher in root than shoot.

Vetiver has been used for rehabilitation in oil contaminated sites as it helps in biodegradation of oil where the petroleum hydrocarbons are more. A greenhouse study has been conducted to study that the vetiver has the potential to degrade the crude oil sites in Venezuela. The results of this experiment indicated that vetiver grown in this contaminated soil have only slight effect but preferable in slightly contaminated soil (Brandt et al. 2006). Heavy metal contamination caused by lead (Pb) in the soil was studied in pot experiments, where vetiver can tolerate heavy metal like lead under EDTA application where EDTA is known to be the chelating agent, where increase in the Pb from the root to shoot was found (Chen et al. 2004).

Vetiver was used in treating coffee effluents in India (Nagesh Rao 2011). Vetiver can be considered a hyperaccumulator of Pb and Zn; heavy metal accumulation is high in roots than in shoot. Vetiver can tolerate salinity up to 12.0 dSm^{-1} pH up to 10.5 and soil ESP up to 55 (Patra 2011).

10.2.5 Carbon Sequestration

Vetiver due to its high biomass production and root growth has been proposed to be a potential candidate for terrestrial C sequestration (Lavania and Lavania 2009). Genotypes most suitable for C-sequestration have been developed (Lavania 2011). In India, estimates were made, based on field studies, the C sequestration potential of vetiver. They showed that vetiver sequesters 15.24 Mg C ha^{-1} $year^{-1}$ in shoot and roots, much higher than that for lemongrass with 5.38, palmarosa with 6.14, and trees with 2.92. In addition the benefit/cost ratio of vetiver, 2.3, is higher than that of rice, 1.97. It was estimated that vetiver cropping could sequester 150 Tg C per year in India, which is nearly 46 % of C emissions in India (Singh et al. 2014). As weather plays an important role in the biomass production and consequent sequestration of carbon, Prakasa Rao et al. (2011) have proposed modeling possibilities for vetiver based land use in west coast of Karnataka, India in relation to climate change. CO_2 emissions from and C sequestration by vetiver were studied in Thailand and vetiver was found to be good in the C sequestration (Taranet et al. 2011).

10.2.6 Other Uses of Vetiver

The other uses of vetiver are: harvested vetiver leaves is an input of agriculture-related activities viz. forage, mulching, composting. In the state of Karnataka, India, vetiver is planted along the field boundaries and cut every 2 weeks or less for use as fodder. Vetiver was found to have relatively higher structural carbohydrates as compared to native grass and rice straw. On the other hand, it also had optimal levels of crude protein, considered to be enough to maximize intake and digestion of the vetiver forage. It was concluded that vetiver may be used as ruminant feed if it is mixed with other good quality feed and forages (Anon 1990).

Vetiver leaves are used as a medium for mushroom cultivation since it contains cellulose, hemicellulose, lignin, and crude proteins. Many investigators have been successful in cultivating Oyster, shiitake, and straw mushrooms using vetiver as the medium for their growth (Chomchalow and Chapman 2003). Nootkatone compound in vetiver roots were able to disrupt termite behavior and physiology as a consequence of direct physical contact, ingestion, or exposure to the vapors. They also found that ingestion of wood treated with vetiver oil or nootkatone causes the progressive death of the protozoa living inside the termite gut, ultimately results in a progressive decline of its colony through starvation, as these termites rely on the protozoa for the digestion of their wooden food (Maistrello and Henderson 2001). In New Zealand, Greenfield (2002) observed that fungal attacks on the vetiver-mulched plants have virtually disappeared and there seem to be little, if any other pest action around the host plants. Thus vetiver mulch seems to have natural fungicide to stop the growth of the fungi which attack the crop plants.

The distilled roots are used for making handicrafts (Fig. 10.3) and the spent biomass is used for making paper and hardboards. In India, dried tops of vetiver have been used since ancient times for making makeshift huts (Fig. 10.4) and cabins as they provide cooling effects during the summer (Lavania 2003a).

Fig. 10.3 Vetiver provides alternate livelihood options to the small farmers of Western Ghats region, India. Handicrafts made from vetiver roots provide extra income to the farmers of the region

Fig. 10.4 Vetiver dry leaves are used to make temporary hutments for the farm workers in the Western Ghats area, India

The studies carried out at the Forest Research Institute, Dehra Dun, India revealed that pulps suitable for making strawboards can be made from vetiver by digestion with lime (Anon 1976). Kuhiran and Punnapayak (2000) described the process of producing ethanol from vetiver leaves by simultaneous saccharification and fermentation technique using *Trichoderma reesei.* Nimityongskul et al. (2003) reported that vetiver grass ash (VGA) can be a cement replacement material as a new building material specifically for the rural areas of the developing countries.

This paper presents the results of a study on large scale cultivation and distillation of vetiver adopting improved agronomic and distillation practices in the fields of nine farmers which was extended to an area of 250 ha.

10.3 Material and Methods

10.3.1 Site and Climatic Data

The study area is located in Uttara Kannada district of Western Ghat region in Karnataka state, India. Uttara Kannada district lies between 13.9220° N to 15.5252° N latitude and 74.0852 ° E to 75.0999° E longitude and covers an area of 10,291 km^2.

The area has a tropical climate with a well-defined south-west monsoon season between June and October, with an average annual rainfall of 4,016 mm. Maximum and minimum temperatures range from 25 °C to 32 °C and 19 °C to 26.3 °C respectively.

The soils of the area are moderately shallow, or moderately deep to very deep, somewhat excessively drained or well-drained, gravelly clay soils, associated with ironstone crust on the surface. The soils have low or medium base saturation, and low or medium cation exchange capacity. They are strongly to medium acidic, have medium to high organic carbon content. Constraints for agriculture are low base saturation, low nutrient status, crusting and steep slopes and moderate or severe erosion (Shiva Prasad et al. 1998).

Thirty farmers were selected in three villages (Uttara Koppa, Kachhodi, Kolegeri) of the region based on their land holdings and economic background. Majority of the farmers are illiterate (86.6 %) with an average land holding of 1–3 acres (76.6 %) and their economic status is very poor. Improved agronomic practices (3.2) and distillation methods (3.3) were tested in nine farmers' fields and were compared with the conventional practices of the region. The soil samples of the farmers' fields were analyzed (Table 10.1). The soils are acidic with low to medium fertility.

10.3.2 Agronomic Practices

The improved agronomic practices tested were as follows. Two varieties viz., local and Gulabi (developed by CIMAP) were evaluated and similar agronomic methods were followed for both the varieties. Vetiver nursery was raised in paddy fallow fields in March-April 2009. In June, 2009 shoots of the nursery plants were cut 25–30 cm above the ground and clumps were dug and split into slips (stem with

Table 10.1 Initial soil analysis of the sampling plots

Sampling plots	pH (1:2.5)	C (%)	Available N (kg/ha)	Available P (kg/ha)	Exchangeable K (kg/ha)
1	5.2	2.8	200.9	4.5	168.0
2	5.3	2.8	140.1	6.7	179.2
3	5.4	3.7	140.7	6.7	184.8
4	5.2	2.6	118.1	13.4	95.2
5	5.3	2.5	120.1	6.7	179.2
6	5.0	2.5	116.6	17.9	140.0
7	5.5	2.8	141.8	11.2	184.8
8	5.3	2.7	142.4	11.3	102.3
9	5.5	2.8	132.2	6.7	128.3
Mean	5.3	2.8	139.2	9.5	151.3

The analysis reveals that the soils are acidic with low to medium fertility
C organic carbon, *N* nitrogen, *P* phosphorus, *K* potassium

some portion of roots intact on them). Land was ploughed and the slips were planted in pits made at a spacing of 30×30 cm. A slip is placed vertically in each pit. Fertilizers were applied to supply N: 60 kg ha^{-1}, P_2O_5: 30 kg ha^{-1}, K_2O: 30 kg ha^{-1} and 5 t ha^{-1} farm yard manure was applied (Patra et al. 2004). Around 100,000 slips ha^{-1} were planted. Irrigation was not essential since it was grown as a rain fed crop. After 3 months of planting, weeding was performed.

Harvesting of roots was performed in April, 2010. The stem was cut at a height of 15–30 cm and the clumps were uprooted. Harvesting was done by manual digging with the help of spades wherein the entire clump is uprooted. The soil adhering to the roots was removed and roots were separated from shoots with the help of a knife. The roots were washed gently with running water in order to remove the adhering mud. The conventional agronomic practices included non-standardized plant population, plant spacing and fertilizer application.

Biometric data (mean root length/plant, shoot height/plant and no of tillers/plant) were recorded. The biomass yields of shoots and roots were recorded.

10.3.3 Distillation Method

In the conventional method of distillation, approx. 150 kg of vetiver roots were distilled for 72–80 h in a cylindrical still made of mild stainless steel of around 150–200 kg capacity. The unit is fixed to the ground with a provision of furnace to burn firewood to heat the water in the still directly (Fig. 10.5). The roots are placed on a mesh and water is filled below it. The still is connected to the condenser through a vapour line. The water is boiled and the steam vapour passes through the roots, vapourizing the oil which gets condensed in a coil type condenser by cooling water. The steam generation in this type of unit is as low as 15–20 kg/h. The condensate oil mixture is then separated through a series of oil separators (4–5 nos.) which are interconnected. The oil recovery is calculated as % of roots (w/w).

In the improved method of distillation, approx. 300 kg roots per batch are steam-hydro distilled for 18–20 h in cylindrical still made of high quality stainless steel cum steam generator of 500 kg capacity (Fig. 10.6) The cylindrical distillation still is placed on a square inbuilt boiler and calandria having smoke pipes which reduces the boiling time of water which results in higher steam generation rate of 120–150 kg/h and lesser fuel consumption (20–30 % of the conventional method). The still is placed on a specially designed furnace having fire grate, flue ducts and fire door for proper control of the firing and draft. The furnace is connected with a chimney of optimum height to maximize the air draft. A compatible stainless steel shell tube type condenser is used to get higher condensation capacity for cooling of the vapours which minimizes loss of oil. The condensed oil- water mixture is then passed through a specially designed stainless steel oil separator. The separator has an inbuilt baffle to maximize the retention time of the mixture thereby minimizing loss of oil along with the outgoing water from the separator. The oil recovery by the improved method was also recorded.

Fig. 10.5 Conventional distillation unit in a farmer's field in Western Ghats area, India. The traditional method of distillation of vetiver roots yield low essential oil yields, take more time and consume more fuel rendering the process less economical and environment friendly

10.3.4 Recycling of Biomass

In order to recycle the bio-wastes generated during cultivation and distillation of vetiver, vermicomposting of the wastes has been studied. Around 100 kg per batch of agro-wastes, which comprised of vetiver plant wastes and dry leaves from the forest trees are charged to the vermicompost pit with dimensions of 20 ft × 6 ft with 1.5 ft depth (1 ft below the ground level and ½ ft above the ground level). Cow dung available in the farm is mixed with water and this slurry is sprinkled in the pit. When the heat evolved during the decomposition of the material has subsided (15–20 days after heaping), 200 earthworms of species *Eudrilus eugeniae* were introduced in to each pit. Appropriate moisture (about 70 %) was maintained and the biomass was turned and mixed once in a week. Vermicompost produced is passed through a 2 mm mesh and stored in gunny bags. An average of 300 kg per pit per batch of vermicompost was produced by this method. Samples of the vermi-compost were analyzed for nutrients by standard method (Puttanna and Prakasa Rao 2002).

Fig. 10.6 Improved distillation unit and oil separator established in a farmer's field in Western Ghats area, India. The unit helps to recover more vetiver oil, consume less time and fuel making the process more economical and environment friendly

10.4 Result and Discussion

Soil characteristics – Initial soil analysis of the sampling plots are given in the (Table 10.1).

The initial soil analysis of the sampling plots reveals that soils are acidic with low to medium fertility.

10.4.1 Biometric Characteristics

The biometric parameters of the plants harvested in different sampling plots of conventional and improved agronomic methods are presented in the (Table 10.2). The mean root length was 43.18 cm and 47.58 cm, plant height was 113.42 cm and 118.3, no of tillers/plant was 41 and 46 respectively. The better growth with the improved agronomy is due to a better variety (cv. Gulabi) and better agronomic practices. Thus, introduction of improved agronomy offers increased productivity in this region.

Table 10.2 Biometric data of vetiver in the different sampling plots from three villages

Sampling plots	Mean root length/plant (cm)		Mean shoot height/plant (cm)		Mean no. of tillers/plant	
	Improved agronomy	Conventional agronomy	Improved agronomy	Conventional agronomy	Improved agronomy	Conventional agronomy
1	50.0	45.5	119.3	109.4	40	37
2	47.6	46.8	119.8	115.0	45	40
3	46.3	42.8	112.2	109.0	49	45
4	45.0	42.3	114.3	109.6	45	39
5	45.3	41.0	115.7	111.8	45	40
6	46.4	40.6	120.3	115.1	51	44
7	48.9	40.7	119.1	114.6	46	41
8	49.5	43.4	123.6	119.7	48	44
9	49.3	45.6	120.8	116.6	47	39
Mean	47.58 ± 0.52	43.18 ± 0.53	118.3 ± 0.82	113.42 ± 0.75	46 ± 0.70	41 ± 0.60

The root length, shoot height and number of tillers are found to be higher with improved agronomy

Table 10.3 Above ground and below ground biomass of the sampling plots of vetiver

Sampling plots	Mean root biomass/plant (g)		Mean shoot biomass/plant (g)		Mean root: shoot ratio (%)	
	Improved agronomy	Conventional agronomy	Improved agronomy	Conventional agronomy	Improved agronomy	Conventional agronomy
1	28.0	26.0	288.6	270.3	9.70: 90.3	9.61: 90.39
2	30.0	24.3	356.3	294.6	8.41: 91.59	8.24: 91.76
3	30.0	28.3	314.0	284.0	9.55: 90.45	9.96: 90.04
4	30.0	26.0	318.1	282.0	9.43: 90.57	9.21: 90.79
5	32.4	28.0	296.6	270.6	10.92: 89.08	10.34: 89.66
6	30.0	26.0	324.0	238.6	9.25: 90.75	10.89: 89.11
7	30.1	24.0	342.0	280.3	8.80: 91.20	8.56: 91.44
8	29.4	24.0	312.0	264.0	9.42: 90.58	9.09: 90.91
9	32.5	28.0	290.3	276.3	11.19: 88.81	10.13: 89.97
Mean	30.26 ± 0.42	26.06 ± 0.36	315.76 ± 4.27	273.41 ± 3.04	9.63 ± 0.42: 90.37 ± 4.27	9.55 ± 0.36: 90.45 ± 3.04

The root and shoot biomass are found to be higher with improved agronomy. However the root: shoot ratio in both the methods remains constant

10.4.2 Biomass Evaluation

Biomass evaluation was done in three randomly selected plants from nine plots. These plants were manually dug and biomass observations were taken on the separated shoots and roots. The biomass in shoots and roots was higher with improved agronomic methods than with the conventional methods (Table 10.3). This may be due to better biometrical values with improved agronomy as shown in Table 10.2.

Nearly 10 % of the plant's fresh biomass is located in the roots. The root: shoot ratio, rather, remained constant in both the varieties. The higher biomass with the improved agronomy can support higher vetiver oil production in the region.

10.4.3 Distillation of Essential Oil

Nine batches each of 100 kg roots have been distilled in conventional and field distillation units. The comparative essential oil recoveries from the conventional and improved methods of vetiver oil distillation are presented (Table 10.4). The oil yields in the two methods are 17 kg/ha and 25 kg/ha respectively.

The improved method of distillation gave 25–30 % higher oil recovery than the conventional method, besides reduced firewood consumption (150 kg as against 600 kg in conventional method) and man days (3 man days as against 18 man days in conventional method) and also distillation cost is reduced to 50 % in the improved method of distillation. The higher essential oil recoveries were also possible due to the improved design of the unit and oil separator. The essential oil recoveries are presented in the (Table 10.5). The improved oil separator reduced the oil losses and improved the recoveries by 22.6 %.

10.4.4 Vermicompost Production

On an average, 300 kg/pit/batch of vermicompost is produced. The analysis of vermicompost is presented (Table 10.6). The analysis reveals that it is rich in organic carbon and nutrients.

Table 10.4 Oil recovery in using conventional and improved distillation methods

Batch	% oil recovery	
	Conventional method	Improved method
1	0.8	1.2
2	0.7	1.3
3	0.6	1.2
4	0.9	1.1
5	0.7	1.3
6	0.8	1.1
7	0.9	1.2
8	1.0	1.3
9	0.7	1.1
Mean	0.78	1.2

About 50 % higher vetiver oil yield is obtained by the improved distillation method than the conventional method

Table 10.5 Essential oil yield from conventional oil separator and improved designed oil separator

Batches	% of oil in traditional oil separator	% of oil in improved designed oil separator	% increase in oil recovery
1	1.0	1.2	20
2	1.1	1.3	18
3	0.8	1.0	25
4	0.8	1.0	28
5	0.8	1.0	25
Mean	0.9	1.1	22.6

The improved method of distillation gave 25–30 % higher oil recovery than the conventional method

Table 10.6 Nutrient composition of vermicompost produced from vetiver waste

Vermicompost	Production of vermicompost kg/pit	pH	C (%)	N (%)	P (%)	K (%)
Pit -1	980	7.6	18.9	1.3	0.14	0.28
Pit -2	1,020	7.0	15.2	1.1	0.11	0.34
Pit -3	980	7.7	34.4	1.6	0.10	0.20
Pit -4	1,040	7.1	24.3	1.6	0.15	0.24
Pit -5	1,000	7.5	21.3	1.4	0.12	0.31
Mean	1,004	7.4	22.9	1.4	0.12	0.22

The average vermicompost production is around 300 kg/pit/batch and the analysis of the compost reveals that it is rich in organic carbon and nutrients
C organic carbon, *N* nitrogen, *P* phosphorus, *K* potassium

Table 10.7 Economics of the major and secondary crops of the region

Particulars	Paddy (Rs.)	Arecanut (Rs.)	Cashew (Rs.)	Vetiver-conventional method (Rs.)	Vetiver-improved method (Rs.)
Cost of cultivation/ha	33,125.00	70,750.00	36,500.00	1,05,750.00	89,500.00
Gross income/ha	45,000.00	1,20,000.00	90,000.00	1,70,000.00	2,12,500.00
Net income/ha	11,875.00	49,250.00	53,500.00	64,250.00	1,23,000.00

The net returns have increased by about 50–60 % with the intervention of improved vetiver agronomic and distillation technologies

10.4.5 Economics of Cultivation

The major crops of the region are paddy, areca nut, and cashew. Vetiver is cultivated as a secondary crop. The monetary returns/ha of major crops and vetiver were studied. The return from vetiver cultivation is more than the traditional crops (Table 10.7). Based on the results obtained in the selected farmers' fields, the improved agronomic and distillation technologies were adopted in 250 ha of the project area.

The improved methods reduced the cost of distillation by half, increased the oil recovery by 22 % thus increased the net returns by more than 50 %.

10.5 Conclusion

Large scale cultivation and distillation of vetiver was tested in a western Ghats region of Karnataka state, India using improved agronomic and distillation technologies. The improved agronomic methods along with recycling of agro-wastes have significantly increased the yield levels of vetiver root. Also, the improved method of distillation helped farmers to realize higher oil recoveries thereby producing higher oil yields. Based on the data obtained in farmers' fields, the improved methods tested in a large area of 250 ha of the region resulted in an increase of more than 40 % vetiver oil yields compared to the conventional methods; nearly 6 tones of vetiver oil was produced in the project area. Thus, we have shown for the first time that large scale cultivation and production of vetiver oil adopting improved agronomy and distillation is feasible that would improve the livelihoods of farmers while helping to maintain ecological sustainability in this region.

Acknowledgements We thank Council of Scientific and Industrial Research (CSIR), New Delhi, India and National Bank for Agriculture and Rural Development (NABARD), Bangalore, India for funding the research.

References

Aibibu N, Liu Y, Zeng G, Wang X, Chen B, Song H, Xu L (2010) Cadmium accumulation in *vetiveria zizanioides* and its effects on growth, physiological and biochemical characters. Bioresour Technol 101(16):6297–6303

Andra SS, Datta R, Sarkar D, Saminathan SK, Mullens CP, Bach SB (2009) Analysis of phytochelatin complexes in the lead tolerant vetiver grass [*Vetiveria zizanioides* (L.)] using liquid chromatography and mass spectrometry. Environ Pollut 157(7):2173–2183

Anon (1976) Vetiver. In: Wealth of India, vol X. Publication & Information Directorate, Council of Scientific and Industrial Research, New Delhi, pp 451–457

Anon (1990) Fodder value of vetiver grass. Vetiver Newsl 4:6

Babalola O, Jimba JC, Maduakolam O, Dada OA (2003) Use of vetiver grass for soil and water conservation in Nigeria. In: Proceedings of the third international conference on vetiver and Exhibition. Guangzhou, China, October 2003, pp 293–309

Bhushan B, Kumar SS, Tanuja S, Lalith S, Hema A (2013) *Vetiveria zizanioides* (Linn.) Nash: a pharmacological overview. Int Res J Pharm 4(7):18–20

Brandt R, Merkl N, Schultze-Kraft R, Infante C, Broll G (2006) Potential of vetiver (*Vetiveria zizanioides (L.)* Nash) for phytoremediation of petroleum hydrocarbon-contaminated soils in Venezuela. Int J Phytoremediat 8(4):273–284

Chen H (2000) Chemical method and phytoremediation of soil contaminated with heavy metals. Chemosphere 41:229–234

Chen Y, Shen Z, Li X (2004) The use of vetiver grass (*Vetiveria zizanioides*) in the phytoremediation of soils contaminated with heavy metals. Appl Geochem 19(10):1553–1565

Chiu K, Ye Z, Wong M (2006) Growth of *Vetiveria zizanioides* and *Phragmities australis* on Pb/Zn and Cu mine tailings amended with manure compost and sewage sludge: a greenhouse study. Bioresour Technol 97(1):158–170

Chomchalow N, Chapman K (2003) Other uses, and utilization of vetiver. In: Proceedings of the third international conference on vetiver and exhibition, Guangzhou, China

Demole EP, Holzner GW, Youssefi MJ (1995) Malodor formation in alcoholic perfumes containing vetiveryl acetate and vetiver oil. Perfum Flavor 20:35–40

Eboli J, Rogério P, Truong P (2011) ICV5 fifth international conference on vetiver, Lucknow, India

Greenfield JC (2002) Vetiver grass: an essential for the conversation of planet earth. Infinity Publishing, Haverford

Grimshaw R (2008) The vetiver system. First national Indian vetiver workshop. The Vetiver International, Kochi, pp 21–23

Jain SK (1991) Dictionary of Indian folk medicine and ethnobotany. Deep Publication, New Delhi

Kuhiran M, Punnapayak H (2000) Leaves of vetiver grass as a source of ethanol. An abstract of a poster paper presented at ICV-2, Phetchaburi

Lavania UC (2003a) Vetiver root – oil and its utilization. PRVN Tech. Bull. No. 2003/1, ORDPB, Bangkok

Lavania UC (2003b) Other uses and utilization of vetiver: vetiver oil. In: Proceedings of the third international conference on vetiver and exhibition, Guangzhou, China

Lavania UC (2011) Vetiver grass model for long term carbon sequestration and development of designer genotypes for implementation. In: ICV5 fifth international conference on vetiver, Lucknow, India

Lavania UC, Lavania S (2009) Sequestration of atmospheric carbon into subsoil horizons through deep-rooted grasses—vetiver grass model. Curr Sci 97(5):618–619

Lucena L, Leão P (2011) ICV5 fifth international conference on vetiver, Lucknow, India

Maistrello L, Henderson G (2001) Vetiver grass: useful tools against Formosan subterranean termites. Vetiverim 16:8

Nagesh Rao HR (2011) Introduction of vetiver system in coffee effluent treatment. In: Icv5 fifth international conference on vetiver, Lucknow, India

Nimityongskul P, Panichnava S, Hengsadeekul T (2003) Use of vetiver grass ash as cement replacement materials. ICV-3 held in Guangzhou, China. In: ORDPB's Thailand's technical papers for the presentation at ICV-3, pp 148–159

Nöffke R (2011) The vetiver system for infrastructure stabilisation in Africa with special references to road batter side slope protection & sand dune stabilisation in Madagascar. In: ICV5 fifth international conference on vetiver, Lucknow, India, 28–30 October 2011

Ohloff G (1994) Scent and fragrances: the fascination of odors and their chemical perspectives. Springer, Berlin/Heidelberg, p 172

Panklang P (2011) The promotion of vetiver grass cultivation for soil and water conservation by farmer's participation. In: The ICV5 fifth international conference on vetiver, Lucknow, India

Patra DD (2011) *Vetiveria zizanioides vis-à-vis* rehabilitation of degraded and contaminated soils-some case studies. In: The ICV5 fifth international conference on vetiver, Lucknow, India

Patra DD, Chand S, Singh A, Anwar M, Lal RK, Singh S, Bahl JR, Khanuja SPS (2004) Agrotechnology of vetiver (Vetiveria zizanioides L & Nash.). J Med Aromat Plant Sci 26(4):784–789

Prakasa Rao, EVS, Gopinath, CT, Khanuja SPS (2008) Environment economics and equality aspects of vetiver in South India. First Indian National Seminar on Vetiver, Kochi, pp 21–23

Prakasa Rao EVS, Goswami P, Ramamurthy V, Naidu LGK (2011) Climate change vis-à-vis soil system modelling possibilities for Vetiver based land use in West coast of Karnataka, India. In: The ICV5 fifth international conference on vetiver, Lucknow, India

Puttanna K, Prakasa Rao EVS (2002) Vermicomposting distillation residues of some aromatic crops. Fertil News 48(2):67–68

Quang DV, Schreinemachers P, Berger T (2014) Ex-ante assessment of soil conservation methods in the uplands of Vietnam: an agent-based modeling approach. Agric Syst 123:108–119

Randloff B, Walsh K, Melzer A (1995) Direct revegetation of coal tailings at BHP Saraji Mine, Australian Mining Council Environment workshop, Darwin, Australia

Rao RR, Suseela MR (2000) *Vetiveria zizanioides* (Linn.) Nash – a multipurpose ecofriendly grass of India. Paper presented at ICV-2 held in Cha-am, Phetchaburi, Thailand, 18–22 January 2000. In: Proceedings of ICV-2, pp 44–48

Rao KPC, Cogle AL, Srivastava KL (1992) ICRISAT annual report 1992, Andhra Pradesh, India

Roongtanakiat N, Tangruangkiat S, Meesat R (2007) Utilization of vetiver grass (*Vetiveria zizanioides*) for removal of heavy metals from industrial wastewater. Sci Asia 33:397–403

Rotkittikhun P, Chaiyarat R, Kruatrachue M, Pokethitiyook P, Baker AJ (2007) Growth and lead accumulation by the grasses Vetiveria zizanioides and Thysanolaena maxima in lead-contaminated soil amended with pig manure and fertilizer: a glasshouse study. Chemosphere 66(1):45–53. Epub 2006 Jul 10

Shealy CN (1998) The illustrated encyclopedia of healing remedies. Brideg water book Co; PMCid:PMC105742

Shiva Prasad CR, Reddy RS, Sehgal J, Velayutham M (1998) Soils of Karnataka for optimizing land use. NBSS Publ. 47b 1998, National Bureau of Soil Survey and Land Use Planning, Nagpur, India, p 111

Singh KK, Maheshwari JK (1983) Traditional phytotherapy amongst the tribals of Varanasi district, U.P. J Econ Tax Bot 4:829–838

Singh SK, Juwarkar AA, Kumar S, Meshram J, Fan M (2007) Effect of amendment on phytoextraction of arsenic by *Vetiveria zizanioides* from soil. Int J Environ Sci Technol 4:339–344

Singh M, Neha G, Prakasa Rao EVS, Goswami P (2014) Efficient C sequestration and benefits of medicinal vetiver cropping in tropical regions. Agron Sustain Dev 34(3):603–607. doi:10.1007/s13593-013-0184-3

Smyle J (2011) Confronting climate change: *vetiver system applications*. In: ICV5 fifth international conference on vetiver, Lucknow, India

Taranet P, Wattanaprapat K, Meesing I, Nopmalai P (2011) Carbon sequestration and carbon dioxide emission in vetiver grass cultivation areas. In: Vetiver and climate change. The fifth international conference on vetiver, Lucknow, pp 8–9

Thwaites C (2010) Haitian Vetiver: uprooted? Perfum Flavor 35(5):22–23

Truong P (1993) Report on the international vetiver grass field workshop, Kuala Lumpur. Aust J Soil Water Conserv 6:23–26

Truong PNV (1999) (12) Vetiver grass technology for mine rehabilitation. Office of Royal Development projects Board, Bangkok, Technical Bulletin

Truong P (2011) Global review of contribution of VST in alleviating climate change disasters. In: The fifth international conference on vetiver, Lucknow, India

Truong PNV, Baker D (1998) The role of vetiver grass in the rehabilitation of toxic and contaminated lands in Australia. In: Proceedings of the international vetiver workshop, Fuzhou, China

Truong P, Truong N (2011) ICV5 fifth international conference on vetiver. Lucknow, India, 28–30 October 2011

Van TT, Truong P (2011) Vetiver system for natural disaster mitigation in Vietnam: some lessons learned after a decade of application. In: The ICV5 fifth international conference on vetiver, Lucknow, India

Van TT, Truong P, Pinners E (2008) R&D results on unique contributes of vetiver grass applicable for natural disaster mitigation in Vietnam: the vetiver system. First National Indian Vetiver workshop. The Vetiver Network International, Kochi, India

Wilson R (1995) Aromatherapy for vibrant health and beauty. Penguin Putnam Inc., New York

Yang B, Shu WS, Ye ZH, Lan CY, Wong MH (2003) Growth and metal accumulation in vetiver and two *Sesbania* species on lead/zinc mine tailings. Chemosphere 52:1593–1600

Index

E. Lichtfouse (ed.), *Sustainable Agriculture Reviews*,
Sustainable Agriculture Reviews 17, DOI 10.1007/978-3-319-16742-8

The manufacturer's authorised representative in the EU is Springer Nature Customer Service Centre GmbH, Europaplatz 3, 69115 Heidelberg, Germany. If you have any concerns regarding our products, please contact ProductSafety@springernature.com

Printed and bound by CPI Group (UK) Ltd, Croydon, CR0 4YY
07/08/2026
02189673-0001